THE FIRST BRITISH ARMY 1624-1628

The Army of the Duke of Buckingham
(REVISED EDITION)

Laurence Spring

'This is the Century of the Soldier', Fulvio Testi, Poet, 1641

HELION & COMPANY

Helion & Company Limited
Unit 8 Amherst Business Centre
Budbrooke Road
Warwick
CV34 5WE
England
Tel. 01926 499 619
Email: info@helion.co.uk
Website: www.helion.co.uk
Twitter: @helionbooks
Visit our blog http://blog.helion.co.uk/

Published by Helion & Company 2024
Designed and typeset by Mary Woolley, Battlefield Design (www.battlefield-design.co.uk)
Cover designed by Paul Hewitt, Battlefield Design (www.battlefield-design.co.uk)

Text © Laurence Spring 2024
Photographs and illustrations © as individually credited
Cover artwork by Giorgio Albertini © Helion and Company 2024
Pen and ink art work by Maxim Borisov © Helion & Company 2024

Every reasonable effort has been made to trace copyright holders and to obtain their permission for the use of copyright material. The author and publisher apologize for any errors or omissions in this work and would be grateful if notified of any corrections that should be incorporated in future reprints or editions of this book.

ISBN 978-1-804514-49-8

British Library Cataloguing-in-Publication Data.
A catalogue record for this book is available from the British Library.

All rights reserved. No part of this publication may be reproduced, stored in a retrieval system, or transmitted, in any form, or by any means, electronic, mechanical, photocopying, recording or otherwise, without the express written consent of Helion & Company Limited.

For details of other military history titles published by Helion & Company Limited contact the above address or visit our website: http://www.helion.co.uk.

We always welcome receiving book proposals from prospective authors.

Contents

Introduction		iv
Acknowledgements		ix
1	Officers and Lesser Officers	13
2	The Rank and File	33
3	Dressed to Kill	54
4	Arms and Armour	67
5	Provisions	93
6	Quartering	103
7	Discipline and Pay	114
8	Tactics	126
9	Life and Death in the Army	146
10	Mansfeld and Morgan	156
11	The Spanish War	174
12	The Île de Rhé Expedition	194
13	The Siege of La Rochelle	220
14	Homecoming	233

Appendices:

I	Regiment 1624–1628	241
II	Pay	265
III	Clothing, 1625	270
IV	Instructions given to Cecil, 1625	276
V	Martial Law	278

Bibliography   285

Introduction

The concept of Britain dates back to Roman times, but it was not until King James VI of Scotland's accession to the throne of England in 1603, that Britain was founded in the modern sense. James bestowed upon himself the title of King of Great Britain, uniting Scotland, England, and the dominions of Wales and Ireland under one crown. Unfortunately for James both England and Scotland hated the idea of a Great Britain, but the first faltering steps had been taken. It was also at this time the 'British flag' was introduced, which flew above the castles and forts throughout the land.

Nevertheless, military historians see the origins of today's British army as being established in 1660 with the restoration of the monarchy. However, during the early part of the seventeenth century there had been three armies known as the 'British army', when regiments from England and Scotland united to form an army. Moreover, in 1627 an Irish regiment was also raised.

Therefore, it can be said that the first British army was raised in Britain between 1625 and 1628 and is remembered for the disastrous campaigns against Spain and France, and has been overshadowed by political events in Parliament and the repercussions of Charles' taxation policies during the early years of his reign. However, despite the lack of money to finance it, the army is one of the best documented of the seventeenth century. It has its similarities with later ones, its recruits being called 'scum', like those referred to in the days of the Duke of Wellington, and pre-dating the First World War, the men were also described as 'Lions'. From the archives we not only know many of the names of these lions, but also some of their former professions and even their ages and descriptions. We also know what clothes they wore, the food they ate and what became of some of them once the army had been disbanded. This gives a detailed insight of an army in the early seventeenth century, and also gives a clear insight of a soldier's lot in the English Civil War as well as the Thirty Years' War.

The second British army was raised in 1631 and was commanded by the Marquess of Hamilton, which is often called an English army by historians, but is referred to by Robert Munro who wrote an account of his regiment's service in Danish and Swedish service, as a British army since it was composed of English and Scottish regiments. It was raised to support the Swedish intervention in the Thirty Years' War, but the majority of the soldiers died of plague within a year.[1] With the outbreak of the Irish Rebellion in 1641 Scottish and English regiments were once more united to put down the rebellion. There was even a tax imposed for the 'British army in Ireland'. Finally with the Restoration the British army that we know today was formed, although it would not be until the 1680s that an Irish regiment was to be found among its establishment.

Unlike Elizabeth I, King James tried to keep Britain out of the Dutch Revolt (1568–1648) and later the Thirty Years' War (1618–1648), which engulfed Europe. Many believed that it was only natural that Britain should join in the fight against the Catholicism, particularly since James' own daughter, Elizabeth, and her husband, Frederick, had been deposed as Elector Palatinate early in 1623 by the Holy Roman Emperor, because he had accepted the Bohemian throne. The English might have hated James for being Scottish, but they were not going to let a member of the British royal family be deposed by a foreign power. In 1620 estimates to raise an army to assist Frederick were compiled, but it proved too expensive and so only Sir Horace Vere received a commission to raise a regiment to assist James' son-in-law.

However, James did allow several colonels who had received their commissions from Frederick to discreetly raise their regiments in Britain, as long it did not upset the Spanish, since he was negotiating to marry his son, Henry, and after his death in 1612, Charles, to Maria Anna of Spain. During these negotiations there was to be no anti-Spanish propaganda and James had to promise that he would restrain the English privateers who were harrying Spanish ships. In return the king of Spain promised a dowry of £600,000 and that they would not interfere in Irish affairs.

However, there was fierce public opposition to the match, including a parliamentary motion in November 1621, declaring that Charles should marry a Protestant princess. Despite this, James dismissed this criticism declaring it as part of his royal prerogative and signed the marriage contract. However, the Spanish delayed signing the agreement and so in February 1623, the Duke of Buckingham and 22-year old Charles journeyed to Spain to speed up negotiations, which by now had lasted about nine years. When they arrived in Madrid, they found that Maria Anna had no intention of marrying Charles nor any Protestant prince and that the whole proposal had been an attempt to keep England from intervening in any foreign wars.

Humiliated, Charles and Buckingham returned to England, but public opinion saw it as another example of God's providential care over England, in that war with Spain during the Elizabethan Age had been marked with economic prosperity. However, when negotiations with Spain began, this boost to trade had begun to decline and was seen as the 'vengeance of God', because England had abandoned a just war.[2]

---

1 The list of Scottish Officers in chief, 1632' in Robert Monro, *Monro His Expedition with the Worthy Scots Regiment* (called Mac-Keyes Regiment) levied in August 1626 (London: William Jones, 1637) part I appendix.

2 Cogswell, Thomas, *The Blessed Revolution* (Cambridge: Cambridge University Press, 1989), pp.284, 287.

INTRODUCTION

With the collapse of the Anglo-Spanish marriage, the now elderly James finally bowed to pressure, not only from the public, but also Prince Charles and the Duke of Buckingham and reluctantly agreed to raise an army of 12,000 men to be commanded by the mercenary Count Ernst von Mansfeldt. Mansfeldt's army set sail early in 1625 with orders to reconquer the Palatinate for James' son-in-law, Frederick. Parliament granted James subsidies to raise this army, but they fell far short of what was required.

Early in 1625 James died and was succeeded by Charles, who was determined to avenge the snub he had received from the Spanish monarchy. He decided to raise another army for a war on Spain. The war should have been popular with the public, but Charles was not a diplomat like his father and alienated his first parliament, who voted him just two subsidies of £56,000, before he dissolved it. However, over £626,000 was needed for the war with Spain and to fill this financial black hole in Charles' budget he turned to forced loans and other means to raise money. On 4 November 1625, the Privy Council sent out letters for the Justices of the Peace to appoint collectors of a loan from Charles' 'good and loving subjects', which was to be repaid in 18 months. The amount to be loaned was based on the last subsidy, that is, if a man owned £100 of land, then he would loan £100 in money.

However, many saw it as a voluntary loan and so did not have to pay it. On 28 November 1625, Godhelp Cooper of Weybridge in Surrey, who was charged £20, wrote to Sir George More informing him that he could not pay this amount, but 'it is (under favour) as I conceive'. However, Charles saw any non-payment of loans as an insult to his royal person. True, Charles did later modify the loan so that the gentry could lend him what they could afford, but as the Earl of Clarendon wrote in his *History of the Great Rebellion* that:

King Charles I (Author's Collection)

> Very many gentlemen of prime quality… were, for refusing to pay the same [loan] committed to prison with very great rigour and extraordinary circumstances and could it be imagined that these men would meet again in a free convention of Parliament without a sharp and severe expostulation and inquisition into their own right, and the power that had imposed upon that right.

THE FIRST BRITISH ARMY 1624-1628 (REVISED EDITION)

Even George Abbott, Charles' first Archbishop of Canterbury, preached against these loans. Charles has been criticised for not disbanding the army after the raid on Cadiz in 1625. Unfortunately for Charles he could not afford to keep the army, neither could he afford to disband it because he would have to pay the soldiers their arrears of pay.

By the time of his coronation on 2 February 1626 Charles had already dissolved one Parliament and England was braced for a possible Spanish invasion. When no invasion materialised, Charles, or rather Buckingham, decided to declare war on France. He increased the size of the army by raising an Irish and a Scottish regiment. This was far from popular, and a Mr Knightly argued in Parliament on 8 April 1628, that 'Shall we trust those Irish and redshanks with the king's service? They are [only] fit to die in ditches'.[3]

The Duke of Buckingham (Author's Collection).

The invasion of France was also a disaster and like in 1940 the British army had to be evacuated from the beaches of France. But in 1627 there was no Dunkirk spirit in England. Criticism of Charles and especially the Duke of Buckingham grew, 'Since England was England', Denzil Hollis wrote, 'it received not so much dishonourable a blow'.[4] Charles and Buckingham tried to relieve Rochelle again in 1628, but with the assassination of the Duke and the surrender of La Rochelle, any further attempt on France was abandoned. In 1628 Charles was finally able to disband the army, so ending the history of the first British Army. He had called three parliaments in as many years and it would not be until 1640 that he would call another one, when war loomed with Scotland.

Although at the time Buckingham was strongly criticised for his actions between 1625 and 1628, it was not until the relaxing of the censorship laws after the First Civil War that some officers of these campaigns still felt the need to publish news sheets openly attacking Buckingham and Charles over their foreign policies.

---

3   BL Harl ms 4,771 Diary of the proceedings of Parliament, 17 Mar to 27 May 1627 f.69.
4   Denzil Hollis to Sir Thomas Wentworth 19 November 1627 printed in Knowler, William (ed) Letters and Dispatches of Thomas Earl of Strafforde with an essay towards his life by Sir George Radcliffe (London, 1739) vol. 1 p.42.

Foreword to the 2nd editon

I have been researching early seventeenth century warfare for over 30 years now and so having the opportunity to write a second edition of this book not only gives me the chance to incorporate additional information I have found since 2016 when the first edition was published, but also to introduce colour plates. Although this book relates to the British Army of 1624–1628, in many ways it the can be used as a guide to those armies raised during the Civil Wars. By studying a notebook of the Parliamentarian officer, Jeremy Baines, which is preserved in the British Library, he took notes from Francis Markham's *Five Decades of Epistles of Warre,* which were published in London in 1622. Baines' notebook is also a surprisingly modern book, with phrases that would be recognised well into the twentieth century, like 'Halt! Who goes there?' when being approached by a stranger. In writing this second edition I have extracted several of these passages from this notebook to give the reader a better understanding of the organisation of a regiment of the early to mid-seventeenth century.

Moreover, Cecil and Buckingham are known to have used the Dutch method of drawing up a regiment in the field, which was also used by both sides during the Civil Wars. When it comes to hand-to-hand fighting most military manuals do not describe the mechanics of fighting a melee, although Johan Jacobi von Wallhausen does include several plates of soldiers in combat in his books on both cavalry and infantry tactics. However, this is the exception rather than the rule because most manuals describe the motions of how a captain should draw up his company and teach his soldiers how to shoulder his pike or musket et cetera. Nothing is written about what to do when it came 'to push of pike', or musketeers 'clubbing their muskets'. Nor do they describe how a commanding officer was to feed and clothe the soldiers under an his command. Using accounts of the campaigns in Spain and France therefore, I hope to rectify this.

In writing this book I have for the most part used archival sources and contemporary printed books which I hope will give the reader a vivid account of a seventeenth century army. To help the reader I have used the modern spelling of words, so 'laughful money' becomes 'lawful money', but I have kept some words the same, such as 'causeth', to give a flavour of the language at the time.

THE FIRST BRITISH ARMY 1624-1628 (REVISED EDITION)

March with your rest in your hand

Acknowledgements

In writing this book I have for the most part used archival sources and contemporary printed books which I hope will give the reader a vivid account of a seventeenth century army. To help the reader I have used the modern spelling of words, so 'laughful money' becomes 'lawful money', but I have kept some words the same, such as 'causeth', to give a flavour of the language at the time.

I would like to thank the staff of the National Archives both in England and Scotland, the British Library and the various county records offices, especially those of Essex, Kent, Hampshire, the Isle of Wight, Surrey, Lancashire, Wiltshire and the House of Lords without whose help this book would not have been possible.

THE FIRST BRITISH ARMY 1624-1628 (REVISED EDITION)

March, and with your musket carry your rest

1

Officers and Lesser Officers

In theory a regiment mustered 10 companies, with a colonel, lieutenant-colonel, major and seven captains, 10 lieutenants and 10 ensigns, the latter carried the company colours and no matter how weak a regiment or company was it always appears to have had its full complement of officers and lesser officers. Fortunately for Charles he did not have to worry about any cavalry regiments, because Buckingham took just one or two troops with him to the Île de Rhé. As in other armies at this time the officer corps was made up of the nobility and the gentry who not only saw it as their right to receive a commission, if they decided to pursue a career in the army, even though they had little or no experience in warfare.

According to tradition when a monarch was not present in person then the aristocracy would command the army or be given the highest positions within that army. However, Lawrence Stone has argued that there was a steady decline in the numbers of aristocracy becoming soldiers, with three quarters of the peers seeing service up to 1540, but by the time of Charles I this had declined to a fifth, although Charles had increased the number of noble families. On the other hand, Roger Manning has argued the opposite that there was a steady increase in the numbers of peers serving in the army throughout the seventeenth century, with 59 out of 103 English peers and 47 out of 85 Scottish peers seeing military service in 1625 alone. The Irish had 43 out of 63 peers serving at this time, which was the largest number of peers from the three kingdoms. As one would expect this rise in military peers would reach its peak during the English Civil Wars, only declining in Scotland after 1660 and in England and Ireland in 1685.[1]

Among this aristocracy was Robert Devereux (3rd Earl of Essex) who in 1619 received a commission as a colonel in the Bohemian service even though he had not seen any military action, although since his father, the 2nd Earl, had commanded an army in Ireland it was believed he would have inherited some of his father's martial qualities, or what Thomas Hobbes describes as a 'foolish superstition'.

---

1   Lawrence Stone, *Crisis in the Aristocracy* (Oxford: Oxford University Press, 1965) p.133; Roger B Manning, Swordsmen (Oxford, Oxford University Press, 2003) p.18. There were also three times as many peers in 1700 as there had been in 1600.

In June 1624 when four regiments of foot were to be raised, John Chamberlain wrote to Sir Dudley Carleton on 6 June about who was to command these regiments:

> The prime competitors are the Earls of Oxford, Essex and Southampton. The fourth place rests between the Lord Willoughby, the Earl of Morton, a Scottish man, and Sir John Borlase. It hath seldom been seen that men of that rank and privy councillors should hunt after such mean places; in respect of the countenance our ancient nobility was wont to carry. But, it is answered, they do it to raise the companies of voluntaries by their credit, which I doubt will hardly stretch to furnish 6,000 men, without pressing; for our people apprehend too much the hardship and misery of soldiers in these times.[2]

Chamberlain was correct in his assumption that Oxford, Essex and Southampton were appointed colonels. However, when Douglas, Earl of Morton was given command of the fourth regiment the other three earls objected, because he was a Scottish earl, 'so easily is the rivalry between these nations revived', recorded the Venetian ambassador. Therefore, Robert Lord Willoughby was appointed to the command of the fourth regiment, although Morton's honour was satisfied inasmuch as he had been offered the colonelcy. He would again be offered the command of a Scottish regiment in 1627.

However, the English commanders did not always get their own way, because later in 1624 the Scotsman, James Viscount Doncaster, received the command of a regiment in Mansfeldt's army. His Lieutenant-Colonel, James Ramsey, Major, Alexander Hamilton, and at least two of his captains, John and William Douglas, were also Scottish although the regiment itself was raised mostly in the London area. Sir Henry Bruce was also a Scot who commanded a regiment during the Cadiz Expedition which was raised in Yorkshire, Middlesex, Shropshire, Cornwall and Surrey.[3] It was this willingness for the Scots to command regiments of a different nationality rather than their own which has given rise to an exaggerated number of Scottish regiments serving on the continent at this time. Robert Monro in his regimental history of Sir Donald Mackay's Regiment lists 26 Scottish colonels serving with the Swedish army in 1632, but only 12 commanded Scottish regiments. One regiment, Sir John Hepburn's Green Regiment, is sometimes referred to as a Scottish regiment, but it was raised in 1627 from recruits from Brandenburg rather than Scotland.[4]

Moreover, the rivalry between the English and Scottish aristocracy did not end with Willoughby's appointment, because the four English colonels then demanded to know who had seniority between them and that they should be superior to Sir Horace Vere and Sir Edward Cecil who were lower in social status, despite their years of military experience and both having the rank of general.

---

2 Folkestone Williams, Robert *The Court and Times of James the First* (Henry Colburn, 1848) p.458.
3 Hampshire Record Office: 44M69/G5/37/29.
4 The origins of the other regiments were eight Dutch, two English, two Finnish and two Swedish.

OFFICERS AND LESSER OFFICERS

In the end after much discussion, due to his age, the Earl of Southampton was appointed as the senior colonel, although the four regiments would come under the command of the General Sir Horace Vere. Even so they were still to be treated as their title required. However, within a year Southampton had died of 'lethargy' and Oxford of wounds he received in battle.[5]

Essex would return to England to command a regiment in the Cadiz Expedition before returning to the continent once more, but such was the jealousy of an aristocrat's position that when Sir Charles Morgan was appointed commander of all the English regiments in Danish service, Essex resigned his commission in December 1626. The Venetian ambassador saw Essex's reaction as natural after the insult he had received to his dignity.[6]

Therefore, it must have come as a great surprise in December 1624 when Count Ernst von Mansfeldt, an illegitimate son of a minor German nobleman with a reputation of allowing his soldiers to plunder the lands of friend and foe alike, was chosen to command the army which was to retake the Palatinate for James I's son in law, Frederick V, (the Winter King). Mansfeldt was good at persuading prospective employers to grant him money to raise an army for them, only to fall foul of them through his actions. Certainly, the Dutch had vowed to have nothing more to do with him, earlier in the 1620s.

Count Ernest von Mansfeldt (from Philippson's *Geschichte des Dreissigjahrigen Krieg*)

Although Mansfeldt's army practically ceased to exist due to sickness it had long since been abandoned by Charles, who wanted revenge on Spain. In 1625 Charles needed a new commander for the expedition against Spain, but Mansfeldt was still soldiering in Europe and both Sir Horace Vere, and the Earl of Oxford were considered too important in their current positions in the Low Counties to be recalled. Therefore, according to his social standing and having the rank of Lord High Admiral, the Duke of Buckingham was chosen for this command, despite having no military experience. However, at the

---

5   Hinds, Allen B (ed) CSP *Venetian 1623–1625* Alvise Valeresso to the Doge and Senate, 7 June 1624, (HMSO, 1914–1919) p.333.
6   Manning, Roger B 'Styles of Command in the 17th Century English Armies' in *The Journal of Military History*, Vol. 71, No. 3, July 2007 (Lexington, Va, Society for Military History, 2007) p.679. CSP *Venetian* vol. 20 p.50.

last moment Buckingham claimed ill health and then that he was needed for a diplomatic mission to the Low Countries, which led to accusations of cowardice on Buckingham's part. Therefore, the command fell to the Duke's deputy, Sir Edward Cecil, 'which much disgusted the mariners', who had wanted Sir Robert Mansel, 'an experienced sea commander' to take Buckingham's place. However Cecil's commission stated that he would have to be advised by his council of war on any movements he might make, which greatly hindered the campaign and it was a total disaster.[7]

Cecil was not the only army officer to be given the command of the navy. In 1625 the Earl of Denbigh was appointed rear admiral even though he had not commanded a single ship. Despite this total lack of experience, he was promoted to full admiral in 1627, a position that he was totally unsuitable for and the following year he would bring shame upon the Royal Navy when he ordered the fleet to sail back to England without trying to break the French blockade of La Rochelle.[8]

When Charles wanted to launch a new expedition in 1627, this time against France, Cecil, now Lord Wimbledon, was in disgrace after the Cadiz Expedition, so the Duke of Buckingham, finally decided to take command himself. Buckingham had all the wealth and political power he needed, but he was hated by most members of parliament who were trying to impeach him and he still had accusations of cowardice hanging over him. Therefore, as politicians have found, a military victory abroad could turn their fortunes around at home, so what better way could Buckingham choose but to be known as the saviour of Protestantism in France? Therefore, it was essential for him not only to lead the army but win the forthcoming war with France.

To make up for his lack of experience Buckingham took several advisers with him, but even the members of his council of war were severely criticised, not because of their lack of military ability, but because of their social standing or faith, one of whom was the 18 year old illegitimate son of the Earl of Suffolk and several were described as 'violent papists' and two had been saved from the gallows.[9]

Once the senior commands of an army and its regiments had been filled by the nobility then any colonelcies and lesser ranks would then be filled by the gentry. In his *Directions of Warre*, Mansfeldt stated that the colonel and captains could choose their own officers and when Sir Donald MacKay was commissioned to raise a regiment in Scotland on 4 March 1626 he was allowed 'free choice and disposition of the officers superior and inferior in the regiment'.[10]

However, in 1624 James wanted to keep a tight control on the four regiments then being raised, which caused the Venetian ambassador, Alvise Valaresso, to write that the colonels will find, 'Their powers very limited, as

---

7   Rushworth, John *Historical Collections* of Private Passages of State (1659).
8   For Apsley and Denbigh's careers see the *Dictionary of National Biography*.
9   British Library Add Ms 26,051 Journal of the Voyage of Rease [Rhè] 1 May to 7 November 1627 f.16.
10 Mansfeldt, Count Ernest, *Directions of Warre*, (1624); Sir Donald Mackay's Commission printed in John Mackay, *An Old Scots Brigade, being the History pf Mackay's Regiment* (Tonbridge: Pallas Armata, 1991) pp.211–222.

the king desires to nominate the captains himself, which seems something new. They number thirty six, selected out of over ninety, all well qualified, such is the number of persons and their love of soldiers'.[11]

Raising of a new regiment or army could bring swift promotion to an officer and on 6 June 1624 John Chamberlain wrote to Sir Dudley Carleton:

> Here is much canvassing about the making of captains and colonels for this new force that are to be raised to assist the Low Countries. Sunday last was appointed, and then put off till Tuesday; when they, flocking to Theobalds with great expectation, the king would not vouchsafe to see any of them, nor once look out of his chamber till they were gone. But word was sent they should know his pleasure 'twixt this and Sunday.[12]

Among those who were promoted in 1625 was Sir Thomas Fryer, who had been a lieutenant in the Dutch service, but now was promoted to major, while Lieutenant Colonel Sir Alexander Brett had been an ensign.[13]

There are various letters from Charles, who was usually influenced by Buckingham and, to a lesser extend Secretary Conway – to army commanders suggesting or recommending officers, for example on 23 November 1626 Conway wrote to Sir Charles Conway 'signifying His Majesty's pleasure for Mr Fleetwood to have a company' under Morgan's command. For Catholic officers they might seek the patronage of the Earl of Argyll, the Count of Gondomar or Archduchess Isabella since James had given Spain permission to recruit in England. When he came to the throne in 1625, Charles I did try and put a stop to this by recalling these officers, but few appear to have obeyed.[14]

Sometimes royal patronage could be ignored if a regiment or army were overseas, for example when Charles wrote to the Lord Deputy of Ireland on 17 June 1628 he 'requested' that Sir John Clotworthy be given the next available company or troop, although he would be passed over several times before he finally received a company, but not before Clotworthy had complained to the king.[15]

However, Charles and Buckingham did not always get their own way, because when Colonels Courtney, Bartie, Ramsey, Conway, Spry, Fryer and Morton raised their regiments in 1627 many of their company commanders had commanded companies in their regiments in the Dutch army in 1624. On the other hand, none of the officers in Sir Charles Rich's Regiment in Dutch service in 1624 appear to have been commissioned into his new regiment.

---

11 Hinds, CSP *Venetian 1623–1625* Alvise Valeresso to the Doge and Senate, 21 June 1624, p.353.
12 Folkestone Williams, *The Court and Times of James the First* (Henry Colburn, 1848) p.458.
13 HMC Earl of Cowper of Melbourne Hall. p.429.
14 Lyle (ed). *Acts of the Privy Council 1621–1623* Warrants to the officers of Lord Vaux's Regiment, 10 April 1622, p.191. Loomie, Albert J 'Gondomar's Selection of English Officers' *English Historical Review* vol.88 pp.574–584. Vaux's Regiment was taken over by Colonel Sir Edward Parham. TNA SP 104/167. Chambers arrived at Hampton Court on 9 December 1625.
15 CSP Ireland 1625–1632, The King to the Lord Deputy for Sir Hugh Clotworth, 17 July 1628, p.366.

Given the need to raise money it is surprising that Charles did not introduce the purchasing of commissions, a method used in the Dutch army at this time, despite Maurice of Orange trying to prohibit the practice in favour of experience.[16] However the English regiments at least appear to have been able to appoint their own officers, at least up to about 1635. Colonels could name their successors, although there seems to have been some financial benefit in appointing an officer within a regiment. In 1622 Sir John Ogle reckoned that a major's rank could be purchased for £300 and a lieutenant colonelcy for £600. A year later when Sir Robert Sydney (Lord Lisle) decided to retire from military service he had two offers for his regiment, one from Sir Charles Rich who offered him £2,000 plus £300 a year for life and Sir Edward Harwood who was the regiment's lieutenant colonel but could only afford £1,500. Nevertheless, Harwood was Sydney's choice, and he was given the colonelcy of the regiment, which he would hold until his death in August 1632.[17]

When Frederick Henry succeeded his elder brother Maurice as the commander of the Dutch army in 1625, he tried to close this loophole in the English regiments by having a colonel put forward the names of five or six officers to fill the vacancy, so that he could choose the best candidate, which was rejected by its officers.

Even though he was commanding a regiment in the Dutch service at this time, Harwood also commanded a regiment during the Cadiz Expedition; but despite leaving his belongings to his family, as well as the British royal family and his fellow officers, he did not say what was to become of his regiment, although his lieutenant colonel, Sir Henry Herbert, who was also the executor of his will, did take over the regiment, but not until December 1632.

Patronage within a family or county could also set a young nobleman on a career as an officer. On 23 June 1627 Sir Ralph Hopton wrote to Lord Conway on behalf of a Lieutenant Hall, 'a near kinsman and one that I have very good proof of both for his honesty and ability'. However, no officer by the name of Hall appears in the list of officers in the regiments raised for the Île de Rhé Expedition and some officers did not necessarily command companies from their own county. Captain Peter Hone commanded a company raised in Middlesex in 1625 and from Warwickshire in 1627.[18] Unfortunately, too many officers who were appointed by patronage had very little or no military experience, while other more capable officers were passed over for promotion. It was these gentlemen appointed by patronage that Cecil complained about in September 1625, when he wrote:

---

16   It would not be until 1683 that the purchasing of a commission would be introduced into the British Army.
17   Dalton *Life and Times of General Sir Edward Cecil* (London, Sampson Low, Marston, Searle and Rivington, 1885) vol. 2 p.25; The National Archives SP 84/101 Sir John Ogle to Sir Dudley Carleton, 4 May 1622, ff.165–166; TNA SP 84/111, Lord Lisle to Sir D. Carleton 31 January 1623, ff.55–56.
18   TNA SP 16/67/99, Sir Ralph Hopton to Secretary Conway, 26 June 1627.

> The captains I have found here and that your Lordship hath again recommended, I could take exceptions, because I know many of them have not been soldiers and although they have taken pains to eat well and lie well, yet I fear I shall see but little from them… I brought down with me, choice men, but those that were so recommended to me, I neither could nor durst return.[19]

In 1628, Sir Charles Morgan endorsed Cecil's sentiments when he complained that most of his officers are from, 'Gray's Inn, Lincoln's Inn or Middle Temple, where they have learned to play the mauvais garcon that they can hardly be made fit to know what belongs to command, but in time I hope to bring them to better experience or else I will show them the way to break their necks'.

He also demanded the names of his absentee officers and NCOs should be, 'nailed to the gallows, being themselves officers and partly the cause of the running away of the rest. In all the thirty seven years I have been in service I was never troubled with such a confusion as these four regiments have put me to'.[20]

In his petition, Lieutenant Thomas Dymoke of Bingley's Regiment complained about his company commander during the Île de Rhé Expedition, whose 'simplicity and having never seen the wars and continual sickness make him incapable of doing [any] service till the day of the assault'. Unfortunately for Dymoke's captain he had recovered enough to be killed in the retreat of Buckingham's Army.[21]

Once the vacancies within a regiment or army had been filled by patronage, then other officers could apply for the remaining positions. These were usually those gentlemen without any fortune, but who had a great deal of military experience. One of the officers who lost out because of patronage was Lieutenant John Felton, who after his captain died petitioned to take over the company, but Buckingham chose another officer instead. While in 1627 Captain Sitthorp who had commanded a company during the Cadiz Expedition, complained that 'other people were promoted over his head'. However, it was not just in England that this happened. Lieutenant Roger Oram who was 'a man of good reputation in these wars having at his father's death quit his lieutenant's place because the company was given away over his head'. He retired from the Dutch service in protest. Despite writing several military manuals Henry Hexham spent 42 years in the Dutch army and only reached the rank of captain.[22]

Hexham's extremely slow climb up the ranks was by no means unique. In 1629 Edward Yates had spent 40 years in the army, only becoming a major in 1628, while Captain Peter Hone's ensign petitioned on his behalf in 1626 to become the major of Essex's Regiment. Hone had joined the army at the age of 15 and had already served 50 years, although his own petition says 42

---

19   TNA SP16/6/36 Cecil to Conway, 8 September 1625.
20   Beller 'The Military Expedition of Sir Charles Morgan' p.530.
21   TNA SP 16/399/22 Statement of Thomas Dymoke, 26 September 1638.
22   TNA SP 63/42 f.803. He is possible the same Roger Oram who served in Colonel Herbert Morley's Regiment of Foot during the English Civil War. Nimwegen, Olaf van *The Dutch Army and the Military Revolutions, 1588–1688* (Woodbridge: Boydell Press, 2010) p.22.

years. He was present at the capture of Cadiz in 1596 and then in Ireland, before serving with Mansfeldt in 1624 and then Cecil the following year. Unfortunately for Hone, Essex's Regiment was disbanded and in 1627 he was still a captain, this time under Sir John Radcliffe. However, in 1628 he was the lieutenant colonel of Knyphausen's Regiment which was one of the regiments assigned to relieve Rochelle.[23]

Apart from Mansfeldt, Baron Dodo von Knyphausen appears to have been the only other foreign officer in the army at this time. Born in 1583 at Lutetsburg in East Frisia he had served in the Dutch army before commanding a regiment under Mansfeldt during the early 1620s and was captured at Dessau Bridge in 1626. Why he was commissioned to command a regiment in the army of the Earl of Lyndsey, who had succeeded Buckingham after his death, after his release is unknown, but the Venetian ambassador, Alvise Contarini, wrote that, 'reports varies about his abilities'. Nevertheless, he later became a field marshal in Swedish service and was killed in 1636.[24]

Whether appointed by patronage or merit it was important that an officer did not lose the respect of his fellow officers or men, because as Robert Munro points out, 'When officers and soldiers conceive an evil opinion of their leaders, no eloquence is able to make them think well of them thereafter… [and they] are despised by their followers'. A song written between 1625 and 1628 refers to Buckingham's army stating that, these officers were brave in the alehouse who 'can fight like soldiers tall' but were cowardly on the battlefield and 'lie drunk in the mire where our foe shall fall'. The song continues about the soldier having no pay and broken promises then finally runs away, although hoping that Charles will give them better song for them to sing.[25]

Once the officers had been appointed, they would be paraded in front of the regiment to receive their commissions so that all would know their rank, but what made these gentlemen take up a career in the army, while others decided just to serve as a volunteer as part of their education? In theory to be classed as a gentleman one had to live off their income he received from his tenants who rented land on his estate. However, for many this was not always possible and so they had to seek additional income. One of the few trades that was seen as honourable for a gentleman was soldiering. It is obvious from his autobiography that Thomas Raymond enjoyed the finer things in life, but as he saw it there was 'no other way to make a fortune, being a younger brother… I buckled myself to the profession'. He enlisted in the Dutch service and would rise to the rank of colonel, but whether he ever made the fortune he desired is unknown.[26]

When his father died, Sir Walter Waller received an annuity of just £140, while the majority of the estate went to his elder brother, Sir William, the

---

23 SP 16/136/47 Petition of Major Edward Yates, 19 February 1629; SP 16/34/83 Petition of Captain Peter Hone, August 1626.
24 CSP Venetian vol. 21 Alvise Contarini, to the Doge and Senate, 7 August 1628 pp.215–216.
25 Robert Monro, *Monro His Expedition with the Worthy Scots Regiment...* (London: 1637), part II p.174; Historical Manuscripts Commission 12th Report Appendix 11 pp.547–548. Although the song is undated it refers to a soldier returning from Mansfeldt's army and the 'new king' so therefore must have been written between 1625 and 1628.
26 Thomas Raymond, *Autobiography of Thomas Raymond* (Royal Historical Society, 1917) p.43.

OFFICERS AND LESSER OFFICERS

future parliamentary general, who had served as a volunteer in the Bohemian Army in 1620. This may have been the reason why he decided to take up soldiering and become a captain in Sir Charles Rich's Regiment of Foot in Mansfeldt's Expedition and later in the Cadiz Expedition. Unfortunately, his military career was cut short, because he died in Ireland from the disease which swept through Cecil's army.[27]

There were many younger sons to be found in the officer corps at this time who were unlikely to inherit their father's estate, like Sir Edward Cecil who was the third son of Sir Thomas Cecil and Sir Horace Vere who was the youngest of four sons born to Geoffrey Vere. However, there were exceptions, Colonels Sir Edward Harwood, Sir John Borlase and Robert Bertie were all eldest sons, so it is unclear why they decided to assume a military career. Although, the eldest son of Patrick Turner, whose father wanted him to become a minister, it appears to have been boredom that made the 17 year old James Turner join the army. Hearing of the exploits of Gustav II Adolph and that Sir James Lumsden was raising a regiment in Scotland for the Swedish service, he records, 'a restless desire entered my mind, to be, if not an actor, at least a spectator of these wars which at that time made so much noise over the world'. Fortunately for Turner he became an ensign in the regiment.

According to the Earl of Clarendon it was the plight of James I's daughter, Elizabeth, in her battle to regain her husband's throne, that persuaded Ralph Hopton, who was the only surviving son of the wealthy landowner Robert Hopton, took up arms. He 'had always born an avowed and declared reverence to the Queen of Bohemia and her children, whom he personally and actively served in their wars… and for whose honour and restitution he had been a zealous and known champion'. Elizabeth also appears to be the reason why Robert Monro took up arms. In his account of the campaigns in Germany, he called her the 'jewell of her sex and the most resplendent in brightness of mind', but despite Hopton's and Monro's, and many others' efforts the 'Winter Queen' would die in exile.[28]

Many also probably recognised the sentiments of the German Heinrich von Uchteritz when he wrote:

> Ever since childhood I… have had the urge to try something noble, according to my station in life, and thereby to maintain and increase the praise of my ancestors and kinsmen had earned through knightly virtue. For this reason, I chose war as the most fitting means for my position and when the opportunity presented itself, I had in mind to seek my fortune in it.[29]

---

27 Adair, John, *Roundhead General* (Kineton: Roundwood Press, 1973) p.30.
28 Dictionary of National Biography. Clarendon quoted in C. E. H. Chadwyck-Healey *Sir Ralph Hopton's Narrative* p.xviii; *Monro His Expedition with the Worthy Scots Regiment* p.2; T. Thompson, (ed) Sir James Turner, *Memoirs of his own life and Times* (Edinburgh: 1829), pp.3–4.
29 Alexander Gunkel and Jerome S. Handler (eds) 'A German indentured servant in Barbadoes in 1652' in *The Journal of the B.M.H.S.* vol.33 (1970), pp.91–92.

Uchteritz would join the Royalist Army during the Third Civil War and would be captured at the battle of Worcester (3 September 1651) and was transported to Barbados along with other Royalist prisoners.

Since the first biography in the seventeenth century, the biographers of George Monck have described him as a 16 year old volunteer in the Cadiz Expedition. However, he is not listed among the officers and since the muster rolls for Devon have not survived we cannot verify whether he was a 'lesser officer', as non-commissioned officers were known at this time, a private soldier or was there at all. What we do know is that in September 1626 Monck killed the undersheriff of Devon when he tried to arrest his father for debt and so had to flee England. Fortunately, Sir Richard Grenville was able to secure a commission as an ensign for him in Colonel John Burrough's Regiment, although Monck would have to travel through France to join the regiment on the Île de Rhè. After serving in the expedition and being unable to return to England, he soldiered in the Low Countries, finally returning to England in 1640 to take part in the Bishops' Wars. He fought in the Civil Wars on the Royalist side before changing sides during the Commonwealth and was instrumental in bringing about the Restoration in 1660.[30]

Monck was not the only officer to flee justice, Robert Munro of Foulis left his pregnant English wife, Mary Haynes, taking her fortune of £600 with him to pay his debts. In 1625 after seven years living in poverty, she heard that he had an estate of £900 per annum and had married Marjorie McIntosh, a Scottish or 'northland woman'. It was only after Mary Haynes petitioned the king. For various crimes including the 'filthy crime of adultery', that he was ordered to appear at the tollbooth in Edinburgh. When he failed to appear, an order was issued for his arrest but in 1626 to avoid justice he obtained a captaincy in Sir Donald Mackay's Regiment. He later became a colonel in the Swedish service and is believed to have died at Ulm in 1633 after a wound in his foot became infected. It is believed that he never returned to Scotland and what became of his two widows is not known.[31]

For those who decided to take up arms there were no military academies to teach them how to be an officer. In 1639 Robert Ward complained that gentlemen are:

> Admitted for their wealth's sake into captainships, which neither have courage, skill nor delight in Arms themselves, nor discretion to command others… How can we hope to make our people soldiers, where they be under blind and ignorant guides, *If the blind lead the blind*. Had we but all these abuses corrected, me thinks we might make our English trained bands, parallel [to] the best soldiers in Europe.[32]

---

30 *Dictionary of National Biography* and Maurice Ashley, *George Monck* (London: Jonathan Cape, 1977).
31 Hume Browne, F (ed) Register of the Privy Council of Scotland 2nd Series (Edinburgh, 1901), vol. I pp.303–304, 377–378, 660–662; John Mackay, *An Old Scots Brigade being the history of Mackay's Regiment* p.206.
32 Robert Ward *Animadversions of Warre*, pp.30–32 His italics.

OFFICERS AND LESSER OFFICERS

Although he was criticising the officers of the trained bands when he wrote this passage the same could be said of those in the army as well. One of the ways a gentleman could learn how to be an officer was to serve in the ranks as a 'gentlemen of a company'. According to Henry Hexham, gentleman volunteers were to be found mainly in a general's or colonel's company. Hexham states that:

> A gentleman's duty at his first entrance into the profession of a soldier… is to stand sentinel for a month to learn the first degree of a soldier, that he may be the better able to command others when he is advanced… A gentleman also is to go round with the captain of the watch or his fellow gentlemen either in the field or in garrison to give the corporal of the guard the [pass]word and to charge the sentinels to look well about them.[33]

Rather than following an apprenticeship as a gentleman volunteer, an officer might purchase one of the many military books that were published at this time, like John Bingham's *The Tacticks of Aelian or the Art of Embattailing* which was published in 1616 and described how soldiers were to be equipped and drawn up in company formations as well as arms movements. Unfortunately, these publications usually fell short of telling the prospective officer how to look after their men and how to conduct a regiment in battle.

These books also described the virtues of each rank. According to Francis Markham, a captain should be a 'gentleman both of blood and quality', while a colonel was to be 'a man of wonderful experience and knowledge in wars'. On the other hand, a general should have all 'virtues' and shun all 'vices',[34] although in practice this advice was probably ignored.

If there were no vacancies for an officer's position then he might become a reformadoe, that is, an officer without a command, which also meant that he would lose the financial benefits a company or troop had to offer, but on the other hand as an officer he could enjoy the status of his rank, without the outlay of money necessary to subsidise a company. Alternatively, he might have to drop a rank or two, for example the lieutenant colonel (Ferrar) and major (Donne) of Sir William Courtney in the Dutch service in 1624 served as his major and a captain in his regiment in 1627, because these two positions had been filled through patronage. On the other hand, if a captain were a knight, then he could also claim to have a higher status than a plain captain, inasmuch he was allowed to have three servants instead on two.[35]

For officers from the lower classes a too speedy a rise up the promotion ladder could bring ridicule, as a Captain Weddell found to his cost during the Île de Rhé campaign. Weddell was promoted to the rank of colonel of a regiment of 500 mariners, but the journal of the campaign records that 'his honour was so small continuance as it only served to make him ridiculous'.

---

33 Henry Hexham, *The First part of the principles of the Art military* (1642) p.3.
34 Francis Markham, *Fives Decades of Epistles of Warre* (London, 1622) pp.134, 163, 198.
35 TNA SP 84/121 Memorandum on the reduction of 11 regiments with particulars, nd 1624, f.314; SP 63/246/1029 Instruction for the officers at musters, 1628.

Not even the support of Buckingham could save him and his regiment, which had also been ridiculed by the soldiers was disbanded.[36]

It is usually said that it was only the New Model Army which was raised in 1645 introduced the concept of a commoner becoming an officer. This is not so, among those from the lower orders of society to make the leap from common soldier to officer, was Sydenham Poyntz. He ran away to join the army and rose through the ranks to become a captain in Wallenstein's Army and eventually commanded an army for the Parliamentary cause during the English Civil War. William Ware had been an 'impressed' soldier but 'by his endeavours obtained an ensign place'. However, this was after 'above 20 years' service, but it would be wrong to suggest that all the gentry were fast tracked through the officer corps, Edward Lord Conway's brother had 'been twenty years a captain', when he asked Frederick V if he would issue a major or lieutenant colonel's commission to him.[37]

What did the soldiery think of the promotions of the lower classes? Unfortunately, no letters survive recording their feelings, but the novelist Hans von Grimmelshausen records in his *The Adventurous Simplicissimus* that when the hero of the book asked a sergeant why the gentry are promoted rather than the more deserving common soldiers, the sergeant relies,

> It is not true that nobly born officers be better respected by the soldiery than they that beforetime have been but servants. And what discipline in war can ye find where no respect is? Must not a general trust a gentleman more than a peasant lad that had run away from his father at the plough-tail and so done his parents no good service? For a proper gentleman rather than bring reproach upon his family by treason or desertion or the like, will sooner die with honour. And so tis right the gentles should have the first place.[38]

Certainly, an officer had to be wealthy enough to clothe, feed and pay his company or regiment if necessary, since the government rarely paid its soldiers on time. In a petition to Lord Conway, dated 8 May 1626, several officers, including Conway's own son, wrote, that since they had landed in Ireland 'we have lived on our private means, we are very poor in consequence and must have relief'. Whereas Sir Henry Spry petitioned the Council saying that he had lost £3,000 in the King's service financing his regiment.

Lieutenant Francis Musgrave had to pawn his 'poor estate' so that he could serve in the Cadiz and Île de Rhé Expeditions, where at the assault

---

36 BL Harl ms 6807, 'Miscellaneous tracts including charges by the Earl of Essex against Sir Edward Cecil and a journal or diary of the most martial passages happening at and after our landing at the Isle de Ree 1627.'

37 See *DNB*, Sydenham Poyntz and A. T. S. Goodrick, *The Relation of Sydnum Poyntz* (Royal Historical Society, 1908); *Privy Council June 1630–June 1631* Petition of William Ware dated 20 February 1631, p.233; TNA SP 16/179/20, 'Petition of William Ware, Ensign to Captain Thomas Broughton, 1630'; TNA SP 63/267 'Edward Lord Conway to Frederick V Elector Palatinate', f.25.

38 Hans von Grimmelshausen, *The Adventurous Simplicissimus* (Lincoln: University of Nebraska, 1962), p.36. Grimmelshausen's role in the Thirty Years' War has been debated since his novels were published. However, most historians agree that due to their contents he must have served in the army.

OFFICERS AND LESSER OFFICERS

on the citadel he was killed while leading the forlorn hope. On the other hand, a Major Watkins of Sir Thomas Morton's Regiment is said to have married into a good estate, but by the time of his death in October 1627 his widow was said to be destitute, presumably he had spent her fortune on his company.[39] Moreover the officers were the last to receive their arrears of pay. On 22 May 1630, Evan Watkins, who had served as the major of Colonel Harwood's Regiment during the Cadiz and the Île de Rhè Expeditions was still owed £205 three shillings and three pence.[40]

One way an officer could supplement their pay was to claim for 'dead pays' which during Elizabethan times seems to have been 10 percent of a company, although by the 1620s a captain could claim for 13 fictitious soldiers, so that according to Lord Conway it 'reduces a company of 100 to 87'. Whereas a captain of horse could claim for 10 'dead pays' for a troop of 100 men.[41]

During the early 1620s the Dutch government began to crack down on these corrupt practices by encouraging the soldiers to denounce their officers, but in October 1623 Francis Wrenham feared that this would, 'encourage the soldiers to accuse their officers in the point of musters'. Certainly in 1620 the soldiers of Colonel Andrew Gray's Regiment deserted in droves because they blamed their officers for embezzling their pay even though the Bohemian government appears to have been at fault. By the time of the English Civil War presenting false musters was likely to get an officer cashiered if he was caught.[42]

On the other hand, officers might receive more than one commission and would leave one company to raise another either in a different service, like Lieutenant John Eaton who commanded in a company in the Dutch and another in the Venetian service. Eaton then returned to England to command a third company and took part in the Cadiz, Île de Rhè and La Rochelle Expeditions. A muster of the forces in Ireland in 1625 found that of the 209 infantry officers, 29 were absent, as well as five of the 30 cavalry officers. These absentee officers would leave a junior officer or a sergeant to command their original company while they were getting paid for commanding both companies and even regiments. Amongst the officers who were guilty of this practice was Sir Edward Cecil himself who had commanded a regiment in the Dutch service since the 2 May 1605 and would continue to command it until his death January 1632, but he also commanded a regiment in the Cadiz Expedition and as well as being the commander of the expedition itself.[43]

---

39 TNA SP 16/120/59 Petition of Isabel Musgrave to the King, 14 November 1628; SP 16/154/4 Petition of Jane Watkins, nd c.1629.
40 TNA SP 63/42 f203 Watkins had received £385 7s 4d.
41 BL Add ms 69907 f.26 Lord Conway to the Lords, nd [c1626], TNA SP 16/521/31 Estimate of the pay by the month and year for 10,000 soldiers, 3 April 1625.
42 HMC 13th Report Appendix II Portland ms p.112; TNA SP 80/3 Carpenter to Secretary of State, 17 November 1620, ff.258–259.
43 CSP Ireland 1625–1632 pp.42, 625. After Cecil's death it was taken over by Philip Pakenham and it would not be until 5 March 1665 when it was finally disbanded, along with the other English regiments in Dutch service when they refused to swear the Oath of Allegiance to the Dutch monarch. However, the soldiers from these regiments would go on to form the 3rd Regiment of Foot, which was also known as the Buffs.

However, not all abandoned their companies, Sir Ralph Hopton, who was a lieutenant colonel under Mansfeldt had refused a commission in the Île de Rhé Expedition and it was not until his company ceased to exist that he returned to England. He declared that, 'the miseries we suffered in the last journey (though I could hazard myself willingly enough) make me afraid to have charge of men where I have any doubt of the means to support them'. Nevertheless, Hopton went on to become a Royalist general during the Civil War and is described by the Earl of Clarendon as a 'man of great honour, integrity and piety of great courage and industry and an excellent officer in any army for any command but the supreme, to which he is not equal'.[44]

An officer's absence was seen as a major cause for ill discipline among the soldiers, which prompted the Privy Council on 12 December 1625 to write to the colonels of each regiment to 'take care that all the officers of companies and captains who have none other commands in the army do remain with their companies except there be some good reason to the contrary'.

Naturally, the Dutch government was not happy to release the English officers serving in its army and refused permission for a Mr Killigrewe and Mr Gibson to join Mansfeldt's army in 1624 and wanted to recall those that had returned to England as soon as possible. Therefore, on 1 June 1626, the Privy Council after receiving a petition from the 'colonels and other officers' reluctantly gave permission for them to return to their other companies.[45] Captain Sheffield Clapham, who was then in the Danish service, even went so far as to petition Secretary Conway on 24 March 1627, 'that he may continue to hold his company in England so that he may have something to return to'. Although with the shortage of money, on 19 January 1627 the Privy Council ordered that the officers and NCOs who were not with their companies would not receive their pay.[46]

However, there was a danger through their absence that they might lose their positions. In 1631 Colonels Sir Thomas Fryer, Robert Ferrar and Philip Hackluit petitioned the Privy Council that they had lost, 'both present and future hopes of preferment' in the Dutch service and 'humbly crave' that the Privy Council 'grant unto each of them the several fees of 13 shillings 4 pence a day as you have already granted to others for the like service'. They may have been among the 15,000 officers and soldiers serving in the Dutch service who were discharged in 1629 because the United Provinces could no longer afford to pay such a large army, although many of them found employment in other armies such was the demand for officers and soldiers. Nevertheless, the officers lost their investment in their company or regiment. In fact, very few officers found wealth, fame or honour before their careers

---

44 TNA SP 16/136/99, Sir Ralph Hopton to Secretary Conway, 23 June 1627; Earl of Clarendon *History of the Great rebellion* vol.3 p.312.
45 Lyle ed. *APC 1625–1626* A letter to Lord Viscount Wimbledon, 12 Dec 1625 p.267; BL Add ms 72422 Trumbull papers of the council of war and the Muster Master General, 1624–1635, f.1.
46 TNA SP 101/46, Holland Newsletters, 1623–1626 f.40; SP 16/5818, Captain Sheffield Clapham to Secretary Conway, 28 March 1627.

were cut short either by death or having to retire due to the effects of wounds, illness or old age. [47]

To support these officers were the lesser or inferior officers, who were composed, in order of seniority, of sergeants, corporals and lance passidores. The latter rank being the seventeenth century equivalent of a lance corporal. Sources suggest that there should be at least three sergeants per company, but some surviving muster rolls at this time show there were just two. According to Francis Markham one of their duties was to teach the soldiers their drill and therefore should be 'valiant, expert, vigilant and diligent' and that a 'good sergeant is an admirable benefit and if he live and execute his place well any long time, no man deserveth advancement before him'.[48] However, some officers were reluctant to promote sergeants especially if he covered up their incompetence, but other sergeants had a meteoric rise. Captains Robert Dish and John Manley had been sergeants in the Low Countries before being commissioned into Cecil's army for the Cadiz Expedition. While others like Degorie Collins who was a sergeant in Sir Henry Killigrewe's Regiment under Mansfeldt in 1624, became an ensign for the Île de Rhé Expedition, being promoted lieutenant in Captain Thomas Gates' company on 13 November 1627, and taking over the company the following year. Thomas Drake was also a sergeant in 1625 and was promoted to the rank of ensign during the Île de Rhè Expedition. In November 1627 after his lieutenant and captain were killed, he received command of the company. Unfortunately, their petitions do not record how long they served as a private soldier, corporal or sergeant nor their social status.[49]

Certainly, the high mortality rate among the officers almost certainly played its part in their rapid promotion, but others were not so lucky, including John Leeds, who was described as a 'poor gentleman and sergeant', and had served in Sir Edward Conway's Regiment in the Cadiz Expedition; but his company was reduced into another and all the officers were discharged, so after 19 months service all he had to show for his pains was 10 shillings. On the other hand, William Hassell, a sergeant in Captain Edward Spring's company, had his military career cut short by being 'disabled' so he was forced to return home from the Île de Rhè campaign.[50]

Francis Markham recommended that a company should be divided into four squadrons, which would give four corporals per company, but Mansfeldt and Henry Hexham suggest only three, which is the number of corporals mentioned on the surviving muster rolls for 1625 and 1627. Each squadron was further divided into 'comrades or fellowships' whose members would look after each other.[51]

---

47 TNA SP 16/205/28, Petition of Colonels Sir Thomas Fryer, Robert Ferrar and Philip Hackluit, 1631; Roger G. Manning, *Apprenticeship in Arms* (Oxford: Oxford University Press 2006) p.49.
48 Markham, *Five Decades of Epistles of Warre,* pp.69, 72.
49 TNA SP 16/38, Petition of Captain Degorie Collins to the Council, f.655 March 1629; SP 16/138/35, Petition of Captain Thomas Drake to the Council, 5 March 1628.
50 TNA SP 16/65/66, Petition of Sergeant John Leeds, May 1627; SP 16/79/74, Pass for Sergeant Hassell to return home September 1627.
51 Henry Hexham, *The First Part of the Principles of the Art Military Practiced in the Warres of the United Netherlands.*

Unfortunately, no petitions for individual corporals appear to have survived, but according to Francis Markham a corporal:

> Ought to be a man carefully chosen out, and induced with valour, virtue, diligence and experience; he ought to be of reverend and grave years, thereby to draw on respect, but withal of a sound judgment; for experience without it is but a jewel in the sea, which neither adorns itself nor others, he is to be a cherisher of virtue and a lover of concord, for he is said to be the father of his squadron and must therefore love them and provide for them as for his natural children.[52]

Markham is the only writer to suggest a corporal be of 'grave years' or elderly, while Mansfeldt devoted almost four pages to the corporal's duties in his *Directions of Warre*, also referring to the corporal as the 'father' of the squadron, because it was his task to receive the men's rations and to see they were bivouacked probably. Hexham also states that the corporal should make sure that the new recruits were not bullied by the veterans and train them in the arms drill and the various positions, although Francis Markham suggests that this was one of the sergeant's duties.

When the company was drawn up for battle the corporal's position was to be on the right flank of his squadron in the first rank and on the left flank would stand the lance passadore. If the corporal was the 'father' of the squadron then the lance passadore was the mother, because he would carry out the corporal's duties in his absence. According to Francis Markham:

> [He] ought to be elected out of the most sufficient of all the meaner rank of persons, and to be a careful, obedient and active spirit of a subtle and wise disposition… and be void of turbulent qualities, lovers of concord and enemies of mutiny and sedition.[53]

Markham continues:

> If they be new levied soldier and have never seen the wars, then there will be some difficulty in the selection, and a man can hardly discharge what he doth not know. But… it is hard if in a number of five and twenty men, who may not be found which have either seen or had some small taste of the wars, yet suppose they have not, it then behoveth the captain to look into their natural inclinations and their aptness, willingness and love unto military exercise and from thence to choose out such as he finds quickest of apprehension, fullest of care, vigilance, valour and observation.
>
>> However, an officer should avoid choosing 'a drunkard, ruffian or profane person'.

Therefore a newly levied soldier with the right talents might find himself quickly on the promotion ladder and if a lance passidore lost his rank for drunkness then he would probably endeavour to regain it because as

---

52 Markham, *Five Decades of Epistles of Warre*, pp.65–66.
53 *Ibid.*, p.62.

Markham continues, there is nothing more 'disgraceful or grievous to a soldier to be pulled back or cast from authority… [so] the very shame of dishonour, will so inflict and torment him, that it will enforce him to endeavour admendment'.[54]

Another rank which appears in a company is the drummer, or trumpeter in the case of a troop of horse, which was the voice of the officers since their orders could not be heard over the din of battle. Mansfeldt specified there should be two drummers in a company, but certainly Lord Cromwell's Regiment had three drummers in the colonel's and lieutenant colonel's companies, although the other companies probably had two.

According to Markham, the drummer was to be, 'Every way fitting for his place, must besides the exquisiteness and skillfulness in his Art and instrument, and the rudiments of marshal discipline, be also a good linguist, and well served in foreign languages: for by the carrying of messages, he must commerce and have to do with people of sundry nations.'[55] Mansfeldt added that when a drummer is with the enemy, he is to 'take as much notice as he can of all such things as concerneth the enemy' to tell the commander when he returned. For this reason, drummers were often blindfolded when being led through the enemy's lines. Such was his importance drummers were paid five shillings per week, the same as a corporal.[56]

Rather than training a recruit to drum, drummers were recruited, which the Privy Council acknowledged was not an easy task. In December 1624, the Council authorised the captains of Mansfeldt's Regiments were to impress sufficient drummers for their companies. The same was true for the regiments raised for the Cadiz and Île de Rhè Expeditions. On 20 May 1627, the Privy Council ordered His Majesty's drum major 'to press 10 drummers' each for Sir Charles Rich and Sir Alexander Brett's Regiments, while another order, dated 31 August 1627, required several captains 'to impress two able and sufficient drummers (but none of them are to be taken out of the trained bands)'.[57]

Some contemporary illustrations also show fifers, but they do not appear in the rank structure, so must have been recruited for their entertainment value rather than any military role.

An important role within a company was the clerk because it was their job to keep the company's paperwork up to date. According to Francis Markham, they had to:

> … keep the rolls and muster books containing the names and surnames of the whole company and these he shall have in sundry and divers manners, as in one book or roll according to the arms and weapons which they carry, sorting every several arms by themselves, as first all the officers in their true ranks, then all the

---

54 Ibid., p.62.
55 Ibid., p.59.
56 Count Mansfeldt's *Directions of Warre* (1622) p.18.
57 Lyle APC 1625–1626 A letter to the Mayor and Aldermen of Southampton, 23 August 1625, p.137; Lyle APC 1627 Warrants for impressing drummers, 24 May 1627 and 31 August 1627, pp.295, 513.

pikes and short weapons, then the muskets and harquebusiers or bastard muskets if there be any.

In another book or roll all their names according to their squadrons having the officers of every squadron first, then the gentlemen, and lastly the common soldier, and in the third book or roll all their names as they march in their own particular Battalia and according to the dignity of their places; so that when the muster master or captain shall come to make a general call of the company he may deliver them the first book.[58]

Markham suggested the clerk should make several copies of these books, so that one could go to the muster master, another to the captain and to his lieutenant, while he should keep a fourth copy for himself. The clerk was also to give a list to a sergeant and a corporal so that he knew what soldiers were under his direct command.

Markham also suggests that only seven soldiers' names be written on each page:

> … that he may better insert any exchange, defect, absence, death or other faults which may procure a check, that so an even reckoning may be kept between the captain and the soldier'.

As well as the paperwork, the clerk was also the company's paymaster:

> He shall receive from the treasurer or under treasurer, all the pay belonging to the captain and soldiers, and see faithfully and truly distribute to every man his due according to the captain's directions, and he shall keep a faithful audit between the captain and all others, clearing every reckoning without doubt or disorder, he is to receive from the victualler all proportions of victuals and to keep a due account of the prizes thereof, he is to receive all provaunt apparel, with the value of the same, and he is to receive all manner of munition and arms which is necessary for the whole company, and of all these he shall keep true records, fetch out the certificates and keep all reckonings even between his captains and all officers… [He] ought to be both the company's physician and the company's merchant, for he ought by information to the captain and by his direction to provide all things necessary to them, both in sickness and in health…
>
> If any man shall happen to be slain, or otherwise to depart out of this life, the clerk of the band shall administer upon his goods, and making a true inventory or praysure thereof, (after his debts and defaulcations are paid) shall be answerable for the rest to his next of blood, or else such on whom by will he had formerly bestowed it. He shall also keep a true note of the time of his death, whereby a certificate may be made to the muster master, and also that neither the Prince may be deceived in his pay, nor yet the victualler over reckon the captain in the multiplicity of their victuals. He is to see that such as are wounded be duly dressed by the surgeons, and if that any be taken prisoner, he is to awaken his captain's memory touching their ransom.[59]

---

58   Francis Markham, *Five Decades* pp.54–55.
59   *Ibid.*, pp.55–56.

OFFICERS AND LESSER OFFICERS

However, Markham warned:

> To conclude if he be a good clerk, he is an excellent member both for captain and soldier, for it is impossible that any captain should thrive if he have an evil and unconscionable clerk, for the way are so infinite by which he may deceive him.[60]

Although Francis Markham's *Five Decades of Epistles of Warre* was published in 1622, it seems to have been a popular book because quotes from it appear in a notebook written by Jeremy Baines, who was a major in Colonel Melve's Regiment of Dragoons in 1643, before being promoted to Lieutenant Colonel in Colonel Samuel Jones' Regiment of Foot later that year.

---

60 *Ibid.*, p.56.

THE FIRST BRITISH ARMY 1624-1628 (REVISED EDITION)

Unshoulder your musket

2

The Rank and File

According to Cecil, a good soldier was made up of three parts, 'Courage, strength and obedience'. Courage being 'the foundation of war', strength was the need to endure the rigours of campaign and obedience to obey orders.

Contemporary sources usually describe soldiers either as mercenaries or volunteers. Volunteers from Britain could be found in the ranks of most European armies at the beginning of the seventeenth century, including the Dutch, Spanish and even that of Russia. The concept of a 'recruiting sergeant', was as much a part of life in the seventeenth century as it would be in later centuries. Officers and NCOs often returned to Britain to persuade men to volunteer to fight for the Protestant or Catholic Cause. As early as June 1621 orders were sent to all English ports to allow any soldiers travelling to Europe to serve 'either on the one side or the other' to leave 'without any let or impediment' and the port books record a steady flow of volunteers either travelling alone or in small groups to join a particular regiment. For a gentleman it was as much a part of his education 'to trail a pike' in a war, as it was to go to university. Among those who were 'educated' in this manner were the future Parliamentarian generals Sir William Waller and Sir Thomas Fairfax, who volunteered to fight in the Low Countries or try and recapture the Palatinate for King Charles' sister, Elizabeth and her husband Frederick.[1]

When a foreign power was given permission to raise a regiment, each of its captains were allocated a number of counties to raise their companies and sometimes several officers were commissioned to raise their companies within the same county, usually 'by beat of the drum only'.[2]

However, many of these recruiters did not care who they enlisted. In April 1622, the Spanish General Spinola, complained that there were 'many heretics' in his army from Britain. Later that year Lord Vaux's Regiment mustered 1,700 Englishmen, but when the regiment arrived in the Low

---
1   BL Harl ms 3638; BL Kings ms 265, Transcripts of military and naval papers, chiefly temp. Charles I, f.81.
2   For example, when Sir Jacob Astley raised his regiment in 1624 each company commander was allocated five or more counties. Astley. Lt Col John Talbot, Major Chambers, Captains Earley, Astley, Wroughton, Goodrick and Howell could raise their companies in Surrey as well as other counties.

Countries many deserted, claiming that they had volunteered to fight for the Dutch, not against them.[3]

For the lower class, some might volunteer because of 'youthful enthusiasm', by wanting to get away from their humdrum lives to seek adventure and wealth, although the latter was rarely achieved. Others might join to escape a strict master, like the 16 year old Sydnam Poyntz who recalls that he enlisted in Mansfeldt's army in 1624 having been:

> Bound an apprentice that life I deemed little better than a dog's life and base. At last I resolved with my self thus, to live and die a soldier would be as noble in death as life, which resolution took strong root in me, that not long after… [I enlisted as] a private soldier.

In 1632 two other apprentices who had run away to join a regiment being sent to Russia were not so fortunate and were ordered to be returned to their master, who no doubt severely punished them.[4]

However, to the population of the Low Countries or Germany many saw these volunteers fighting for their, or their enemy's, cause, as mercenaries. Whether these volunteers also saw themselves as mercenaries is not known for certain, but many believed they were fighting for a noble cause. However, this enthusiasm would probably soon wear thin with the rigours of campaign.

Mercenaries were seen as not caring what cause they fought for, as long and they were paid; as one mercenary – Carlo Fantom (who served in the English Civil War) – allegedly boasted, 'I care not for your cause but your half crowns and handsome women'. Since they were paid to fight, mercenaries were also seen as prolonging a war. Despite this many generals saw mercenaries as a necessary evil because he was a trained soldier.[5]

However, there was another type of recruit, the conscript, or 'pressed man', who usually had no position in society. It has been estimated that during the 1620s 50,000 men were pressed into service, 28,000 alone being between 1624 and 1627. This is compared with 100,000 men during entire reign of Elizabeth I (1558–1603). They were impressed not only for Mansfeldt's, Wimbledon's and Buckingham's armies but also the English regiments in Ireland as well as those in the service of Denmark and the Low Countries.[6]

There was a long procedure of raising an army. First, the Privy Council would order the Lord Lieutenant to raise a certain number of men. He would

---

3 Albert Loomie 'Gondomar's Selection of English Officers in 1622', in *English Historical Review* vol. no 88, no. 348 Jul 1973 p.577; *CSP Venetian 1621–1623* Girolama Lando Ambassador to England to the Doge and Senate, 11 February 1622, p.233; *CSP Venetian 1623–1625*, Valaresso to the Doge and Senate, 21 June 1624, pp.354, Valaresso to the Doge and Senate, 28 June 1624, p.363.

4 Sydnam Pontz, *A True relation p.45* TNA PC2/42; Lyle *APC*, May 1632–April 1633, f.136. Poyntz implies that he joined Mansfeldt's army but does not give any details and he is not mentioned in the muster roll for the London recruits. However he does give information about Lord Vaux's Regiment, so he may have joined this regiment instead.

5 Quoted in Charles Carlton, *Going to the Wars* (London: Routledge, 1994) pp.20–21.

6 Stephen Stearns, 'Conscription and English Society in the 1620s' *Journal of British Studies* May 1972 pp.4–5; Anthony Fletcher, *Sussex, a County in Peace and War* (Chichester,:Philimore, 1980) p.193.

then pass the order onto his deputy lieutenants, who would allocate a certain number of recruits to be raised in the county's administrative areas, which were usually known as Hundreds, Lathes, Rapes and so forth. They would then inform the high constable of these areas how many recruits were to be raised within their administrative area, who would in turn inform the constables how many men were to be raised in their parish. The number could vary with the size and population of the parish. The constable would then choose who were to be selected and then send them to the high constable, who would then take them to the rendezvous appointed by the deputy lieutenants.

Writing during the latter part of Elizabeth's reign, Barnaby Rich said, about choosing the recruits within a parish, 'we disburden the prison of thieves, we rob the taverns and alehouses of tosspots and ruffians, we scout both town and country of rogues and vagabonds'. This was no idle boast, being officially sanctioned by the Tudor government, because the Council of State had ordered the sheriffs and Justices of Nottinghamshire, 'to apprehend idle persons and masterless men' for service in the Low Countries. No doubt this order was not unique to Nottinghamshire.[7]

In theory the Stuart government had a more enlightened approach ordering that only, 'fit and able men' should be enlisted, and that 'unnecessary persons that now want employment and live lewdly or unprofitably' should be avoided. However, the parish constables continued to choose those who were unemployed or masterless men, who were a burden to the parish. This would have the benefit of not ruining their parish's internal economy, while ridding themselves of those who were likely to claim poor relief. Anti-social behaviour could also lead to someone being impressed into the army. In 1624 Matthew Nicholas hoped the two recruits supplied by the parish of Winterbourne for Mansfeld's Army, 'will behave better in war than they did in peace'. With the army being made up with these type of men it is not surprising that Lord Herbert of Cherbury, pre-dating the Duke of Wellington by two centuries, called the recruits which he saw as 'the mere scum of our provinces'.[8] Prisoners who were conscripted into the army could expect a royal pardon for their crimes and so escape the hangman's noose, although it would be a mistake to class all the recruits as prisoners, in 1627 only seven out of the 250 recruits raised in Norfolk were from the county gaol.

On 30 April 1627, Thomas Crosfield, a fellow of Queen's College, Oxford records in his diary, the recruits were taken during the night, which was not unique to Oxfordshire or to the 1620s, because in 1632 Matthew Davy of St Andrew's parish in Holborne, London complained that he had been 'taken out of his bed by the constables of that parish to serve as a soldier'. On 18 July

---

7   HMC Salisbury vol. XV p.60.
8   TNA SP 16/72/48 Secretary Coke to Secretary Conway, 31 March 1627; SP 14/181/15 Matthew Nicholas to Edward Nicholas, 3 January 1625; Lyle *APC 1623–1625,* Minute of the Privy Council, 27 April 1627, pp.249–50, Edward Cherbury, Baron Herbert of, *The Expedition to the Isle of Rhe* (London, Whittingham and Wilkins, 1860) p.46. Certainly, criminals and outcasts were favoured. On 5 June 1629 47 prisoners 'condemned of felonies' from Newgate and Bridewell Gaols, including Elizabeth Leech were ordered to be delivered to the Swedish Ambassador for service in Gustav II Adolph's army, CSPD 1628–1629, p.568.

1632, the Privy Council ordered his release, but not because they disapproved of the way he had been recruited, but because there was not a warrant to raise any additional troops at that time.⁹

When Major Leigh conducted a survey of the recruits at Plymouth in 1625, he found that of the recruits from London four were blind, one was a minister, another was mad, while another was a 'simpleton'. Twenty six were described as 'aged', i.e. too old and two were either lame or 'imperfect in arms' and a further 200 were described as 'defective', although what their defects were is not recorded. A further four were foreigners or strangers to the area, including a Dutchman who did not speak English and had only been in London to see a play. Another recruit was just in London for 'legal business' when he was conscripted. These examples were by no means unique, one recruit from Norfolk was described as 60 years old and 'imperfect in his arm'.¹⁰

Personal gain might also result in a man being impressed. One Norfolk constable had bought some land from a 'fool' for £5 when it was worth £30 and so wanted to rid himself of the person involved. On the other hand, blackmail could also be used, Elizabeth Rawlins who was a widow had to pay 26 shillings to the constable of Marlborough, not to take her son. While others who had been impressed were able to pay to be released.¹¹

In Somerset, at least, there appears to have been faction fighting between the gentry, which caused William Rogers to be pressed for the Cadiz Expedition which was intended 'to bring some disgrace upon Sir John Stowell'. It was claimed that Rogers 'was pressed out of private spleen rather than for service of his Majesty'. Fortunately for Rogers he was released from military service, although this would mean that another man would have to be impressed to replace Rogers. Stowell also appears to have used the same tactic against his opponent's faction.¹²

These abuses were not confined to just the recruits of 1625. In 1627 of the 147 impressed in Norfolk, 136 men were to be chosen for service, including several 'strangers' who had the misfortune of staying in a parish when they were conscripted to make up its quota.¹³ However not all recruits were the 'dregs of the people' as Sir John Saville called them, adding that:

> I have taken some pains for sending away of these soldiers which are to go from here to see them well apparelled, well chosen, well provided for by the county so as they may be an example hereafter to other shires for 100 of them are in such worth 200 of such as are ordinarily pressed.¹⁴

---

9   Boas, Frederick, (ed) *The Diary of Thomas Crosfield* (Oxford University Press, 1935); TNA PC2/42 Acts of the Privy Council, May 1632–April 1633 f.155.
10 Stearns, 'Conscription and English Society in the 1620's pp.8, 9.
11 TNA SP 14/185/21 Deputy Lieutenants of Wiltshire t the Earl of Pembroke, Lord Lieutenant of Wiltshire, ?6 March 1625; Stearns, 'Conscription and English Society p.9.
12 Lyle *APC 1627,* Letter to the Judges of the Assize for Somerset, 27 July 1627, pp.453–44.
13 *State Papers relating to Norfolk* (Norwich: Norfolk and Norwich Archaeological Society, 1907) p.69. One of the two recruits from Henstead and one of seven from Laundish were described as a stranger, pp.66, 67.
14 TNA SP 16/59/57 Sir John Saville to the Duke of Buckingham, 4 April 1627.

Moreover, with the constant call for soldiers to serve in the various armies the sources of masterless men would quickly dry up and so the constables would have to make their way up the social ladder for recruits. Bachelors who had no dependents were preferred over married men because they would not leave any family members to claim the parish's poor relief. On the other hand, a Buckinghamshire Deputy Lieutenant requested one man to be excused from military service because his mother would have no other income, while some constables claimed that they had just married men in their parish.

Sometimes the parish constables left it up to the household to choose who they could spare. In 1624 two Dorchester constables went to the house of Robert Snookes and finding him at work with four 'very able men' and a boy they asked which one he could spare. Snookes pointed to the young boy who was then carried off to be a soldier. Age was also not a barrier to another Dorset boy when 'an extraordinary stout' miller was replaced by a boy of 15 'as little as [he was] young'. However, not all impressed men went quietly, in 1625 John Goodwyn of Leicester, assaulted the local official and used 'very unreproachful speeches' against the mayor. He was tried and for punishment he was sentenced to become a soldier.[15]

Others might petition for their servant's release from the levy, like Bartholomew Fromonde, who on 13 December 1624 petitioned Sir Nicholas Carew of Surrey, 'because I have need of him, he is a very simple fellow and nothing fit for a soldier'. A family's hardship might also be a factor in trying to have a recruit rejected. On 23 May 1627, the Earl of Montgomery wrote to Sir George More, also of Surrey, on behalf of William Elisander of Egham because 'his mother is lame and his father aged and he is the only support' to his parents. Whether these two men were exempted from service is not known, but if they were then replacements would have to be found to meet the county's quota.[16]

On the other hand, one Hampshire resident was terrorised by her former servant who believed that she had not done enough to protect him from being impressed into the army. He 'swore he will be even with me' and after three days with the army he deserted. She petitioned Sir Thomas Jervoise on 19 May 1627 that she was certain:

> He will do me some injury if it lay in his power for he hath been two or three nights about my house and I conceive it can be for no good to me, he knowing my husband hath been gone from home ever since a month before Easter and I have now but a weak household about me.

She asked for her former servant to be apprehended, but unfortunately Jervoise's actions are not known.[17]

---

15 Thomas Home Cogswell, *Divisions, aristocracy, the State and Provincial Conflict* (Manchester: Manchester University Press, 1998) p.40.
16 Berkshire Record Office D/ELL/C140, f.141; Surrey History Centre 6729/6/90, Letter from the Earl of Montgomery to Sir George More, 23 March 1627.
17 HRO 44M69/G5/48/76, Eliza Arthur to Sir Thomas Jervoise, 19 May 1627?

On arrival at the rendezvous the county's deputy lieutenants wrote the names, and sometimes their parish and profession, on an indenture. The indenture for those raised in Ipswich for Mansfield's army is just one example:[18]

Name	Age	Stature	Hair	Profession
Thomas Bacock	32	Low	Brown	Mason
Nath. Gildston	33	Middle	Brown	Carpenter
Jacob Reynolds	25	Middle	Brown	Joyner
James Tarrant	24	Middle	Brown	[not given]
George Balls	30	Middle	Brown	Smith
Thomas Barwick	30	Middle	Brown	Shoemaker
Ambrose Payne	40	Tall	Fair	[not given]
Robert Adds	22	Tall	Brown	Baker
Roger Tremant	24	Middle	Reddish Brown	Carpenter
Thomas. Brown	30	Middle	Fair	Gardiner
Robert Forster	50	Tall	Grey	Sherman
John Camplin	20	Middle	Black	[not given]
James Spain[?]	34	Middle	Fair	[not given]
Michael Starthope	20	Tall	Brown	Taylor
John Cook	40	Middle	Black	Sawyer
John Wilbe	33	Tall	Brown	[not given]
Thomas King	20	Middle	Fair	[not given]
Michael Dixon	26	Tall	Reddish	[not given]
James Hareman	40	Reasonable	Fair	[not given]
Thomas Colbie	40	Tall	Black/Brown	Glover
Thomas Bett[?]	40	Reasonable	Black/Brown	Sawyer
Thomas Martin	25	Reasonable	Brown	[not given]
Edmond Welsam	25	Tall	Brown	[not given]
John Payne	25	Middle	Brown	[not given]
William Clen	20	Middle	Brown	Smith

There were three copies of this indenture, one was sent to the Privy Council, another remained with the deputy lieutenants and a third was given to the conductor who would then present it to the captain whose company they were to form. At this rendezvous, the recruit would be given a coat and 'press money', sometimes also known as 'arram anglice prest'. This would later be known as the 'king's shilling', but during the 1620s it could be as little as four pence. In Suffolk's case, at least, the recruits were told, 'here is press money

---

18   TNA SP 14/179 f.34 Indentures for the delivery of recruits levied for Mansfeldt's Army, 1624.

to serve his Majesty, and we charge you thereby to be ready upon one hour's warning, when so ever you are called for, upon pain of death'.[19]

In 1624 the Warwickshire indenture show the profession of its 200 recruits as follows:

Trades	Number	Percentage
Shopkeepers		
Food and Guest trades, including Butchers, millers, brewers	17	8.5
Clothing & textile trades	20	10
Building, Wood & Metal processing trades, Inc blacksmiths, carpenters, bricklayers	22	11
Transport trades, such as carters	1	0.5
Paper trades, inc. papermakers & binders		
Educated, that is, students & teachers		
Agriculture such as labourers.	105	52.5
Other trades, such as servants, cooks, basket makers, Ostlers	10	5
Unknown or unreadable	9	4.5
Yeoman	2	1

The large percentage of labourers is not unique, because of the 100 men raised in Oxfordshire in 1624, 55 were labourers and in June 1627, 50 of the 100 men from Derbyshire were labourers, with 15 miners forming the second largest group followed closely by those in the clothing and textile industry.[20]

However, in 1624 of the recruits from Somerset only 17 percent were labourers and 14 percent were either weavers or cutters. The largest majority, 40 percent, are described as 'grooms'. The precise meaning of this profession has been lost, but possibly was related to the cloth trade, i.e. cleaning the cloth, rather than household servants. Among the other 29 percent, one is described as a 'gentleman'.[21]

The 1624 Ipswich indenture shows that of the 25 conscripted, the majority (12) were between 20 and 26, seven were in their 30s and five were 40 and only one was 50. Sixteen are described as of middle or reasonable stature, and only one as low stature, the remainder are described as tall. These age groups

---

19 BL Add ms 39,245, Muniments of Edmond Wodehouse; *The manuscripts of Rye and Hereford corporations, Capt. Loder-Symonds, Mr. E.R. Wodehouse, M.P., and others (13th Rep., app. iv)*, Royal Commission on Historical Manuscripts, 31 (London: HMSO, 1892) p.181.
20 TNA SP 16/72/48 Secretary Coke to Secretary Conway, 31 March 1627.
21 TNA SP 14/178 Indentures for the delivery of recruits levied for Mansfeldt's Army, 1624.

are similar to those volunteers who went to fight on the Continent between 30 December 1628 and 31 July 1629, whose ages are known:[22]

Age	21–25	26–29	30–35	36–39	40–45	46–49	50–55	56–59	60
No.	187	56	66	22	21	10	11	3	1

Presumably, there were younger soldiers but anyone under 21 was considered to a minor and therefore was not listed in the port books.

However, there might be a long delay between the time the recruits were summoned and leaving to go to join the army, because a conductor was not always on hand to take charge of them. On 18 November 1624, the deputy lieutenants of Suffolk were ordered to raise 150 men for Mansfeldt's army, which they managed to assemble in four days. They arrived at Sunbury on 23 November, but it was not until Christmas Eve that they were again summoned to the rendezvous because an officer was expected to arrive the follow day to conduct them to the army. When Lieutenant Woodhouse finally arrived, he refused to take those he considered unsuitable for service, so substitutes had to be quickly found. On 26 December, the lieutenant found that one recruit was missing, and so refused to wait for this individual and left ordering the deputy lieutenants to send the levy to Ipswich where other recruits were gathering once the missing man had been found.[23]

It was the same for other counties who were ordered to raise their quota by 30 November 1624, but it was not until mid-December that the deputy lieutenants received orders for them to be ready to be handed over to the conductor. A further week or so passed before the recruits actually left their county. The order from the Privy Council made no allowance for the recruits of the northern and western counties to march to Dover, all were told to be at the port on the same day. During this time, it cost the county eight pence per man per day to look after these recruits. This 'coat and conduct money' was to be deducted from the county's subsidy.[24]

Some recruits made a will before they left, including John Biddle of Horsham in Sussex, 'being a pressed soldier in Count Mansfeldt's voyage' gave a verbal will before his departure that his property should go to his wife and eldest son.[25]

When it came to raising recruits in 1625 and 1627 it was a different matter. Counties were ordered to appoint their own conductors so that the recruits were to be sent to the rendezvous as quickly as possible. On 6 May 1625, the Privy Council ordered Northamptonshire to raise 200 soldiers and to see that they arrive at Plymouth by 20 May, or just 14 days in all. This also meant that

---

22   TNA E157/14 Register of soldiers taking the oath of allegiance and register of licences of persons to pass beyond the seas, 1628–1629.
23   Stearns, 'Conscription and English Society in the 1620s' p.3, 17; HMC *13th Rep., app. iv*, p.181; BL Add ms 39,245 Muniments of Edmond Wodehouse.
24   Anthony Fletcher, *Sussex a County in Peace and War* p.193.
25   Captain John Glen who had served in the Cadiz expedition before leaving for France with Buckingham made provision in his will that Thomas Pendlebury of London received the £50 which he had lent Glen. The remainder of Glen's estate was to go to his wife.

there was no one to examine the recruits for their suitability to serve in the army, but this did not give a recruit much time to desert if he so wished.[26]

According to James Bateson, who ran away after stealing some items from his master, he found himself in the company of two 'decoy ducks', who persuaded several men to join the army with them, but they would then desert after receiving their levy money. Abuses like this prompted the deputy lieutenants of each county to order that all runaway soldiers be sent to gaol until they could be tried at the quarterly Assizes. In July 1627 John Barton, William Beach and George Hanshaw who had enlisted in April 1627 as part of the county's quota were brought before the Kent Assizes charged with desertion. They were stupid enough to return home where they were arrested and sentence to death. Fortunately for them they were reprieved and sent back to the army. However, the fate of two deserters from Wapping and two from Hammersmith, who were also sentenced to be hung, this time by the Middlesex Sessions in 1627, is not known.[27]

In 1625 the recruits from Leicestershire were given a rousing send off and were given an extra penny for beer. However, this did not prevent 16 men deserting along the way, some bribing their conductor, John Everard, to release them. Bribery was only possible for the wealthier recruits, who usually paid between 40 to 50 shillings to get themselves released from military service, although bribes of up to £10 were not unknown.

To cover up their crimes the conductors would quickly have to find replacements. Among the replacements Everard found was one who was 'troubled with a bloody issue [i.e. a discharge]' while another had been injured by a scythe which had caused 'his guts [to] fell out of his belly'. On the other hand, one wonders what ailment caused one substitute to have 'issues' from both his breasts. In all a total of 26 of Everard's recruits were found to have had some disability on their arrival at Plymouth and so had to be discharged, but Leicestershire had to find replacements for them.[28]

It was not just Everard's recruits that were unfit for service. On 12 June 1625 Sir John Ogle complained that 'the men [are] very unfit by reason of age, impotency and sickness. Nothing had changed two years later when Sir John Saville wrote to the Duke of Buckingham on 4 April 1627 referring to some of the recruits as 'the dregs of society' Sir George Blundell wrote on 1 May 1627 '200 pressed men from Hampshire arrived at Portsmouth, but such base rogues that he sent 120 of them back'.[29] It seems in these two cases at least it was the fault of the county where they were raised, rather than the conductor, since they were probably trying to get rid of those who were a burden to the parish.

---

26 HMC Lord Braye of Stanford Hall, Rugby Manuscript p.112.
27 J. L. Rayner & G. T. Crook, *Complete Newgate Calendar* (Navarre Society Ltd., 1926) vol. 1 p.189; Assizes of Kent during the reign of Charles I, pp.51–52; *Middlesex Session Rolls 1627*, (Middlesex County records, 1888) vol. 3, pp.13–17.
28 TNA SP 16/22/5, Everard JP of Leicestershire to the Council 1 March 1626; Cogswell *Home Divisions, aristocracy, the State and Provincial Conflict* pp.41–42; TNA SP 14/181/11 Sir John Hippisley to Buckingham, 3 January 1625.
29 *CSPD 1625*, Ogle to Secretary Conway; *CSPD 1627*, Blundell to Buckingham.

On 31 July 1627, the Privy Council warned the Lord Lieutenants of these frequent abuses by conductors and that the deputy lieutenants should take 'more than ordinary care' in selecting the conductors and keep 'a special eye' on them. If a conductor was caught abusing his position, then he was to be brought before the Privy Council to answer for his offence. It was also suggested that the captains themselves should choose the recruits from the county levy and there is some evidence that when a conductor was a serving officer, he did dismiss unsuitable recruits. However, these measures were to no avail and the abuses continued.[30]

It would be wrong to assume all conductors were corrupt, in 1628 a man was tried at the Sussex Quarter Sessions for trying to bribe 'the King's Majesty's officers with money to permit him to escape after he had received his press money'. Some recruits managed to escape themselves, and on 7 January 1625 Sir Thomas Dutton complained to the Lord Chamberlain Pembroke about the recruits from Wiltshire, that, 'before they left their own county, 30 of them ran away and since I have lost 20 more'. In 1625 of the 2,500 recruits which were sent to Plymouth, about 200 had run away or died on their journey and in 1627 80 of a party of 400 had also run away.[31]

The recruits were expected to march between 12 and 15 miles a day and receive eight pence per day travelling expenses. At night they would be quartered with the civilian population, who received six pence a day for each soldier's 'supp and lodging' and two pence for their breakfast. The conductors of the recruits were to receive four shillings per day. Each county would eventually be reimbursed by central government, but in 1625 the deputy lieutenants of Essex pointed out to the Privy Council that during the reign of Queen Elizabeth it had cost £322 2s 8d to conduct 400 recruits to the place of rendezvous, but now it cost £567 13s 4d. A conductor might also try to make money by not paying all the money owed to the recruits. In 1625 a Mr King who was the conductor of a party of Gloucestershire recruits, 'detained their pay one day and a half which amounteth to fifteen pounds'.[32]

It was not just quartering of the recruits that the local population had to worry about, but their behaviour as they progressed through the countryside. On 26 December 1624, the Privy Council wrote to the Earl of Montgomery, the Lord Lieutenant of Kent, ordering him to assemble the county's trained bands because the recruits travelling to Dover:

> March not in company and orderly upon the highways as they ought to do but straggle up and down the county and not only spoil and take what they list [*sic* = wish?] but do also terrify the poorer sort of inhabitants, molest and offend all that

---

30 Lyle *APC 1627* A Minute of letters to the Lord Lieutenants of several counties, 31 July 1627, pp.456–457, the same to same, 10 August 1627 p.481.
31 Fletcher, *Sussex, a County in Peace and War*, p.193; Sterns, 'Conscription and English Society in the 1620s, p.15; TNA SP 16/4/160, Report of Captain Edward Leigh, July 1625.
32 BL Add ms 39,245 Muniments of Edmond Wodehouse f139; Lyle *APC 1627* p.98; TNA SP 16/53/57 Robert Earl of Sussex to the Council, 9 February 1627; SP 16/4/160 Report of Captain Edward Leigh, July 1625.

pass upon the highway… This disorder comes chiefly by the negligence and fault of the captains and officers that have the conduct of them.[33]

Even if the recruits reached their rendezvous, many were too exhausted and in a poor condition to continue. Of a party of 200 men from London, only 152 were described as 'serviceable', seven had died en route and the rest were 'defective'. On 19 January 1625 so many recruits had either deserted, fallen sick or died on their journey to Dover that the Privy Council was forced to order a further 2,000 men to be impressed.[34]

Furthermore, when recruits arrived at their rendezvous there was no guarantee that there would be anyone to receive them. On 25 March 1627 Sir John Coke wrote from the Tower of London about the recruits assembled there:

> Last night there came 50 soldiers from Northampton, proper men well clothed, but the conductor have left them at random and no officers at all to take charge of them… [and] without present officers the soldiers will disorder and run away.

In total about 946 recruits had gathered at the Tower of London, but within a week 137 had deserted.[35]

Scotland

It has been estimated that between 1626 and 1632 a total of 25,000 Scots served in the Danish and Swedish Armies, which was about 10 percent of the adult male population of Scotland.[36] Recruitment followed a similar pattern to England. When Sir James Leslie received a commissioned in 1626 to raise a regiment for Mansfeldt's army, he was to use all lawful means necessary to complete his regiment. These 'lawful means' were probably similar to those imposed by the Privy Council of Scotland on 3 April 1627 for raising recruits destined for Danish service:

> That all Egyptians [gypsies], strong and sturdy beggars and vagabonds, idle and masterless men wanting trades and competent means to live upon and who in that respect are unprofitable burdens to the country… should be apprehended… and to deliver them to such captains, officers and sergeants as shall make first acquisition for them.[37]

---

33 Lyle *APC 1623–1625* Privy Council to the Earl of Montgomerie, the Lord Lieutenant of Kent, 26 December 1624 p.409.
34 SHC LM/Cor 4/50, Copy of a letter from the Privy Council to the Lieutenants of the County of Surrey, 19 January 1625.
35 HMC Cowper vol. 1 pp.299, 301.
36 Geoffrey Parker, 'Soldiers of the Thirty Years' War' in Konrad Repgen, (ed.) *Krieg und Politik* (Munich, R Oldenbourg, 1988) p.305.
37 F. Hume Browne, (ed.), *Register of the Privy Council of Scotland 2nd Series* (Edinburgh, 1901) vol. 1 pp.565–566.

This brought in a flood of complaints that the recruiting parties had dragged:

> Diverse lawful subjects out of their beds, has taken others from the plough and some in their travelling about the country; and their insolence is now come to this height that no single man dare travel in the country, attend their labour in the fields.

In May 1627, this practice was outlawed but each parish had to draw up lists of all masterless men over the age of 20 who were to be enlisted when required.[38]

Unfortunately, very few of these lists survive, but those which do record various information, like Patrick Grey of the Old Town of Schewas who was 'thirty years of age, unobedient to Kirk and King and troublesome to his neighbours', or Robert Rich of Darforky who is 'offensive to Kirk and King's laws', while two others from the parish of Tarves were 'inclined to great vice'. Several men from the parish of Jedburgh were described as 'idle, neither [of] trade, calling nor competent means'. The eldest man on these lists was 50 year old Alistair Diker of Logiret, who had lived in the parish '20 and more years, had no calling nor means'. Another 50 year old was John McInkaid who 'has no calling, but gives himself out for a sorcerer and charmer', who is a 'refracture to the Kirk'.[39]

On 28 February 1627 when the King of Denmark commissioned Robert, Earl of Nithsdale to raise a regiment of 3,000 men in Scotland he called upon James Lord Ogilvy for help. On 16 August 1627 Ogilvy informed Nithsdale that he had sent Proclamations to Forfar, Dundee, Couper, Arbrott, Montross, Killemuer and had written to:

> All the Justices of the Peace and ministers to whom the enrolling of the idle and masterless men were committed desiring them not to suffer any captain or their officers (except such as has warrant from your Lordship) to take or apprehend any of the idle men within their parishes and… [he had] also written to the ministers to send me the rolls already made, with advertisement how many of them are already apprehended and when they were taken.[40]

However, men's names on lists were one thing, actually turning them into recruits was another. On 28 June 1627 George Hamilton informed Nithsdale that:

> I have been hitherto as careful and diligent in levying men as any man could be, but as I wrote lately unto your Lordship the news of my warrant to press being divulged, hath made all the idle and suspected people go upon their keeping.

---

38 *Ibid.*, pp.603–605.
39 *Ibid.*, pp.689–693.
40 Sir William Fraser, *The Book of Carlaverock, memoirs of the Maxwells, Earls of Nithsdale* (Edinburgh, 1873) vol. 2, pp.93–94.

By 7 July Hamilton had raised 'some forty [men] all of them, except one or two, volunteers'.[41]

When Douglas, Earl of Morton raised his regiment in August 1627 for Buckingham's Army he resorted to publishing a recruiting pamphlet called *Encouragement for the Wars of France* which was aimed at the 'heroic noblemen, gallant men and courageous soldiers' to enlist in his regiment. On 25 September, the Burgh of Edinburgh responded by ordering a proclamation to be sounded in the Burgh by the sound of the drum, 'to all persons of whatsoever rank or degree' to enlist. Those wishing to do so were to present themselves to the magistrates and 'enrol their names betwixt the day and date hereof and the third of October'.

In return for their service, they would be made burgesses of the town. Unfortunately, only 12 men enlisted, seven being described as 'indwellers', three as tailors, one as a mason and the profession for the other person is described as 'post'. Morton's Regiment was said to be filled with a 'great number of base and worthless men', who quickly deserted giving an 'evil example to the others'. Fortunately for the deserters of Morton's Regiment they were to be returned to the Colours, whereas those apprehended while serving in other Scottish regiments in Danish or Swedish service could expect the death penalty.[42]

Ireland

What of Ireland? In 1635 the Venetian Vicenzo Gussoni wrote, 'As a rule levies are raised in Ireland for the Spaniards'. Irishmen had served in the Spanish Army during the reign of Queen Elizabeth, particularly in Sir William Stanley's Regiment, which had been raised by Elizabeth, but later changed sides. Unfortunately for Stanley, few Irish recruits were able to join his regiment which by March 1595 contained just 232 Irishmen men, the remaining 318 were Walloons.

In 1605 an 'Irish Regiment' was again raised by Spain under the command of Henry O'Neil. The recruits came from Irish refugees after the Spanish invasion of Ireland in October 1601, which ended in defeat at Kinsale early in the following year. The Spanish used several independent Irish companies to bring it up to strength, but it was not until the reign of James I and much of Charles' that Spain was given free rein to raise Irish recruits. With this freedom to recruit the regiment of Henry O'Neil (now Count of Tyrone), mustered 102 officers and 1,078 men in April 1621 and by March 1622 it had increased to 151 officers and 1,322 in 12 companies. The Irish made excellent soldiers and the regiment was given the honour of being called a Tertio, a title given to the best units in the Spanish Army.

---

41 *Ibid.*, pp.98–99.
42 Marguerite Wood, *Extracts from the Records of the Burgh of Edinburgh* (Edinburgh, Scottish Burgh Records Society, 1882) p.35; Hume Browne, *Register of the Privy Council of Scotland* vol. 2, pp.34, 35, 40, 90, 285.

Of course, there were strict instructions for captains raising recruits in Ireland for Britain's long-standing enemy, Spain, including that the recruits should be gathered together in groups of no more than four or five men. They were to travel to the port in a peaceable manner and not bear arms, nor be a financial burden on the country or the population. Furthermore, the captain was to keep a list of the names of the men he had recruited, where they lived and their route to the port. This list was to be regularly updated, 'or once every fortnight at the furtherest' and presented to the commander of the nearest garrison. Once a captain had raised his quota of recruits, they were to leave Ireland 'without delay' and not wait for their fellow captains.[43]

However, with the outbreak of war with Spain in 1625 Charles forbad any further levies being raised and so recruits had to be raised discreetly, but these levies could not meet the demands of Tyrone's Tertio which would see a steady decline it its strength. Even so, on 15 June 1629 Dermot O'Sullivan received a commission to raise an Irish company.[44]

With the Peace of Madrid on 15 November 1630 between Britain and Spain, Spain could once more enjoy the recruiting grounds of Ireland, insomuch that on 13 January 1632 the Count of Tyrconnell received a commission to raise a new tertio and there would be other Irish tertios raised, including those of Owen Roe O'Neill, which was considered by the Cardinal Infant as the best Irish tertio and argued that it should not be disbanded when there was to be a reduction in the number of Irish tertios.[45]

Spain was not the only power to recruit for its army in Ireland, as early as 1609 Captain Richard Bingley was commissioned to raise between 200 and 300 Irishmen to serve in the Swedish Army, which seems to have been the first of many. However, the Swedish King, Gustavus Adolphus distrusted the Irish, but this seems to have been because of their Catholic religion rather than any act of treachery.[46]

By the 1630s France also had its Irish regiments, but they were less numerous than they would be under Louis XIV. It would also be members of Colonel Walter Butler's Irish Regiment who assassinated the Imperialist General Albrecht von Wallenstein at Eger in 1634.

In 1627 Sir John Bingley suggested raising a regiment of 2,000 Irishmen for Buckingham's Army, because the Irish were:

Acquainted with skirmish, nimble and skilful shot, bold and valiant that can endure labour, hunger, sparse diet and course lodging. They are healthful [healthy] and seldom fall sick… When our English soldiers fall sick as they ever do after their first landing, they will be healthful and be of great strength to our army… [and] Ireland might well spare them, for it swarms with idle men.

---

43 SP 63/236 Conditions dated June 1622, f.179.
44 Edward de Mesa, *The Irish in the Spanish Armies of the Seventeenth Century* (Woodbridge: Boydell, 2014) p.41.
45 *Ibid.*, pp.27–28, 31.
46 *CSP Ireland 1608–1610*, Chichester to the Lords of Council, 13 July 1609.

Bingley added that they should recruit only be 'honest men, if not it will be dangerous'.⁴⁷ With the need for manpower Charles' Privy Council readily agreed and Sir Piers Crosby was chosen as the regiment's colonel. Irish recruits were also used to bring Bingley's Regiment up to strength, which had been raised from the remnants of Cecil's Army that had landed in Ireland after the Cadiz Expedition.

On 20 August 1627 Lord Esmond remarked that Crosby's and Bingley's Regiments, were 'proper men and like to do service [as] I have seen'. They sailed from Waterford on 29 and 30 August 1627, 'furnished with brave and well promising men'. Nevertheless, Lieutenant Thomas Dymoke of Bingley's Regiment complained of his company, 'the pain I took with those raw rebellious and untutored Irishmen was excessive not one in a hundred speaking or understanding English or any discipline, yet I made them fit for service'. Despite this only seven men of his company survived the Île de Rhè Expedition.⁴⁸

Although the recruits were meant to be volunteers, Edward de Mesa states that there is some evidence that conscription was used, but does not elaborate upon this statement.⁴⁹ On 29 April 1627 Sir George Hamilton wrote to Lord Deputy of Ireland that the 'papists in the North objected to their idle swordsmen leaving the country or entering the king of Denmark's service', but whether this was because of conscription or not is not recorded. Certainly, the local authorities in Ireland saw the raising of Crosby's and Bingley's Regiments as a good way of getting rid of possible rebels, among whom were Patrick O'Mulhallen and Daniel O'Neil who had taken part in the Woodkerne's Rebellion, but for their distinguished service during the Île de Rhè and at La Rochelle they received a pardon.⁵⁰

The Rendezvous

It was only when the recruits reached the appointed rendezvous and swore the oath of allegiance that they officially became soldiers and the responsibility of the government. Robert Monro in his *Expedition of the Worthy Scots Regiment* he described Colonel Donald Mackay's Regiment being sworn into Danish service. The regiment was drawn up in formation and after the King of Denmark had inspected it, the regiment paraded before him. The king then praised the regiment, and the senior officers kissed his hand. Once these officers had returned to their positions:

---

47 SP 63/244/626 Sir John Bingley reason for raising 2,000 Irish Soldiers, 30 March 162, ff.182–183.
48 TNA SP 16/399/15i, Statement of Thomas Dymoke's former military service, 26 September 1638.
49 de Mesa, *The Irish in the Spanish Armies of the Seventeenth Century* p.40.
50 *CSP Ireland 1625–1627*, p.227 Lord Deputy to Lord Viscount Killultagh, 29 April 1627; *CSP Ireland*, Charles I to the Lord Deputy, 30 March 1628, p.428; TNA SO1/1, Signet Office, January 1627–December 1629.

> The oath was publically given both by officers and soldiers being drawn up in a ring by conversion… The oath finished, the Articles of War read and published by a bank of the Drummer Major, and his associates.
>
> The regiment then marched off by companies to their respective quarters.[51]

Unfortunately, the oath that Wimbledon's or Buckingham's soldiers had to swear is not known, but probably included obedience to their officers, to uphold the Protestant religion and abide by the Articles of War. Francis Markham states there was no standard oath for the soldiers, but records that once it had been read out to them an officer would say:

> Brethren and Companions in Arms, you have heard the commandments of our king, containing the principal laws of the field, and the Oath which every soldier should take; All you that intend faithfully to follow the same, let him now either openly refuse to be a soldier or else hold up his hand and say after me: 'All these Articles which have been read, we hold scared and good, and will valiantly and truly fulfil the same. So help us God and his divine Word, Amen.[52]

Once these formalities were out of the way then the soldiers could march off to war.

Wives and Families

However, the burden of conscription almost certainly had a devastating effect on the soldiers' families since they might be the only bread winner. On 8 June 1627, the Deputy Lieutenants of Dorset complained to the Privy Council that they had with difficulty impressed their quota of recruits, with the 'lamentable cries of mothers, wives and children', who petitioned for their release.[53]

With their menfolk away, the soldiers' families had little choice but to throw themselves on the mercy of their parish's poor relief, like Janet Frodsan of Wigan in Lancashire who:

> Being a poor woman and left with four small children, the eldest but twelve years of age, the second nine years of age, the third seven years of age and the youngest but one year old and having my husband taken to be in the King's service for a soldier to my undoing for ever… and having no other relief nor sucker but begging to relieve my wants and keep my children from famishing and many peoples' hearts are so hard that I cannot get them half [the] food [they need] to keep them from famishing…. [who] sits with hungry bellies which often grieves me and [I] knows no way how to relieve my wants but pays five shillings for a poor cottage to sit in and [I] am no way able to get it.

---

51 Monro, *Monro His Expedition with the Worthy Scots Regiment,* Part I p.4.
52 Markham, *Five Decades of Epistles of Warre,* p.42.
53 SP 16/66/41, Deputy Lieutenants of Dorset to Council, 8 June 1627.

She ends her petition, 'for God's sake make and set down some order to relieve my wants'.[54]

Janet Frodsan was not alone in her struggle, at the beginning of 1626 Ann Tilly of Chiverscoton petitioned the Warwick Quarter Sessions because her husband, Richard, had been pressed as a soldier and she and her small child was left without any means of support. Ann was granted an allowance of just six pence per week which was to be reviewed after a year. What became of her is unknown.[55]

Another possibility was to accompany their husbands on campaign. The trail of women and others who followed a regiment could be immense. In 1630 one German cavalry regiment at Langenau mustered 368 cavalrymen, 600 horses plus 66 women, 78 girls, 307 stable boys and 24 children and in 1632 another regiment mustered 2,014 soldiers in Bamburg who were accompanied by 916 women, 521 children and 560 horses. To a certain extent being an island, Britain could control the number of women with a regiment, as was the case during the Napoleonic Wars when three to six women per company were chosen by lot to accompany their husbands. However, on 28 April 1627 George Tucker wrote from Gravesend, that 'divers women' had arrived with their husbands 'who will not suffer them to be taken from them. You would be pleased that a general pass may be granted for all such women as shall go with the soldiers to Denmark'.[56]

It is not known whether any women and children accompanied the regiments with Cecil or Buckingham. Certainly, none are mentioned in the ships' returns for either the Cadiz and Île of Rhè Expeditions, but it would be extremely unusual for an army to be without any camp followers. Neither do any appear in the muster rolls of Mansfield's army, but Ann Owldery aged 20, Sara Halliday, aged 26, Jane Reynolds, aged 24, and Ellen Howlden aged 40, are known to have accompanied their husbands, but unfortunately history does not record what became of them.[57]

The absence of women in the regimental returns for Cecil and Buckingham's Armies might be due to a similar order dated 15 July 1631 by the Privy Council:

> Whereas there are women, maidservants and boys whose names cannot be particularly expressed, who are to pass with the troops now going beyond the seas under the command of our very good Lord Marquis Hamilton. These are therefore to will and require you forthwith on sight thereof to permit and suffer such women, maidservants and boys quietly to embark themselves and to pass with the said forces for which his Lordship shall give you warrant.[58]

On the other hand, in March 1623 Lady Daringall Easton was given permission to take with her a 'chambermaid, a cook, maid, and two other

---

54 Lancashire Record Office QSB/1/30/37, petition of Janet Frodsan of Wigan.
55 S. C. Radcliffe and H. C. Johnson, (eds), *Quarter Session Order Book* (Warwick, 1935) p.28.
56 George Tucker to Sir John Coke, 28 April 1627 in HMC Cowper, p.303.
57 TNA E157/13, Register of licences of persons to pass beyond the seas, Apr–Dec 1624.
58 TNA PC 2/41 Acts of the Privy council, June 1631–April 1632, f.102.

servants' when she left England to join her husband who was serving in Lord Vaux's Regiment, while other women seem to have travelled alone to join their husbands either a few days or weeks afterwards. Since they knew where their husbands were stationed there must have been a certain amount of correspondence between the soldiers and their wives. Certainly, in her petition, dated about 1640, Anne Nugent who wanted to join her husband, who was a captain in Colonel Preston's Irish Regiment in the Netherlands, had received 'several messages from her husband to repair unto him'. Unfortunately, none of these letters appear to have survived and on rare occasions soldiers might be given leave to fetch their wives. Fortunately for Anne Nugent her brother-in-law was able to escort her to her husband. For an illiterate soldier sending word home to his wife was almost impossible and so she would have to remain in ignorance until he returned home again or was informed of his death by a former comrade.[59]

Women who followed the army can be divided into three types; soldiers' wives, whores, who were unmarried but performed the same duties as a soldier's wife, and prostitutes. 'Whores' might attach themselves to a soldier for a campaign or two, although a late sixteenth century German engraving called *The Mercenary's Whore* warned:

> If you're not into gluttony and boozing
> I don't want to follow you for long
> If I stay near you for any length of time
> I'll certainly let you have the French disease.[60]

Since the 'French disease' or syphilis appeared in Europe in the late 1490s shortly after the discovery of America it is often said that it came from the New World, but there is no evidence for this. In theory according to the Articles of War prostitutes were to be whipped from the camp, but this seems to have been largely ignored, even though they could also help spread syphilis throughout the army.[61]

Whatever their function women were considered essential to an army. According to Sir James Turner:

> Women are great helpers to armies, to their husbands, especially those of the lower condition... They provide, buy and dress their husband's meat, when their husbands are on duty or newly come from it; they bring in fuel for fire, a soldier's wife may be helpful to others, and gain money to her husband and

---

59 TNA E157/15 Registers of soldiers taking the oath of allegiance and register of licences of persons to pass beyond the seas, March 1631–December 1632; Lyle *APC 1623*, A pass for Lady Daringall Easton to go to her husband, 25 March 1623, p.451; Lancashire Record Office DDKE/Box 32/63. On 21 April 1628 the Earl of Warwick gave Henry Rycroft, a soldier a licence to 'go about his needful affairs for the space of three months'; TNA SP 16/475/19, Petition of Anne Nugent, nd [c.1640].

60 Quoted in Keith Moxey, *Peasants, Warriors and Wives* (Chicago: University of Chicago Press, 2004) p.81.

61 Herbert Langer, *The Thirty Years' War* (Poole, Dorset Press, 1990) p.97; Hubsch *Das Hochstiff Bamburg* (Bamberg, 1895) p.120.

herself; especially they are useful in camp and leaguers, being permitted (which should not be refused them) to go some miles from the camp to buy victuals and other necessaries.[62]

However, one source which does mention soldiers' wives are parish records. I. J. Maclean has studied the registers of about 56 garrison towns in the Low Countries and found of the 2,070 entries, 34 percent (711) of soldiers in Scottish regiments married Scottish women or those with Scottish surnames within the garrisons, while 64 percent (1,316) married non-Scottish women. The remaining 2 percent (43) were Scotswomen who married non-Scotsmen. About half of these women were widows, while a third of the men were widowers. A survey of Scottish officers recorded a similar pattern.

Likewise, the register for the English garrison of Brill between 1589 and 1613 records 194 marriages, including 43 brides who are described as widows and 84 had Dutch names. There was even one bigamist marriage, which resulted in both parties being publicly punished. The register also included 443 baptisms, even though the garrison itself mustered just 363 in 1599, although it is unclear whether all these baptisms were for the children of soldiers or English civilians within the town. No doubt a study of other parish registers for soldiers of other nationalities will be just as fruitful.[63]

In his *Expedition*, Robert Monro has nothing but admiration for soldiers' wives:

> In my judgment, no woman are more faithful, more chased, move loving, more obedient nor more devout than soldiers' wives… whose husbands do daily undergo all dangers of body for their sakes, not fearing death itself, to relieve and keep them from danger.[64]

In another incident Monro recalls:

> There was a sergeant's wife… who without help of any women was delivered of a boy… The next day she marched near four English mile[s], with that [baby] in her arms, which was in her belly the night before, and was Christened the next Sunday.[65]

Whether the boy survived or not, Monro does not record, but the memoirs of a German soldier, who has been identified as Peter Hagendorf and who served in the Thirty Years' War, records that four of his six children died on campaign, along with his first wife. Of course, with the high child mortality rate there is no way of telling whether these four children would have

---

62 Turner, *Pallas Armata* p.277.
63 Mathew Glozier, 'Scots in the French and Dutch Armies during the Thirty Years' War' in Steve Murdoch (ed.) *Scotland and the Thirty Years' War* (Leiden: Brill, 2001) pp.132–136; I. J. MacClean *De Huwelijksintekenungen van Schotse Militairen in Nedeland* (Zutphen, 1976); TNA SP 9/95 Register for Brill garrison, ff.41–53, 180–190; SP 84/58, List of Her Majesty's officers at Brill and muster roll, 7 April 1598, f.128.
64 Monro, *Monro His Expedition with the Worthy Scots Regiment* part II p.29.
65 Ibid., p.6.

survived if Hagendorf had followed a different profession, but when he was wounded his wife helped nurse him back to health.[66]

Nor were these women always young, both Lucie Howard, whose husband was 'a soldier of the League,' and Ann Morecott, whose husband was a 'soldier in Dort,' were both 50 years old when then joined their husbands at the end of 1629. While Marie Doughton, the 'wife of Thomas Doughton, soldier in Flanders' was 53 when she sailed on 25 May 1631 to join her husband. Some were not alone, Hester Milborne travelled with her five year old daughter when she joined her husband in 1632. Boys who accompanied their parents might become a servant to a soldier and when they were old enough join the ranks themselves.[67]

While with the army, according to Sir Horace Vere's standing orders dated about 1614, 'The women and boys must march in the rear of all with one of the [provost] marshal's men to keep them together that they do not run among the soldiers' ranks'.[68]

There was a hierarchy amongst the women, a sergeant's wife was higher than a musketeer's, but a musketeer's wife was higher than a musketeer's whore, who in turn was higher than a prostitute. The women's status in camp society would also depend on whether she sat in a baggage wagon or walked. On the other hand, if she tried to rise above her place then this could lead to disorder among their partners who often became embroiled in the quarrel. If she became a widow, a soldier's wife would have to quickly find herself another husband to survive the rigours of campaign and not slide down the pecking order of camp society. No doubt these widows were in great demand from the unmarried soldiers.[69]

However, sometimes the women could be a detriment to military discipline. In August 1633 part of the garrison of Erfurt was ordered to dig some fortifications, but:

> Their women began to wonder where their men had gone and they went looking for them. They brought food and drink with them. Once they had begun to settle in, little work was done afterwards. More pints of beer and wine were drunk than carts of earth moved.[70]

Unfortunately, how many of these women returned home once the wars were over is not known.

---

66 Jan Peters, *Ein Soldnerleben in Dreissigjahrigen Krieg*. (Berlin:Akademie Verlag, 1993).
67 TNA E157/14, Registers of soldiers taking the oath of allegiance and register of licences of persons to pass beyond the seas, 1628–1629; E157/15, Registers of soldiers taking the oath of allegiance and register of licences of persons to pass beyond the seas, March 1631–December 1632; E157/16 Registers of soldiers taking the oath of allegiance and register of licences of persons to pass beyond the seas, December 1631–December 1632.
68 TNA SP 9/202/1/18 Compendium of the Discipline of the Art of War under Sir Horace Vere, 1603–1625.
69 John Lynn, *Women, Armies and Warfare in Early Modern Europe* (Cambridge: Cambridge University Press, 2008) p.67.
70 Quoted in Holger Berg *Military Occupation under the Eyes of the Lord* (Gottingen: Vandenhoeck & Ruprecht, 2010) p.58.

THE RANK AND FILE

Poise your musket

3

Dressed to Kill

Clothing was not only essential to a soldier's health and welfare but was also seen as his right on enlistment, inasmuch as when the clothing for Captain George Tucker's company failed to arrive, he complained to Sir John Coke, that 'the soldiers pressed for the King of Denmark out of London are in mutiny for want of cassocks and other necessaries'. The word coat and cassock are often seen as interchangeable in military contracts at this time, but cassocks were often worn over the doublet, likewise hose could mean breeches or stockings.[1]

Originally the pikemen wore armour which protected their arms and so only musketeers wore coats, but by the beginning of Charles' reign, judging by the quantities involved, all the soldiers appear to have been issued with coats. Moreover, during Elizabeth's and James' reigns soldiers were issued with two suits of clothing a year, a woollen one for the winter and a linen one for the summer.[2] However, by the time Mansfeldt's Army was raised in 1624 soldiers appear to have been issued with just the winter clothing to save money, which meant that their clothing had to last 12 months instead of six.[3]

During Queen Mary's reign each county was ordered to supply a white coat to each of its recruits and Elizabeth's government usually specified either blue or tawny coats. However, under the early Stuart Kings there was no such specification and each county was free to issue coats to their recruits in whatever coloured cloth was available, i.e. in 1625 Captain Thomas Lynsey's company from Bedfordshire had grey cassocks lined yellow, whereas Captain Walter Morton's company from Breckon wore red cassocks lined

---

1   George Tucker to Sir John Coke, 29 April 1627, *HMC Cowper*, p.303.
2   For winter this was, A cassock of Kentish broad cloth lined with cotton and trimmed with buttons and hoops, 14s 6d, a doublet of canvas with white trim, 7s 6d, three pairs of neats leather shoes, 5s, two shirts of Osnabridge and two bands of Holland cloth, 6s 5d, three pairs of Kersey stockings at 20d the pair, 5s, a pair of Venetians [breeches] of board cloth with buttons and lined with white lined, 8s. For the summer, a canvas doublet lined with white linen, 8s 6d, a pair of Venetians of board cloth lined in white linen, 8s, two shirts of Osnabridge and two bands of Holland cloth, 6s 5d, two pairs of shoes of neats leather, 3s 4d, two pairs of kersey stockings, 3s 4d, a hat with a band to the same, 3s 9d.
3   The winter one was to be issued by the end of January and the summer one between the end of July and 24 August; TNA SP 14/181/39, Secretary Conway to the Earl of Lincoln, 10 January 1625; TNA SP 16/522/85 Lord Conway to Council Board, December 1625.

blue.⁴ Moreover, some counties only issued their recruits with just a coat no matter how poorly they were clothed, for example in June 1625 when the recruits from Northumberland arrived at Plymouth they were described as 'for the most part naked, save their coats'. Nothing had changed two years later because despite the Privy Council ordering that the recruits be 'clothed and fitted for our service', they still arrived at the rendezvous poorly dressed.

In theory each county was reimbursed for these coats, but the government paid just four shillings per coat, which had been the practice during the Elizabethan era, although by 1627 it cost 12 shillings to buy the materials for a coat and have them made-up. This resulted in many counties trying to cut the cost in their manufacture so that when Major Leigh inspected the soldiers at Plymouth in June 1625, he found that many of the coats were not fit for purpose. Leigh found the coats of the recruits from the Isle of Ely were worth 9 shillings and were 'lined all through but the arms', and were 'very short and mard with making'. Whereas the coats of the Cardiganshire soldiers were also described as being 'very short', while the coats of the Shropshire recruits were no better being 'worth five shillings, some six shillings and some seven shillings'.⁵

On the other hand, only 10 coats were made at Norwich for the Norfolk contingent the other 340 coats were made by London tailors, while 13 of the 100 recruits from Buckinghamshire were issued with 'old coats' from the county's store and the remainder were issued with new ones, although the county claimed 14 shillings per coat.⁶

Several accounts for making these coats have survived. In 1625 Leicestershire paid £90 for 225 yards of green broadcloth and £18 for 240 yards of yellow cotton to line the coats for its 150 recruits destined for Mansfeldt's Army, plus £15 for thread and making the coats. In addition, a further 16s was paid 'for cloth and cotton for a pattern for a soldier's coat'.⁷

In 1625 the bill for clothing the Hampshire recruits was:⁸

For 2½ yards of broadcloth, six shilling per yard	15s
For bayes for lining of the jacket	3s
For a doublet cloth	3s 6d
For canvas to line the breeches	1s 6d
For a hat	4s
For a pair of stockings	2s 9d
For making a suit	3s 6d
For two bands	1s
For a shirt	2s 5d

---

4 TNA SP 16/4/160, Report of Captain Edward Leigh, July 1625.
5 *CPSD 1625*, Captain William Courtney to Secretary Conway, p.51. Of the 38 recruits from Hampshire which were reported to be a Southampton in August 1627, only three were described as 'well clothed', the remainder needed shirts, shoes, stockings and some even breeches.
6 BL Kings ms 265 letters of the Privy Council to and orders issued by the Commissioners for musters, of Norfolk, f.264: *Verney papers* pp.289–293.
7 TNA SP 16/70 f.64, Accounts for Leicestershire.
8 Hampshire Record Office 44/M69/G5/48/126 Nicholas Prescod's Account of Equipment Provided to the Soldiers, 1625–1627.

For two pairs of shoes	4s 8d
For buttons	4d
For bringing down the hats, stockings and broadcloth From London	4d
For a pair of shoes more	2s 4d
	£2 4s 4d

Since many of the recruits arrived at Plymouth half naked, Charles' government was forced to issue 1,003 breeches, 1,618 shirts, 1,278 pairs of shoes and 1,278 pairs of stockings to these soldiers. In addition, on 28 June 1625 the Privy Council also directed the commissioners at Plymouth to provide a further 2,803 pairs of shoes at two shillings per pair at £208 and six shillings, 396 pairs of breeches at eight shillings per pair at £158 8 shillings and 1,900 shirts at three shillings or £286 12 shillings and 1,800 pairs of stockings at 6d per pair or £120 six shillings and eight pence. Despite these orders on 23 August 1625, Ogle wrote to Lord Conway complaining, that 'clothes of all sorts are wanting to make this body complete'.[9]

However, despite the condition of the soldiers' clothing it was not until 12 December 1625, after their return from Cadiz, that the Privy Council agreed to 'provide every soldier [with] a complete suit of apparel', adding, 'you should use all necessary frugality for the king and the soldier so we would have you carefully provide them with very strong lasting and sufficient [clothing], but [we] leave the price to your own discretion and to furnish or contract for the same'.[10]

The coats appear to have come from various sources because an account from Hampshire in December 1625 also records payments for items of clothing:

85 soldiers of Colonel Bruce's company were issued with suits to the value of 44s 4d, £190 12s 8d
The like for 86 soldiers of Sir James Scott's company at the like rate, £190 12s 8d
40 suits for 40 soldiers billeted at Fareham at 47s the suit, £84 00s 00d
The like for 25 soldiers of St Legers company at 43s the suit £55[11]

Whereas in the Isle of Wight between 15 December 1625 and 6 October 1626 the soldiers received:

Four score [and] eight suits of apparel for the soldiers at 44s per suit
Shoes for them 176 pairs at 2s the pair

---

9   TNA SP 16/4/160, Report of Captain Edward Leigh, July 1625; SP16/5/70 Sir John Ogle to Secretary Conway, 23 August 1625; AO 1/300/1138.
10 Lyle APC 1625 Privy Council to Cecil and other senior officers of the army, 12 December 1625, p.267.
11 TNA AO1/300/1140, N. Prescod, Treasurer of the Loan Money of Southampton, 20 December 1625 to 30 September 1626; TNA E351/287, N. Prescod paymaster for the billeting of troops, 20 December 1625 to 30 September 1626.

Henry Ringwood was paid for 37s 10d for cloth and other things for apparelling, Lieutenant Edward Chadwell and 34s 7d for Sergeant Thomas Thompson's clothing.[12]

Why these suits should vary in cost between companies is not recorded, but it may be that the coats were made from different coloured cloth.

In 1625 Berkshire had issued its recruits with 'grey cassocks lined white with ash coloured buttons', but for its 1627 recruits they were issued with the same coloured coats except with blue buttons. In October 1627, the Western Division of Somerset also supplied its 75 recruits with blue coats, while those of Hampshire in 1627 received grey ones. In addition, the Hampshire records also show they were issued with shirts, stockings, shoes and breeches which came from a different supplier.[13]

Judging from the coats made for the Leicestershire and Hampshire recruits the coats appear to have been made from 1½ yards of cloth, and the breeches one yard. However, the exact design of these coats is open to debate; some contemporary illustrations show soldiers wearing at least two types of coats; one an overcoat with buttoned sleeves that could be undone leaving the sleeves hanging and also a tight fitting coat, with or without buttons on the sleeves. Some soldiers may have had both which was probably the case with Mansfeldt's Army because on 1 December 1624, George Hooker was contracted to supply, '10,000 coats or cassocks to be made large and plain of good strong broadcloth and lined through the bodies with bayes'. Since these men would have already received coats from the counties where they had been raised these must have been the 'overcoat' style of cassock. Hooker also supplied a further 2,250 cassocks for the troops in Ireland.[14]

However not all the soldiers received their new issue of clothing. On 3 January 1626, the Mayor of Dartmouth complained to the commissioners at Plymouth:

> I have according to your order provided shirts for all those sick soldiers which are here amongst us and do hope that other places where the rest are billeted have done the like, but howsoever the poor creatures are so destitute of other apparel to cover their nakedness that the cold pinching them some die and others cannot recover their health.[15]

On 7 January 1626, the Deputy Lieutenants of Hampshire wrote to Secretary Conway that the two companies that were at Southampton, were 'in such a miserable case as that they are not to be lodged anywhere until they be clothed'. Nothing had changed by 25 January when the mayor of Southampton

---

12 TNA E351/286, Army paymaster and treasurer of the forces Sir E Dennis to Secretary Conway, 28 January 1626.
13 TNA SP 16/53/72, Account according to the allowance made by the Council in Queen Elizabeth's time, 9 February 1627; SP 16/82/65 Secretary Conway to George, Earl of Totnes, Master of the Ordnance, 23 October 1627.
14 HLRO HL/PO/JO/10/1/27. Main Series of papers.
15 TNA SP 16/18/99, Commissioners at Plymouth to Council, 20 January 1626.

wrote to Conway, that the soldiers, 'shall be completely apparelled as soon as possible… which I find will amount to 40s each soldier or thereabouts'.[16]

Meanwhile at Plymouth, Colonel St Leger wrote to Conway on 28 January 1626, that although in the past three weeks he had issued warrants for 2,000 suits of clothing from the market towns in Devon and Cornwall, his soldiers still 'lie in bed for want of clothes', adding 'without present monies the service is not to be performed… [and] I fear we must be supplied from London'.[17]

Apart from the coats issued to the recruits by each county, was there any uniformity among the regiments at this time? In 1627 when Phillip Burlamachi was commissioned to supply the clothing to the army, he is known to have purchased cochineal and indigo, which are used for dying cloth red and blue. Unfortunately, the quantities are not recorded.

Certainly, Captain Ogilvy's Company in the Earl of Nithsdale's Scottish Regiment of Foot appears to have worn red coats, while the majority of the men in Captain Leslie's Company, probably also in Nithsdale's regiment, were described as having blue coats; although a William Porter, Ogilvy's company fifer, was to have 'a suit of clothes of any colour'. This is the only reference to a musician's coat, which are often depicted as being in more elaborate coats than the rest of the regiment. Therefore, presumably drummers were dressed in the same fashion as the rest of the regiment, or the coats came from a different source.[18]

Alternatively enough cloth could have been purchased to issue several regiments in the same colour, as happened in 1642 when at least 4,500 grey coats were supplied to the English troops in Ireland. On the other hand, the majority of the regiments of Manchester's Eastern Association were in red coats, which colour was later adopted by the New Model Army. A regiment might also change the colour of its coats each time it was issued with new ones.[19]

The coats may also have been sorted at a collection point, so that each regiment was appointed to have a different colour, as happened in 1642 with Essex's Parliamentary Army, and probably the Royalist army too.[20]

Since many counties formed companies in a regiment, it is possible that as well as a company uniformity, there was a regimental one also when they were supplied with the 'overcoat' style of cassock. There also seems to have been a distinctive pattern between soldiers' and civilians' coats, because on 23 November 1627 James Chidley of Hampshire was charged with possessing a 'blue coat of a soldier' belonging to Captain Drury's company of Sir Charles Rich's Regiment. Chidley was 'bound in' for £10 an enormous sum at the

---

16 TNA SP 16/18/21 deputy Lieutenants of Hampshire to Secretary Conway, 7 January 1626; SP 16/19/88, Commissioners at Plymouth to Council, 31 January 1626. In July 1624 it was reckoned that English soldiers levied for the service with the Dutch cost 45 shillings for clothing.
17 TNA SP 16/19/66, Sir William St. Leger to Secretary Conway, 28 January 1926.
18 Fraser, *The Book of Carlaverock* vol. 2 pp.91, 96, 104.
19 TNA SP 28/1b, Warrants issued by the army committee, 1 August 1642 to 30 September 1642.
20 Peachey, S and Turton, A, *Old Robin's Foot* (Leigh on Sea: Partizan Press, 1987).

time.[21] Unfortunately, the indictment does not say how they could identify it as a soldier's coat.

Robert Swann was also a military contractor, who in 1624 had provided the shirts for the Mansfeldt's regiments, but he does not seem to have received the contract to make the shirts for Cecil's army. Instead, the order went to a London linen merchant who was to provide 10,000 shirts at two shillings and six pence each, which were packed in dry fats and sent to Plymouth. In 1626 Swann petitioned the Privy Council for a new contract on behalf of the 300 to 400 men and women that he employed and that he still had a large quantity of linen left over from the shirts he had made in 1624. Whether he was successful or not is not known, but his petition shows the large number of civilians, including children, that were needed to put an army into the field, whether tailors, spinners, weavers, dyers and other professions who engaged in the manufacture of cloth and leather.[22]

One of the few contracts to survive was issued to Daniel Condy, a Devonshire merchant, on 19 January 1626, who was to make 200 suits, 40 cassocks were to be a 'medley' colour:

> The rest to be deer colour, sage colour, watchett and olive colour. The cloth [was] not to shrink, the cassocks to be lined all with bayes, except the sleeves, which are only to be faced. The suit to be garnished with a gross of thinned buttons…The hose to have cotton within and linen within. The doublet to be of fushian serge or sackcloth with two linings, to be a yard at the least in the waste and some more in the sleeve to be three-quarters or above.

Condy was also to supply every soldier with a hat, two shirts, a pair of stockings, two pairs of shoes, three soles of neates leather, 'size ten or above' and two bands, which were to be 'both for fashion, goodness and largeness to be answerable unto a pattern remaining in our hands'. The seventeenth century shoe size 10 is the modern day equivalent to size seven. The shoes were probably tied together and marked on the sole with its size. In return Condy was to receive £3 per suit to be paid 'a third in hand, another third at the delivery of the clothes and the other third three months…after delivery'. How the five different coloured cassocks were to be distributed is not recorded.[23]

Between 18 January and 9 February 1626, 16 clothiers in Devon were ordered to supply 2,856 suits of clothes for the soldiers ranging from 16 to 800 suits per merchant depending on their capabilities. The Dorset clothiers also issued badly needed clothing to the soldiers on credit, but in March 1626, despite £5,000 being set aside for their payment it was not until the end of 1627 that they finally received their money and some even had to

---

21 R. C. Anderson, *Examinations and dispositions, 1622–1644* (Southampton: Cox and Shorland, 1929) vol. 2, pp.10–11.
22 TNA SP 16/34/112, Petition of Robert Swann,? August 1626; AO1/299/1135 Roll Sir William St Leger Colonel and land forces in the Expedition to Cadiz, 4 July to 31 October 1625.
23 TNA SP 16/148/100.1, Agreement between his Majesty's Commissioners and Daniel Condy, 19 January 1626.

wait until April 1628. In addition, Captain Mason was ordered to supply 500 suits for the soldiers at Plymouth which cost £1,000 and William Stroud was to supply 'shoes and stockings for the soldiers at Plymouth and elsewhere'.[24]

Those soldiers who had landed in Ireland after the Cadiz Expedition also needed clothing and on 31 March 1626 the Privy Council specified to the Lord Deputy of Ireland how to clothe the soldiers there:

> Your Lordship [is] to proceed forthwith to the contracting for suits for them at the cheapest rates you can, but so as the clothes may be good and serviceable. The allowance made here for the soldiers is after the rate of 48s viz. cassock well lined with bayse, a good fustion doublet and a pair of cloth hose both well lined, two shirts, two pair of stockings, two pair of shoes, two falling bands and a hat.[25]

However, by the following year these clothes had worn out, which again brought a flood of complaints. On 22 January 1627, the Privy Council reported that there were 3,800 soldiers who were in a 'great want of hose and shoes', so five shillings was to be allocated per soldier to purchase these items. On 14 March 1627, the Privy Council also ordered each county where the soldiers were billeted to issue each man with 'a shirt, a pair of stockings and a pair of shoes' before they marched to their rendezvous, although on 29 April Captain Mason was ordered to supply hose and shoes for the soldiers quartered in Dorset.[26]

In 1627 the clothing contract again went to Philip Burlamachi, who had, in 1624, clothed the 6,000 English troops serving in the Low Countries and had also supplied them with arms. On 27 March 1627, the Privy Council ordered that Burlamachi was to:

> Make provision of the clothes for the five regiments of foot to be employed in his Majesty's fleet, shortly to put to sea, to the end therefore that the said clothes may be as well for the quality and fashion of them as for the prizes of each suit, such as nay be fit and serviceable for the soldiers and well husbanded for his Majesty; it is thought fit and ordered that Sir John Burlace and Sir John Burroughs knights and colonels confer and advice with the said Burlamachi and agree as well of the fashion as of the rates for to be allowed for each suit and that accordingly the said Burlamachi proceed in providing of the same.[27]

Burlamachi estimated that it would cost to clothe each soldier as follow:[28]

Large cassock and hose, 2 yards and a half at eight shilling	20s
Cotton to line the cassock, 2 yards at 2s 6d the yard	3s [sic]
Lining for the hose either leather or linen	2s 6d

---

24 TNA E403/2669. Privy Seals Dormant 1627–1628; SP 16/23.
25 Lyle *APC 1625* Privy Council to Lord Deputy and Council 31 March 1625 p.402.
26 Lyle *APC 1627* Privy Council to Deputy Lieutenants of Kent, Sussex, Hampshire, Dorset and Berkshire, 14 March 1627. pp.130–131.
27 Lyle *APC 1627* Privy Council minute concerning Phillip Burlemachi, Merchant of London, 27 March 1627, p.162.
28 *Ibid.*, Estimate of the charge of clothing the soldiers, 31 March 1627, p.194.

Making with buttons	4s
10 dozen buttons to each cassock and hose at 2d the dozen	1s 8d
Silk [thread] for buttonholes	31s 10d
Shoes, [of] neats leather, three soles at the pair	2s 10d
Shirts	2s 6d
Band	1s
Large knit stockings	2s
Hat or cap	3s
Total	43s 2d

In addition, Philip Burlamachi was also contracted to supply 400 recruits with 'close cassocks and breeches' on 13 August 1627 at a cost of '27 shillings and 8d per man'. These cassocks appear to have been the same size as those supplied to the recruits raised in Hampshire and Leicestershire in 1625.

When Crosby raised his regiment in Ireland early in 1627 it was estimated that 10 shillings would be enough to clothe a soldier, whereas Bingley's English Regiment which was also stationed in Ireland, was to receive £1 10 shillings per man for clothes. It may be that Bingley's regiment were issued with clothes made from English cloth, whereas Crosby's men were issued with cheaper Irish cloth.[29]

Further demands and issues for clothing came during the summer of 1627. On 3 June 1627 Sir John Burgh wrote to Buckingham 'shirts, shoes and stockings for three thousand new men are wanted'. Since the army was just about to sail the Privy Council had to act quickly and on 21 June it ordered Burlamachi to be paid a further £1,200 for 3,000 pairs of shoes, 3,000 pairs of stockings, and 3,000 shirts 'to be distributed amongst the soldiers… that have most need'. On 27 June Buckingham's army mustered just 5,934 men, which had increased to 7,424 by 20 July, even so this is a considerable size of men lacking clothing.[30]

Despite these issues of clothing, the soldiers in France were said to be in 'extreme want… and that most of them do go barefoot', which, in October, forced the Earl of Holland to provide 'two or three thousand pairs of stockings [himself] that are very scarce for our men that work in the trenches'. The situation was aggravated by some soldiers selling their clothes to the local population.[31]

On 15 October 1627, the Privy Council ordered that payment of £600 be made for, '400 close cassocks and 400 caps and as many pair of stockings for the clothing of the Highland men, which are to come about from Scotland… the said clothes at the rate of 25s for cassocks and hose, and 5s a cap and pair of stockings'. These were for the Highlanders in Captain Machnaiton's and Captain Campbell's companies of Lord Morton's Regiment, who had not reached the rendezvous in time to receive their clothes. This time David Edwards, who received the clothing contract on 24 October 1627, was

---

29 *CSP Ireland 1627* p.240, An estimate of the charge of recruiting, apparelling, arming, 3,000 men out of Ireland, May 1627.
30 TNA E403/2746, Order Book Pells, 1627, f.102.
31 SP 16/83/8, Earl of Holland to the King, 27 October 1627.

ordered to deliver them to Southwark, from where they were presumably checked and sent to the regiment.[32]

In November 1627 Burlamachi was again contracted to supply 6,000 suits, 'viz cassocks, hose of cloth, shoes, stockings, shirts, bands and Monmouth caps… that the soldiers may be the sooner accommodated with clothing winter now already [having] begun'. Despite the thousands of Monmouth caps being issued to English soldiers during the sixteenth and early seventeenth centuries, it was such a common item of head wear that no one thought to describe it. All we do know that it was knitted, milled, blocked and worked with teasels and in 1644 cost between two and four shillings. Suggestions as to its appearance vary between a closely fitted cap to having a brim and a tall crown.[33] Although taxation records suggest that there was more than one design so both descriptions may be correct.

Burlamachi was again ordered to supply an additional 4,000 suits on 28 July 1628 and two days later he was paid £500 for 'stockings, shoes and shirts for 1,000 men of Colonel Crosby's and Sir Ralph Bingley's Regiments'. The shoes were to be supplied by William Stroud who was to receive £1,000.[34]

Unfortunately, neither Hooker's, Burlamachi's or Edwards' contracts have survived, but contracts made during the early part of James' reign state that the clothing 'shall in their several kinds be fully as good as the patterns of every parcel, now provided and remaining in His Majesty's Great Wardrobe in London'.

Once the order had been completed it was to be:

> Brought to such convenient place or places in the city of London as shall be met for the same to be viewed and there shall suffer and permit the persons that shall be appointed by his Majesty's behalf for viewing of the said apparel to see, visit and survey every part and parcel thereof. To the end that they find the same apparel agreeable with the patterns may give their allowance of the transportation thereof, or disallow and reject such part of the same as they shall see to be defective and not answerable to the patterns.

Examiners of the newly made soldiers' clothing are also mentioned in the New Modal Army contract book and dates back to at least Elizabethan times when it was suggested that there should be only one 'reviewer' if he was 'an honest man'. However, since there were at least two examiners still inspecting the clothing during the 1640s it would appear that honest men in this position appear were hard to find.

---

32   Lyle *APC 1627–1628* Minute of the Privy Council, 15 October 1627, p.85; TNA E403/2746 Order Book Pells, 1627, f.168; Edwards' warrant says 6s per cap and stockings; E403/2669, Privy Seals Dormant or Posting Books, 1st Series 1627–1628.

33   TNA E403/2746, Privy Seals Dormant 1627–1628; Stuart Peachey (ed.), *Richard Symonds, The Complete Military Diary* (Leigh on Sea: Partisan Press,1989) p.10; Morris, Robert *Headwear, Footwear and Trimmings* (Bristol: Stuart Press, 2001) p.20.

34   BL Add ms 18,764, Philip Burlamachi, Merchant, Accounts of Money paid to him for advances for the King's Service, 1620–1634, ff.19–28; Lyle, *APC 1627* Privy Council Minute concerning clothing, 2 November 1627, p.125, 162; TNA SP 16/84/7, Order of the Council to Lord Treasurer Marlborough for clothing, 2 November 1627.

Once the examiners were satisfied that all the clothing was correct it was to be:

> Put in several packs…The pack to be sealed with the particular seal of the said persons or some other common seal to be agreed upon by them which packs so sealed shall without restraint of any officers of the Customs House [to be] immediately transported and not opened either here or beyond seas until they arrive at the garrison or place… Whereunto the same shall be consigned and there seen by the commissary of the muster being now the deliverer of the apparel… The same to be opened and the apparel immediately distributed to every company.[35]

During the Civil War, these packs or 'dry fats' as they are also known, contained about 500 coats and were numbered and loaded on to carts to be transported to the army. Unfortunately, it is not known whether the clothing within each bundle was of the same colour, although presumably a numbered dry fat was allocated to a particular regiment.

There are examples of the some of this clothing being stolen or even becoming rotten due to it becoming wet during transportation due to foul weather. Moreover, one of the many abuses committed by the clothiers was stretching the cloth too much or not washing it before it was made into suits so that when it became wet the cloth would shrink. On the other hand, paying the merchants for the clothing was not high on the government's agenda and payments were still being made in 1632.[36] They were also required to supply the clothing very quickly, so may not have had the resources to complete the orders on time. For example, when George Hooker was given his contract, he had only 20 days to complete his order of 10,000 coats, and so, like Robert Swann who is mentioned above, must also have had a large network of artisans. In theory outer garments were made by male tailors who had served a seven year apprenticeship, whereas shirts et cetera were mostly made by women. However, to save money some merchants hired 'foreigners', or rather unskilled labour, who were much cheaper than their professional counterparts. One of the charges brought against Colonel Sir John Clotworthy during the early 1640s, who was responsible for clothing the English Army in Ireland, was that he used unskilled labourers, or 'apple women' who received as little as 4d for making a soldier's suit.[37]

For these women, the cloth was cut out beforehand and delivered to the unskilled workers to be sewn together, which resulted in many of the coats and breeches tearing when the soldier tried to put them on because either they were too small or had not been stitched properly. The soldiers'

---

35 TNA E351/170, U Babington and R Bromley Contract for Clothing 30 September 1602 to 1 April 1606.
36 BL Add ms 18,764, Philip Burlamachi, Merchant, Accounts ff.19–28.
37 HLRO HL/PO/J0/10/1/27. Main Papers, Hooker received the contract on 20 December 1624 and it should be completed by 20 December 1624. In February 1629, some counties were still being reimbursed for the expense they had paid in 1625, although it was just 4s.

complaints also show that despite supposedly having been checked by the reviewers, this poor quality clothing had been issued to the soldiers.

Unfortunately, Burlamacchi's records have not survived, but we know that he employed workers in England and the Low Countries. However, William Calley who supplied suits to the Spanish Army during the reign of James I, records the purchase of 438 pieces of cloth in 1606 from 11 clothiers. He also paid for the drawing and pressing of the pieces of cloth. A piece if Reading cloth cost Calley £18 and £24 for Kentish cloth. The following year 1,710 pieces of coloured cloth were shipped to Gravelines and 1,092 pieces of cloth and 11 bayes were sent to Dunkirk by Calley to be made into suits for the Spanish Army, but the colour of this cloth is not recorded.[38]

There was no cavalry with Cecil's force in 1625, but in 1627 there were one or two troops of horse with Buckingham, which according to the various sources were harquebusiers. In theory these were equipped with a buff coat, back and breast plates, a helmet and armed a carbine and two pistols. However the clothing was usually supplied by those that had raised them, who were to make sure that they were 'well apparelled and furnished with buff jackets, swords, daggers, cases of pistols.'[39] Rather than being supplied from a central sources the troopers appear to have been supplied by individuals, such as when Secretary Conway informed Lord Montjoy on 18 May 1627, that he had supplied four troopers with, 'A headpiece, a back and breast of pistol proof [armour] a carbine of four foot long with a belt a flask leather and flask and a case of pistols with their flasks and a token of good wishes to the journey they have buff coats, scarves and feathers according to the direction'.[40]

Even during the Civil War, of the 265 arquebusiers raised for the Earl of Essex's army in 1642, 202 wore just buff coats and 60 had just back and breast plates with a helmet, with just two having a buff coat and a helmet and just one was without either.[41] Three types of buff coats can be identified during this period, those with buff sleeves, those with cloth sleeves and those without sleeves revealing the cavalryman's coat. What type was issued to Montjoy's troop is not known.

Since each nation's army was dressed in a similar fashion scarves or sashes were adopted, e.g. red for Spain and white for France, but what of Britain? There are two portraits of Cecil at about this time which shows him in a pink or rose red coloured scarf which according to his biographer, Charles Dalton, had originally been orange but had faded to that colour and that it was unlikely that he would be painted in 'Spanish colours'. However, in 1635 the cuirassiers (at least) of the trained bands were ordered to wear a 'scarf of rich crimson taffeta…which is to be two ells length cut through the breadth'. During the Civil War the colour of the scarves for the Royalist army was the same colour as in Cecil's portraits.[42]

---

38 Wiltshire Record Office 1178/325, Account book belonging to William Calley, 1600–1606.
39 Letter from Mr Beaulieu wrote to Reverend Joseph Mead dated 4 May 1627, in Birch, *The Court and Times of Charles I*, p.222.
40 TNA SP 16/63/101, Secretary Conway to Lord Montjoy, 18 May 1627.
41 TNA SP 28/131 part 3. Account of horse raised in London in 1642.
42 Dalton, *Life and times of General Sir Edward Cecil*…pp.360–361; *State papers relating to Musters, Beacons, Ship money etc* (Norwich: Norfolk and Norwich Archaeological Society, 1907) p.208.

Although contracts for clothing during Elizabeth's and the first part of James' reigns including the supply of apparel for gentlemen, no reference is made in the contracts between 1625 and 1628 for any officers' clothing, so presumably they supplied their own clothes. It is said that before he sailed for France Buckingham was often seen dressed 'like a soldier in his plume and buff coat'. However, his personal accounts record he had several suits made with gold buttons especially for the forthcoming campaign. For those gentlemen who decided to 'trail a pike' they may have worn their own clothing instead of regimental clothing that was supplied, a practice condemned by some German newssheets of the day.[43]

---

43 Quoted in Thomas Cogswell, 'Published by Authortie' newsbooks and the Duke of Buckingham's expedition to the Ile de Rhè' in *Huntington Library Quarterly* vol. 67 no. 1 (Berkeley: University of California Press, March 2004) p.2; TNA E351/170 U Babington and R Bromley Contract for Clothing 30 September 1602 to 1 April 1606; BL Add ms 12,528, Sir Sackville Crowe's Book of accompts containing receipts and disbursements on behalf of the Duke of Buckingham, 1622–1628.

THE FIRST BRITISH ARMY 1624-1628 (REVISED EDITION)

Join your rest to your musket

4

Arms and Armour

By the end of James' reign there were three types of infantryman: musketeers, pikemen and calivermen or arquebusiers, although the latter appears to have been found mainly in Ireland since the caliver, a lighter form of musket, was more suitable to the type of warfare there.[1]

In theory pikemen wore a helmet, back and breast plates, a gorget, which protected the neck, and tassets which protected the thighs. Four sets of pikeman's armour with tassets, belonging to the Littlecote Collection (named after the house where they were stored) weighed between 7lbs 8oz and 9lb 11oz, whereas one without tassets weighted 5lb 7oz. Some back plates were fitted with a small hook where the pikemen could hang his helmet on, which according to Markham 'will be a great ease to the soldiers… in the time of long marches', but on most surviving plates this no longer exist. Originally pikemen also wore vambraces, which protected their arms, but these appear to have gone out of fashion at the beginning of the seventeenth century.

There were many armourers who could supply this armour, among them were John Franklyn and William Couch, both of London, who were each paid £475 on 29 July 1625 for supplying 500 back and breast plates, gorgets and headpieces. On the other hand, John Ashton and Thomas Stephens received £1,371 10s 3d for 1,500 back and breast plates with gorgets and headpieces at 18 shillings a set. This armour was packed in 43 dry vats for an extra £21 10 shillings.[2]

Despite surviving examples of pikemen's armour having tassets, none of the payments record these being supplied. It would appear that, in Europe during the 1620s there was a fashion for pikemen's armour being produced without tassets. Among the armour captured on the 28 September 1627 on the French ship *The Saint Esprit* there were 991 breastplates with no fittings for tassets, which was sent to the Office of the Ordnance at the Tower of London. Some of this armour may have been issued to Buckingham's army in 1628 or during the Civil Wars because in 1650 only 260 back and breast

---

[1] Although in Ireland a company of foot was to be composed of 30 pikemen, 10 musketeers, 54 calivers, and six targeteers, *CSP Ireland,* Instructions for officers at musters, 1628 p.347. It was also found in the Spanish Army until at least the late 1630s.

[2] TNA WO 49/55, Debenture book, 1625, ff.95, 96.

plates remained in the stores and are, at the time of writing, on display at the Tower. The 'Toiras' armour as it is known because the name Torias is stamped on the armour after the French marshal, are not the only surviving examples of breastplates having no fittings for tassets. A set of armour preserved at the British Museum which is believed to have come from Antwerp around 1635 is also of a similar fashion although unlike the Toiras armour, which has a plain rim, it has a false plated design with rivets to match.[3]

In his *Duties of a private soldier*, Cecil recommended that a 'buff jerkin' or buff coat, be worn under the armour for added protection, but there is no mention of them in the surviving accounts and the back and breast plates were often lined with leather anyway to give added comfort to the wearer, although this added to the expense of the armour.[4]

To protect the pikemen's head there were various styles of helmets he could wear. The 'English morion', which was similar to the Dutch model with a large brim which could deflect sword cuts. A Spanish morion, which was in a form of a cone with a narrow brim and a German morion, which had a curved brim and a comb on top. In 1622 Francis Markham, recommended that pikemen should wear:

> Spanish morions well-lined within which a quilted cap of strong housewives' linen; for Buckram which is the usual lining is too course and galleth the soldier's head, as also is too stiff and unplyable by which means it will not guilt like the other. The ear plates shall be lined also.

Spanish morions and linings are also included in the accounts. John Cooper and Richard Nash were paid 6s 18d on 23 July 1625 'for 20 pieces of buckram to line the Spanish morions at 4s a piece' and 46s 5d for 18 pounds 13 ounces of 'cruel fringe to border these morions at 3s the pound weight'. A tailor was also paid £5 12d for lining 600 headpieces at 16s 8d per 100 and on 18 April 1627 a payment of 42s was made for 62 yards of russet linen to line 200 Spanish morions, 10s for towe and thread and a tailor was paid 33s 4d 'for lining them'.

There are several surviving examples of helmets having cotton lining, including two at the Wallace Collection in London. One is a red wool and linen cap filled with raw cotton wadding, although this was designed for a burgonet helmet. The other is made of canvas and linen padded with bast.[5] However, English armourers could not supply enough armour to the troops, therefore when Mansfeldt's Army was being raised it was supplied with armour from the Low Countries and in 1625 a further 2,000 sets of pikemen's armour was purchased from the Dutch armourers by Sir William St Leger.[6]

---

3   Guildhall Miscellaneous vol. 2 part 8 pp.334–335, S Ackermann, E. Gatti and T. Richardson 'A seventeenth century pikeman's armour from Antwerp' in *Arms and Armour* 2,000 vol. 7 no. 1, pp.30–39.

4   BL Harl ms 3,838, A volume containing a variety of historical papers, 1600–1690, ff.155–159.

5   Markham, *Five Decades of Warre* p.39; A. V. B. Norman Wallace, *Wallace Collection Catalogue, European arms and armour supplement*.

6   TNA 16/181/70, Sir John Ogle to Sir William St Leger, 15 January 1625.

Despite armour being issued for protection it could quickly become unserviceable, therefore Sir William Haydon, the lieutenant of the ordinance, had to send some armourers to Portsmouth to repair the 'arms of the troops which are unserviceable'.[7] One of these armourers was John Cooper, yeoman of His Majesty's armoury, who on 16 April 1627 received £33 9s 6d for repairing armour of the regiments which had returned from Cadiz. His bill for repairing the armour was as follows:

For 500 of great and small buckles	20s
For seven hides of leather	8s 15d
30,000 of nails of all sorts	4s 10d
For 220 joints for armour	£11
For 18,000 rivets for armour	18s
For 10cwt of black lead	10s
For two gallons of oil	8s 10d
For 600 hooks to hang headpieces there on	£4 4s 8d
150 dozen of tape for bordering of the headpieces	38s 6d
For hampers to pack these things in to be sent into the country.[8]	4s 6d

The leather was probably for lining and straps, whereas the black lead was to paint the armour to prevent it going rusty.

However not all the armour issued was new. There are several payments to armourers altering old back and breast plates to the 'modern fashion' and one payment made on 26 August 1627 to Richard Nash for £40 16s 1d was for:

Altering and repairing of 250 old Almain [German] and Flanders corslets, being before unserviceable… for lining towe and thread for the headpieces. for joints and colouring of the armour, for hides of leather and other necessaries and workmanship… towards the furnishing of the 2,000 men which were to come out of Scotland.[9]

However, the armourers' trade was by no means just a man's occupation, between 1606 and 1639, 10 women belonged to the London Guild of Armourers. Most are described as widows and not only continued to trade after their husbands' death but took on apprentices.[10] Each armourer had their own mark which would be stamped on their armour, such as Widow Blofield who had a ship and the letters MB. Once finished a pistol was fired at the breastplate, which made a small dent in the armour, so that the buyer

---

7   Lyle *APC 1627* Privy Council to John Jepson, Lieutenant of Portsmouth, April 1627 p.209.
8   TNA WO 49/58, Debenture Book, 1627.
9   TNA WO 49/58, Debenture Book, 1627 f.203.
10 These were Widow Foster, Jane Drake, Katherine Harris, Margaret Hall, Widow Moseley, Widow Hodgeson, Widow Blofield, Widow Wilkinson, Widow Cope, and Widow Hayes. Guildhall Library 12079/2. Guildhall library 12085.

knew that it was pistol proof. However unscrupulous armourers could easily forge this proof mark with a hammer.

The weapon from which the pikeman got his name was the pike, which according to Gervase Markham should be 'strong, straight, yet nimble... [and] of ash wood, well headed with steel, and armed with plates downward from the head, as least four foot and the full size or length of every pike shall be fifteen foot besides the head'. Whereas Cecil's *Duties of a private soldier*, suggests it should be:

> No more than sixteen feet. The head should be broad pointed like a Spanish sword and a thickness and temper strong and enduring. The checks of tough iron and well riveted because when they fail the pike ceases to be serviceable and thereby a man is lost. Lastly it should have a foot of iron, as well to preserve it.[11]

On 29 July 1625, John Harmer was paid for 3,594 long pikes at 3s 2d per pike and John Edwards was also commissioned to supply 2,000 long pikes, also at 3s 2d. He also received a payment for '200 long pikes of the Spanish fashion, the staves being coloured and armed with sword pointed heads and iron feet at the butt ends, 5s 4d'. What colour these pikes were is not recorded.[12]

The second type of soldier was the musketeer. Up to 1621 there does not appear to have been any specifications for his musket, but on 20 February 1621 the *Act for making of arms of the Kingdom more serviceable* had its first reading in Parliament. This stated that:

> A musket [was] to be four foot long and the bore to be so big that a bullet of the eleven [to the pound] be for it and the bastard musket to be of such a bore that a bullet of fourteen will serve for it... The marks of the man's [gunsmith's] name to be engraven with the year of the lord upon it.

On 26 March 1627 4,176 muskets were supplied by 24 gunsmiths who supplied between 15 to 355 muskets and on 6 November 1627, 22 gunsmiths supplied a further 2,633 muskets 'with tricker [trigger] locks [with a barrel] of four foot in length furnished with moulds' for casting musket balls. This time each gunsmith supplied between 15 to 390 muskets at 18s 6d apiece. One of the gunsmiths was a 'Martina Pope, widow' who supplied 40 muskets.[13]

Since muskets weighted about 12.1kg, the musketeers needed a rest to steady it while they took aim. According to Markham:

> For their right hand they shall have rests of ash wood or other tough wood, with iron pikes in the nearest end and half hoops of iron about to rest the musket on, and double strong strings fastened near thereunto, to hang about the arm of the

---

11 BL Harl ms 3658, A volume containing a variety of historical papers.
12 TNA WO 49/55, Debenture Book 1625, ff.96–97. in 1628 the Leicester Trained Bands were reported as having pikes 'all of one length with Spanish heads according to the modern form': SP 16/10 f.94, Henry Earl of Huntingdon to Conway, 30 November 1625.
13 She may have been the widow of Richard Pope who was paid for 29 muskets on 26 March 1627. TNA WO 49/58 Debenture Book, 1627, ff.79–81; WO 49/59, Debenture Book, 1627 ff.203–206.

soldier when at any time he shall have occasion to trail the same. And the length of these rests shall be suitable to the stature of the man, bearing his piece so, as he may discharge it without stooping.[14]

In July 1625 Henry Rowland was paid £758 12s 6d for 2,610 muskets rests at 9d each. By 12 April 1627, this price had risen to 10d when John Edwards 'his Majesty's pikemaker', was contracted to supply 600 musket rests for £25. Since these two manufacturers of musket rests do not amount to the number of musketeers known to be in the army at this time, presumably the others were issued with ones already in storage.

In his aim to make money Charles had granted the sole monopoly for making gunpowder to John Evelyn who owned some mills in Surrey. On 1 July 1624 Evelyn was contracted to supply 20 lasts of powder a month for a year to the Crown at 8½ pence a pound, one last being 24cwt or just over 2,687.5 pounds of gunpowder. This contract was renewed in subsequent years. However, despite payments being made to Evelyn by August 1625 the Crown still owed him £2,250 and in 1627 he had not received a penny from the Crown for six months. With only 50 lasts of powder remaining in the store the Earl of Totnes wrote to Buckingham on 12 January 1628 informing him not to expect any powder from Evelyn until he had received payment for the gunpowder he had already supplied.

Cecil is known to have been issued with 60 lasts of gunpowder in 1625 and Buckingham with 92 lasts. Even in peacetime Evelyn appears to have struggled to maintain the quantities of powder he was contracted to provide and with the outbreak of hostilities an additional source had to be found. On 3 April 1627, Philip Burlamanchi was paid £7,459 to import 100 lasts of gunpowder or saltpetre 'with all convenient speed he can', from the Low Countries and a further 56 lasts in July 1627. However it was found that English gunpowder was better quality than the Dutch gunpowder and cheaper, although Charles was reluctant to increase the number of powder mills in England, there were several other English manufacturers who produced gunpowder.[15]

With the increased demand for gunpowder the need for saltpetre (one of the ingredients for gunpowder) greatly increased so the saltpetremen were given greater powers to dig for nitrogenous earth, in dovecotes, barns, stables and outhouses and any empty building. However, by April 1628 it was said that these saltpetre men 'had grievously oppressed the people' having 'exceeded the limits' of their commission. Despite being ordered not to enter any occupied house, on 14 February 1630 Sir Francis Seymour complained that 'the saltpetremen care not in whose house they dig, threatening men that by their commission they may dig in any man's house, in any room and

---

14   Gervase Markham, *Souldiers Exercise,* pp.2–3.
15   TNA SP 16/433 Report of the officers of the ordnance, 30 November 1639, f.37; H E Malden *Victoria County History of Surrey* vol. 2 pp.316–317; TNA E 351/2709, Ordnance, miscellaneous P Burlamachi, powder from foreign parts, April–July 1627; E44/431, Agreement between James, Earl of Marlborough, et cetera, and Philip Burlamachi, 3 April 1627, Richard Weston says £14,000, so the £7,459 may just have been a part payment.

at any time… If any oppose them they break up men's houses and dig by force'.[16] Another complaint dated 30 April 1630 recorded that:

> There is no part of their commission which they have not extremely abused. As in digging in all places without distinction, as in parlours, bedchambers, threshing and malting floors, yea, God's own house they have not forborne… in bedchambers placing their tubs by the bedside of the old and sick, even women in childbed and persons on their death beds.[17]

Once the saltpetre and other ingredients for gunpowder, sulphur and charcoal, were mixed it was then stored in barrels until it could be issued. When the gunpowder was issued to the musketeers, they carried it in bandoliers which according to Gervase Markham were:

> About their bodies baldrick wise from the left shoulder under the right arm, he shall carry bandoliers of broad leather…and to this bandolier shall be fastened by long double stings (at least a quarter of a yard in length a piece, that they may with more ease be brought to the mouth of the musket) one large priming charge… and at least twelve other charges of wood, all made of some tough light wood or else of horn, and covered with leather.[18]

Francis Markham suggested 12 or 13 charges which 'must contain powder according to the bore and bigness of the piece by due measure'. Whereas Cecil recommends the musketeer have a 'bandolier of 15 charges'. This powder was poured down the musket barrel, and 'a charge greater than the rest… for pan powder'. Sometimes a musketeer would be equipped with a separate triangular shaped flask. However, since no flasks are recorded as being issued to any of the musketeers at this time, presumably they had a larger charge on their bandoliers as suggested by Cecil.[19]

Although there are several examples of tin being used to make the bandolier containers, they were usually made of wood and either covered with leather or painted. Since many of the bandoliers had 12 charges this has caused them to be erroneously nicknamed by some secondary sources as 'the 12 Apostles', but there is no contemporary evidence for this term.

Also attached to the bandolier belt was a pouch to carry musket balls, and in theory a mould to cast new musket balls, a worm to clear any blockages and screwers so that the musketeer could take the musket apart if needed. However, the bag on surviving examples of bandoliers appears to be too small to carry all these items.

Among the suppliers of bandoliers was John Gace and William Beacham who supplied '263 bandoliers of the middle size' for which they received £32

---

16   TNA SP16/1, Commissioner to Saltpetre men, nd [c.1625]; SP 16/101 f.99 letter to the Duke of Buckingham, dated 21 April 1628; SP 16/166/1, Sir Francis Seymour to Secretary Coke, 14 February 1630.
17   TNA SP16/165, Sir William Russell, Sir John Wolstenholme and Sir Kenelm Digby to the Lords of the Admiralty, f.38.
18   Markham, *Five Decades of Epistles of Warre*, p.34.
19   Markham, *The Souldiers Accidence*, p.3.

17s 6d in October 1625. Another payment was made to them on 12 April 1627 for supplying 2,500 bandoliers for muskets with 'broad girdles or belts at 3s a piece' and 500 with 'middling girdles or belts at 2s 6d a piece', for which they were to receive £437 190s. Another payment was made to them on the same day for £253 6s 8d for 1,000 bandoliers for muskets and 800 bandoliers for arquebusiers or calivers and 4,800 priming irons.

As already stated, calivers were a lighter form of musket and so did not need a rest, but they were less powerful and according to Sir Roger Williams' *A Briefe Discourse of Warre* 'one musket shot does more hurt than two caliver shots'.[20] Apart from the British forces in Ireland by the time of Charles' reign the caliverman had disappeared in favour of the musketeer, but some of Buckingham's soldiers are known to have been armed with them, because one account records that when Charles inspected the men at Portsmouth in June 1627 he ordered all those with defective arms and calivers should be issued with new muskets. However, since they were lighter than the musket, they were useful when it came to skirmishing so were ideal for service in Ireland. In fact, at a muster in 1625, 1,860 soldiers were armed with calivers and just 183 with muskets, and the Earl of Strafford's men in Ireland were still carrying them as late as 1638.[21]

Gace and Beacham continued to supply bandoliers in 1628 and beyond, but apart from the references to 'wooden charges' and 'middle size' belts the payments made no reference to how these bandoliers were made. John Gace was among those who were contracted to supply bandoliers to the New Model Army, which are described as being:

> Of wood with whole bottoms, to be turned within and not bored, the heads to be of wood, and to be laid in oil (viz), three times over, and to be coloured blue with blue and white strings with thread twist and with good belts.[22]

During the 1620s the Privy Council recommended that the bandoliers were to be 'covered with leather'. The twisted tread for attaching the wooden charges to the belt was considered better than leather strips because they were less likely to rot. The amount of gunpowder each charge held is not specified, but the amount of powder poured down the barrel of the musket was supposed to be half the weight of the musket ball, so a ball weighing 1.4 oz had to have 0.7 ounces of powder to propel it towards the target.[23]

One of the drawbacks of bandoliers was that the containers hanging loose on the belt could rattle, which could give away a musketeer's position at night. However, this was not the least of his worries because according to

---

20 Quoted in Henry J. Webb *Elizabethan Military Science* (London: University of Wisconsin Press, 1965) p.98.
21 TNA WO 49/58, Debenture Book, 1627, f.87; BL Harl ms 390, Collection of letters of Joseph Mead f.162; *CSP Ireland* 1625–1632, Sir John King's certificate of musters taken at Michaelmas, 1625, p.42; William Knowler (ed.), *Letters and Dispatches of Thomas Earl of Strafforde* (London, 1739) p.199.
22 Gerald Mungeam, 'Contracts for the Supply of equipment to the New Model Army' *The Journal of the Arms and Armour Society* 1968, vol.6 no.3 (Sept 1968), pp.88, 90.
23 TNA WO 49/79 Debenture Book, 1643.

the Earl of Orrery, they were 'Often apt to take fire… and when they take fire, they commonly wound and often kill him that wears them, and those near him, for… if one bandolier take fire, all the rest do in that collar'.[24]

Moreover, they were easily damaged, so Gace and Beecham were paid £156 8s 4d for 'repairing of 1,120 of old bandoliers for muskets with new charges and strings and for new repairing of 1,800 other bandoliers for muskets without adding of charges… being before unserviceable'. On 24 December 1628 they received an additional £77 11s 8d for repairing 1,080 bandoliers for muskets and 55 for carbines, which had been received into the stores after the army was disbanded.[25]

However, another way to carry their gunpowder was by using pre-filled cartridges, which are known to have been used during the early seventeenth century by both cavalry and infantry, as John Vernon states:

> … if you use cartridges, you shall find in your cartridge case a turned wooden pin which you must take, having cut lengths of white paper something broader then the pin is in length, and roll the paper one the pin, then twist one of the end of the paper, and fill it almost full of powder, then put the bullet on the top of the powder, twisting that end also, then put it into your cartridge case, now when you come to load your carbine or pistol with these cartridges, you must bite off that end of the paper where the powder is, pouring it into your carbine or pistol, then put in that bullet, and some of that paper will serve for a wad after it, and ram home, but always observe that your bullet be not too big, but that if may roll home to the powder, for if there be any distance between the bullet and the powder, it is likely to break the barrel of your carbine or pistol: But if you use a flask, which is my judgement is far better than cartridge, because that many times the trotting of your horse in long march, shaketh out all the powder of your cartridges and thereby causeth you to be unprovided for the sudden charge of the enemy, you must gage your flask and so lade your carbine or pistol with powder and bullet as before, but never prime before you have spanned. Now the quantity of the powder usually required for the loading of either carbine or pistol is half the weight of the bullet, but to avoid the carriage of either cartridge case or flask, there is a new invented spanner which contains some six charges with priming powder, which is more many times then is used in our skirmishes'.[26]

James Turner preferred cartridges (or patrons) to bandoliers because:

> … it is impossible for soldiers especially wanting cloaks… to keep these flasks (though well and strongly made) from snow and rain, which soon spoils them, and so makes the powder altogether useless. Besides the noise of them betray those who carry them in all surprisels, Anslachts and sudden enterprises. Instead of those let patrons be made, such as horsemen use, whereof each musketeer should be provided with a dozen, these should be kept in a bag of strong leather,

---

24 Orrery, *A Treatise of the Art of War* p.31.
25 TNA WO 49/55, Debenture Book, 1625; E407/13, Ordnance Quarter Book, January-December 1625; WO 49/59 Debenture Book, 1627.
26 John Vernon, *Young Horseman or the plain dealing cavalier* (London: Andrew Coe, 1644) p.10.

or the skin of some beast well fow'd, that it be proof against rain, this bag he may carry about his neck in a bandolier, or if the weather be extremely rainy in one of his pockets and in the other a horn with priming powder'.[27]

George Monck also suggests musketeers marching at the front or vanguard of the army:

… ought to have two pair of bandoliers furnished with powder and bullet and in case you have no bandoliers let there be provided for each musketeer… 12 cartridges, which they ought to carry in their right-hand pocket, and 12 bullets apiece in their pockets besides'.[28]

In *Kriegs-Buchlein* published in 1644 Hans Conrad Lavater describes how experienced musketeers threw away their musket rests and kept their musket balls in their mouths and loose powder in their pockets. To load their musket, they would grab a handful of gunpowder from their pocket and pour it into the barrel and then 'taps the butt of the musket twice on the ground so that the powder settles in the barrel. On top of it you let a ball run out of your mouth'. According to Lavater, the impact of the ball on the powder was like the impact of the musket being rammed once with the 'scouring stick', or ramrod. By this method a musketeer 'fires five shots before the other [i.e. an inexperienced musketeer] can do two or three'.[29] Although one wonders how many musket balls were accidentality swallowed doing this, and how many more managed to blow themselves up by keeping their powder in their pockets. Although there is some evidence that soldiers were supplied with breeches with leather pockets.

When firing a musket, gunpowder would quickly build up in the barrel, which would corrode the metal if left unattended, so to clean muskets scouring sticks or ram rods were used and if a musket misfired a worm, which was a sort of cork-screw shaped device, would be attached to the scouring stick to try and extract the charge from the barrel. In 1625 these worms cost 10 pence each.[30]

In 1614 Sir Horace Vere commanded that 'the musketeers are always to have [a] good store of match hung at their bandoliers', although in bad weather the match was likely to get wet if carried this way.

In 1627 this match cost £30 per ton and in June 1628, 15 lasts of powder, 26 tons of match and 10 tons of cast shot were issued to the land forces for the forthcoming campaign. However, a musketeer could get through 3 yards of match every 24 hours, so even on the march, or on sentry duty, musketeers would still be burning match even before they came in to sight of the enemy. To save match musketeers would often extinguish their match leaving every eighth man still alight, but this could lead to a desperate scramble for the

---

27   James Turner, *Pallas Armata* (London: Printed by M.W. for Richard Chiswell, 1683), p.176.
28   George Monk, General, *Observations upon Military and Political Affairs* (London: R White, 1796) p.150.
29   Hans Conrad Lavater, *Kriegs-Buchlein* (Zurich, Johann Jakob Bodmer, 1667), pp.85–86.
30   TNA WO 49/55, Debenture Book, 1625.

other musketeers to light their match if the enemy appeared. One such incident happened at the siege of Newark in 1644 when a detachment of Sir Miles Hobart's Regiment of Foot who had been relieved from guard duty extinguished their match only to be cut to pieces when they were suddenly attacked by Royalist cavalry. Those who were not killed were taken prisoner.[31]

Another disadvantage with the match was that you could see the lighted match in the dark, which gave away a musketeer's position. A Dutch invention, said to have been the idea of the Prince of Orange himself was a tin pipe, or cane, about a foot long in which the match was placed. Holes were made in the pipe to stop the match going out. How successful this was is not known, but on 16 July 1625, Thomas Horsell a London merchant received £536 for various items including '2,000 canes or plate for [match]' at £13 6s 8d. When Cecil's regiments were reorganised, '198 canes for match' were delivered to each regiment.[32] However, this disadvantage could also be used as a decoy at night by commanders using lit match to hide their escape, such as Sir William Waller did after the battle of Lansdown on 5 July 1643.

Not all musketeers were armed with matchlock muskets. After the Cadiz expedition 72 'snaphaunce' or flintlock muskets were returned to the stores. These were often used by soldiers who were assigned to guard the gunpowder stores, where lit match would prove extremely dangerous if not lethal. Later they appear to have been used to arm elite regiments.

Although in theory each musketeer had his own mould to cast bullets, in 1628 only 100 moulds were issued to the 4,000 musketeers that were going to relieve La Rochelle. There is little evidence to suggest moulds were issued previously, certainly a report into the Cadiz expedition found some soldiers' muskets 'were insufficient, some of them so grossly [made] that they had no touch holes' others complained that the musket balls 'did not fit the pieces to which they were assigned and that the moulds for bullets were so disorderly shipped that they could not be found'.[33] When musket balls did not fit down the barrel, according to the Earl of Orrery, the 'soldiers were forced to gnaw off much of the lead, others to cut their bullets', which would slow down the firing process and probably did not do the soldier's teeth much good either.

It is known that the government paid for large quantities of musket shot to be made and almost nine tons of musket shot destined for Mansfeldt's Army had to be melted down and recast because it was too large for the musket barrels.[34] Cecil is known to have taken 20 tons of musket shot and 20 tons of match with him to Spain, and Buckingham took 37 ton 4cwt of musket shot and 75 tons and 15cwt of match with him to France in 1627. Among those who supplied this shot was Joseph Day, a plumber, who on 9 March 1625 was ordered to be paid £168 for 12 tons of lead 'by him upon short warning cast

---

31   TNA SP 16/88/35, Book of accounts of the military company from 1616 to 1627; TNA SP 9/202/1/18 Compendium of the Discipline of the Art of War under Sir Horace Vere, 1603–1625; British Library, Thomason Tracts E 39/8.
32   C. H. Firth, *Cromwell's Army* (London: Meuthen & Co., 1912) p.84; TNA AO1/1833/13, Roll 13 H Lord Viscount of Valentia's account 13 August 1625 to 31 May 1632.
33   John Glanville, *The voyage to Cadiz in 1625* (London: Camden Society, 1883), p.28.
34   Orrery, Roger Earl of, *A Treatise of the Art of War* pp.29–30; TNA E407/13, Ordnance Quarter Book, January-December 1625, payments to Joseph Day, unfolio.

into musket shot for furnishing Count Mansfeldt'. Whether he also supplied Cecil's army with shot is unknown, but on 6 November 1627 Day was again ordered to be paid £160 2s 7d for musket shot at £16 a ton and 10 tons of shot were listed as being aboard several ships bound for La Rochelle. In April 1627 Day was paid £20 for five tons of caliver shot, presumable for the soldiers in Ireland or Buckingham's men who were armed with calivers before the King had them exchanged for muskets.[35]

However, the badge of the soldier was the sword. Modern sources just refer to soldier's swords as either 'hangers' or mortuary swords. True, cheap swords were referred to as hangers in various Civil War news sheets, but in contemporary accounts hangers were the belts that attached the sword to the waist belt and there were various types of swords in the accounts including rapiers, backswords, English, Turkish, Dutch, Venice and French swords. Swords were imported from Europe, but since these swords were made by English cutlers it would appear that these were names for the style of blade rather than their country of origin.

Francis Markham suggests that both pikemen and musketeers should be armed with a 'sword with a basket hilt of a nimble and round proportion after the manner of the Irish…as for the blade, it should be broad, strong and somewhat massie, of which the Turkey or Bilboe are best'. 'Irish' hilts were in the shape of a basket hilt with three bars to protect the hand, and probably are what we refer to as 'mortuary swords'. Markham does not give the length of the blade for musketeers, but for pikemen he says the blades should be 'a full yard and one inch'. Gervase Markham says that the swords should be 'good, sharpe and broad swords (of which the Turkey or Bilboe are best)'.[36]

In April 1627 3,000 recruits who were to be sent to the King of Denmark's army with Sir John Burlacy, 2,200 were issued with swords, 300 with rapiers and 500 with 'Turkey swords.' One account for the soldiers raised for Buckingham's army, records the delivery of 328 Dutch swords, 200 Dutch rapiers and 1,323 had either Venice or French blades. On 13 September 1627 Thomas Cheshire, Robert South, Thomas Tuck, William Cane and John Harman were contracted to supply 1,700 Venice and Turkey backswords with Irish hilts and 300 'ordinary English swords', with Irish hilts for the 2,000 men of the Earl of Morton's Scottish regiment. How these swords were divided in the regiment is not recorded, but backswords appear to be what we know today as 'hangers' which were sharpened only on one side of the blade. The 'Turkish blades' were almost certainly curved.

These swords were made by cutlers, who referred to them as 'long wares', as compared to knives which were 'short wares'. Cutlers had to serve an apprenticeship of at least seven years, but terms of 10 or more years are not unknown. Therefore, like other Guilds they strove to protect their profession insomuch that in June 1615 when one cutler employed a blacksmith, he was

---

35 Parliamentary Archives HL/PO/JO/10/1/27, Main Series; TNA WO49/58, Debenture Book, 1627. f.84; WO 49/59, Debenture Book, 1627, f.101; TNA SP 16/12, Kenrick Edisbury to the Lords Commissioners, 9 November 1628; SP 16/433 f.37. The English army sent against the Scots in 1639 took 80 tons of shot with them.
36 Markham, *Five Decades of Epistles of Warre*, pp.34, 39; Markham, *The Souldiers Accidence*, p.2.

fined 40 shillings for doing so because 'the blacksmith is acquainted with the secrets of this mystery' of making swords.[37]

Early in 1624 the government contracted the Cutlers Company of London to supply 5,000 swords per month, although the minutes do not record for how many months this was to continue. Nevertheless, on 25 April 1624 to make it easier for cutlers to meet this demand the company set about purchasing 576 sword blades per month at a cost of £92 16s 2d. These blades were then sold to its members, including Robert South who purchased 32 sword blades at £5 and Thomas Cheshire, who in January 1625 bought 28 blades at £4 11 shillings. In February 1626, the Guild sold to its members a further 439 backsword blades, 66 Dutch blades, 47 French blades and 12 French Rapiers. This scheme was repeated again the following year because on 26 January 1627 the Guild's account book records the purchase of '26 dozen and 4 Flanders blades' from John Wilkinson at the rate of 43 shillings per dozen. These were subsequently sold to 16 cutlers. The cutlers were given two to three months to pay the Guild, which made a hefty profit from this scheme.

As well as blades the cutlers could also purchase ready-made hilts from halters or hilt makers, so that they could quickly assemble the swords. On average a cutler could supply 20 swords to the Royal Armoury every six days.

Among the Board of the Ordnance accounts are various payments to London cutlers. On 11 July 1625, £1,541 13s 4d was paid for supplying 5,000 swords, unfortunately what type of swords these were in not recorded. However, on 12 April 1627 John Harmer received £183 6s 8d for supplying '500 Turkish swords' at 7s 4d a piece, whereas on 12 April 1627 Robert South and Thomas Cheshire, both London cutlers, received £169 11s 4d for 364 Dutch swords at 7s a piece and 136 Venice swords at 6s 4d each. On 24 April they, along with William Cave, received another £371 3s for supplying a further 504 Dutch swords at 7s 6d a piece and 501 Venice swords at 6s 4d a piece. Thomas Rogers received £41 12s on 27 April 1627 for 112 'backswords at 7s and 82 edge swords' at 6s. Unfortunately, none of the hilts of these swords are recorded, apart from, John Harmer supplied 300 swords with Irish hilts at 7s 4d a piece and 260 rapiers at 6s a piece. Certainly, there are examples with broadswords having Irish hilts, but there are also swords with straight and curved blades with just a single bar to protect the hand and a shell-like guard.[38]

On 27 June 1628 more swords were issued to the army from the stores at the Tower of London, including '2,000 Dutch swords with Irish hilts at 7s 6d, 1,983 ordinary swords at 7s a piece, 1,350 Dutch rapiers at 6s 2d a piece, 350 Venice swords with Irish hilts at 6s 4d' and 450 Dutch swords with Irish hilts at 7s 6d. This is a total of 6,133 swords, some of which appear to have been old swords, which were kept in the armoury at the Tower of London, because

---

37 Charles Welch, *History of the Cutlers Company of London* (London: Privately printed for the Cutlers' Society, 1923), p.54.
38 TNA WO 49/55, Debenture Book, 1626, f.69. The Venice swords are possibly what is now known as Proto-Mortuary Swords and the Turkish swords 'Hangers', whereas the Dutch swords are now known as basket hilt swords. Rapiers were between 47.8' to 52.7' long, hangers 28.8' to 31.9' long, Mortuary swords, 39.3'–43.2', and Broadswords 31' long.

among the accounts we find that Thomas Rogers of London, cutler, received £31 0s 12d 'for making clean of 1,863 arming swords they being before unfit for service'. Thomas Cole of London, cutler, also received £17 13 s 4 d 'for making clean 460 army swords at 4d a sword'. However, cutlers were not the only profession repairing swords, William Saunders, armourer of London was paid £29 10s 8d 'for making clean of 1,722… swords they being unfit for service… at 4d a piece'.[39]

An undated note, found among papers from the early 1630s, records the cost of proofing swords for their resilience. Rapier blades which could endure '3 blows over a table to be valued at three shillings', whereas a rapier which could endure '2 blows over a table and one cross the edge of a door shall yield four shillings'. Swords with shorter blades were also to be tested in the same way, three blows over the table would also cost 3 shillings, or three blows over a table and one cross a door' cost 4 shillings. A rapier and a sword which could go through these tests and pierce buff leather was to cost 6 or 5 shillings, respectively. There was a similar price list for longer and wider swords going through the same tests which would cost between 6 and 13 shillings. Unfortunately, the source does not record whether these tests were an already established or a proposed one which may or may not have been adopted. Certainly, these prices are cheaper than the swords which were supplied during the late 1620s.[40]

At this time there were two sorts of sword belts, the baldric, which went over the shoulder, and the hanger and belt (or girdle) which went around the waist. In 1622 Francis Markham wrote, that:

> Hangers are best if they be side and large; especially when he shall come to receive the encounter or charge of horse, where crouching his body down low with the bending of his pike. The baldric girdle (being loose) is apt to fall (by the poise of the sword) in a troublesome manner before him so as he shall neither readily draw it out nor nimbly use his pike by reason that his sword will hang dangling before him.[41]

During the 1620s, at least, waist belts were preferred not just in England but also in Europe. John Hambleton and Toby Bury were ordered to supply girdles and hangers for the soldiers' swords and in June 1628, 6,000 more 'girdles and hangers' were purchased at 2s a piece.[42]

Writing at the end of the sixteenth century Sir John Smythe lamented the passing of the archer, remembering such victories as Poitiers (19 September 1356) and Agincourt (25 October 1415) and that for every one enemy slain with a bullet 100 were slain with the arrow because an archer could fire quicker than the musketeer since they had 'many more things to

---

39 TNA WO 49/59, Debenture Book, 1627, f.102. The swords were made by Robert South, George Moore and William Cave and the belts by John Hambleton, Toby Bury and William Taylor.
40 TNA SP 16/256 ff.80, 82. Swords two ounces lighter would cost 4 shillings: 4 ounces, 4 shillings and 6 pence: 5 ounces, 10 shillings: 6 ounces, 14 shillings: and 8 ounces, 20 shillings more than the former prices.
41 Markham, *Five Decades of Epistles of Warre*, p.39.
42 TNA WO 49/58, Debenture Book, 1627, ff.47 86, 89, 97, 104, 143.

do in the charging and discharging of their pieces than archers'. However, by Smythe's time the arrows from the longbows could no longer penetrate the armour worn by the cavalry and pikemen and it took a lifetime to develop enough strength to use the bow, whereas being taught how to fire a musket could be done in an afternoon.[43] Nevertheless in September 1627 the Privy Council ordered that some of the newly raised recruits should be armed as archers, but the Derbyshire contingent had sent away their men before the order arrived and Glamorganshire could not supply any archers because of the 'long neglect and disusage of archery'.[44]

However, there appears to have been some archers with the army before this time because 300 pallisadoes tipped at both ends were purchased on 21 April 1627 at a cost of 18d each to protect the archers. A warrant dated June 1627 from Office of the Ordnance records the payment for:

For 200 bracers for archers at 6d a piece	£5 00 00
For 200 shooting gloves for archers at 6d a piece	£5 00 00
For cutting shorter and new making of 588 livery bows, being before unserviceable	£39 4 00

Other payments were made between September and November 1627 for the following:

Paid John Powell, His Majesty's fletcher, £40 for 400 sheaves of arrows, delivered.
Paid William legate of London, budget maker £12 10s for 40 quivers of leather for arrows.
Paid Thomas Redding of London, tailor, £20 for 800 bow cases, 400 being of linen and 400 of wool.
Paid George Hambledon of London, £12 10s for 40 quivers of leather for arrows,
Paid John Jefferson, £39 18s 8d for repairing 500 bows.[45]

Some of these were probably for Macnaught's Highlanders who were part of Morton's Scottish regiment and who are known to have been issued with 200 swords, 300 bows and 6,000 arrows.[46]

However, these archers are not included in an estimate, dated 1627, of how much it would cost to furnish 20,000 soldiers:

10,000 Pikes at 3s	£1,500
10.000 Corselets, without tassets at 19s	£9,500
10,000 Bandoliers at 3s	£1,500

---

43 BL Harl ms 135 ff.16-17, Sir John Smythe's answer to Captain Humphrey Barwike's book on Military affairs, 1595.
44 TNA SP 16/77/2, William, Earl of Devonshire to the Council, 6 September 1627; SP 16/81/46, Deputy Lieutenants of Glamorgan to Council, 15 October 1627.
45 TNA PRO/30/37/1, f.47, Payments, allowances, wages arranged quarterly, 1627, WO 49/58 Debenture Book, 1627.
46 TNA SP 16/73/37, Secretary Conway to Totnes, 4 August 1627.

10,000 Musket rests at 10d	£416 12s 4d
10,000 Muskets at 18s 6d	£9,850
20,000 Swords at 7s 6d	£7,500
20,000 Girdles and hangers at 2s	£2,000

Although the army would never reach this strength it clearly shows that the intended ratio was one musketeer to each pikeman. However, when the army was re-equipped in June of the following year, it is clear that there were to be two musketeers to every pikeman, although General Morgan's force continued to have a ratio of one musketeer to each pikeman.[47]

Despite these payments there was a reluctance for the government to pay these cutlers and armourers, which on 17 December 1627 prompted the Duke of Buckingham to petition the Privy Council on behalf of:

> Several artificers for arms, who had been ordered to provide arms and armour on 24 September but now wanted payment for them which was worth about £4,480 or else permission to sell them. They had received a further order to replenish his Majesty's stores on 27 November.

The Lord Treasurer was ordered to pay these artificers either in full or in part 'as may be to their contentment'.[48]

However, despite all these warrants for the payment of arms et cetera, not all of the soldiers receive their weapons. An undated account entitled 'The total amount of the arms delivered, remaining and wanting in the six companies of the regiment of Sir Edward Conway', records:[49]

---

47 TNA WO 49/59, ff.140–142, Debenture Book, 1627; SP 75/9, Muster of English regiments at Zwolle, 17 May 1628.
48 Lyle *APC 1627* Minute of the Privy Council meeting, 17 December 1627, pp.175–176.
49 TNA SP 16/62, The total amount of the arms delivered, remaining and wanting in the six companies of the regiment of Sir Edward Conway.

	Delivered	Lost or wanting	Remaining
Headpieces	240	73	167
Gorgets	240	77	163
Backs	240	29	163
Breasts	240	28	212
Pikes	240	104	136
Swords	880	249	231
Girdles & hangers	880	234	206
Muskets	240	33	207
Rests	240	80	160
Bandoliers	24	69	170

In 1627 it was estimated that 1,000 sets of pikemen's armour, 1,000 long pikes, 1,000 muskets with bandoliers, rests and bullet moulds and 2,000 swords with hangers and girdles were needed for the soldiers already in France.[50]

Furthermore, on 3 August 1627 the Privy Council informed the Earl of Totnes, Master of His Majesty's Ordnance that there were not enough arms to equip the 2,000 men of the Earl of Morton's Regiment, and as late as October 1627 the soldiers of this regiment who were waiting to embark at Plymouth for the Île de Rhé were only armed as follows:

	Sent	Wanting
Armour	218	782
Swords	500	1500
Muskets	1,000	-
Pikes	0	1,000

Nor did the musketeers have any match for their muskets.

In 1628 an inquiry was held to discover why the soldiers had performed so badly, which found that the pikemen's armour was in such short supply that it had been necessary to purchase sets of armour from the Low Countries, and that 'the quality [was] so defective as… no corslet were able to abide pike proof'. The pikes themselves were 18 inches too short and made of sawn ash which could not endure the rigours of campaign and that many of the muskets had been broken during the expedition. Moreover, it was also found that the weapons in the Office of the Ordnance had been issued too many times and had often been returned broken and not repaired.[51]

As to the officers and sergeants, they were armed differently from their men. For the officers Cecil ordered that 'all captains [were] to have leading staves and a target'. The targets or shield were made by Thomas Stephens, a

---

50 Lyle *APC, 1627,* Letter to the Earl of Totnes, 3 August 1627, p.474, same to the same, 15 August 1627, pp.488–489.
51 TNA SP 16/126/40, Report on offices of the ordnance, 1628.

London armourer, who on 26 April 1625 received £17 for 42 targets, lined with leather and fringed. Most were either russetted or blackened to protect them from rust, but two were highly decorated being trimmed with crimson velvet and gold. Presumably, one was for Buckingham himself.

The 'leading staffs' or partisans for the officers and halberds for the sergeants were pole weapons which could be either elaborately decorated or plain. In 1627, 20 gilt halberds and 20 halberds were delivered to each of the 10 regiments of foot, the best coming from the Low Countries. At the end of March 1628 John Edwards was paid for 40 'white halberds' at four shillings each for the Earl of Morton's Scottish regiment.[52]

Cavalry

There were three types of cavalry at this time; the harquebusier, the heavily armoured cuirassier and the lancer. As already mentioned, unlike the soldiers of the foot regiments whose equipment were supplied by the government, a cavalryman was sponsored by an individual. In all 53 people sponsored a total of 163 cavalrymen, some just equipping one trooper, while others, such as the Bishop of Bath and Wells, equipped two cavalrymen. The Duke of Buckingham supplied eight cavalrymen and the office of William Laud, the future Archbishop of Canterbury equipped 30 troopers.[53]

There are several references to equipping troopers in the Office of the Ordnance's account including six purple velvet covers for saddles and richly embroidering with silver and deep silver fringe, which was supplied by Thomas Smith, saddler for £200. Presumably, these were for the officers and trumpeters of the troop of horse which was to accompany Buckingham to France.

One hundred sets of horseman's armour consisting of back, breast and headpieces, were also supplied to the army along with 61 saddles and 100 cavalry pistols, but whether these were for Montjoy's troopers or to supply the expected influx of French recruits is not recorded.[54]

Although none of Buckingham's cavalry appeared to have been the heavily armoured cuirassiers, on 16 July 1625, 650 suits of cuirassier armour and 1,500 of arquebusier armour arrived at Hull from Flanders, although 147 'pots' or helmets were missing. The order also consisted of 1,500 wheel lock carbines for the harquebusiers, but 80 were missing their spanners which were used to wind the mechanism up so that the carbines could fire. This consignment also contained 1,562 pairs of wheel lock pistols, but again some

---

52 BL Lansdown ms 844, Miscellaneous articles, 1558–1726, f.314; TNA WO 49/58, Debenture Book, 1627, WO 49/59 Debenture Book, 1627, f.38.
53 SHC 6729/4/139 A list of the colonels and principal officers of the seven regiments, nd [1627].
54 BL Add Roll 77175, Account of Richard Granham for the sum of £1,000 spent on horses and provisions for the military expedition to the Island of Rhè in 1627; TNA AO1/217/729, Sir R Graham, account for buying horses and provisions for an expedition to the Île de Rhé under the Duke of Buckingham, 1627; WO 55/1684, Surveys of stores and arms ships returning from Rhè, 1627.

THE FIRST BRITISH ARMY 1624-1628 (REVISED EDITION)

Contemporary Illustrations detailing the armour worn by Cuirassiers.
(from Philippson's *Geschichte des Dreissigjahrigen Krieg*)

of the spanners were missing along with the pistol holsters. A week later 500 saddles arrived from Flanders which were unloaded at London. It is likely that these consignments were to re-equip the trained bands which had been ordered to purchase new weapons et cetera at this time.[55]

However, not all the arms for the cavalry came from Flanders, On 26 May 1627, £70 was paid to John White for 60 carbines for the Earl of Dorset's troop, which did not serve in France. Another warrant issued by the Office of the Ordnance was an agreement with the armourers to alter 40 old sets of horsemen's armour by putting 'bars with screws to the face of the head pieces and make the long bellies [of the armour] fashionable'. On 4 June 1627 George Hambledon was paid £4 'for cleaning and mending 112 gauntlets at 6d a piece', probably for the cavalry. Whether any of these items was destined for Buckingham's cavalry is not known.[56]

The third type of cavalryman was the lancer, which were equipped similar to the cuirassier, but armed with a lance, although this type of cavalryman was rare by the 1620s, a few were still to be found in the county trained bands.[57]

---

55 HMC Braye' Manuscript p.112; TNA SP 16/251 f.115. The arms that arrived at London also consisted of 1,039 corslets, 1,280 pikes, 2,727 and muskets and 23,347 hundredweight of match.
56 TNA AO1/217/729, Sir R Graham, account for buying horses and provisions, 1627; WO 55/1684, Surveys of stores and arms ships returning from Rhè, 1627; WO 49/58, Debenture Book, 1627, f.99; BL Harl ms 429, Journal of the Office of the Ordnance in the Tower of London, Aug 1626 to Feb 1630.
57 Kevin Sharpe, *The Personal Rule of Charles I* (London: Yale University Press, 1992) pp.31–32; BL Add ms 72,422, Trumbull papers of the council of war and the Muster Master General, 1624–1635 f.91. Among those counties who are known to have contained lancers at this time was Lancashire, which had 23 lancers, Monmouthshire four, Northamptonshire 37 and Leicestershire 28.

However, one type of soldier which was increasing in popularity at this time was the dragoon, who were infantry mounted on inferior quality of horses to the cavalry and would dismount to fight. Despite one source referring to Buckingham's cavalry as dragoons there is no evidence that any were present with Mansfeldt, Cecil or Buckingham.

Colours

With the union of England and Scotland James introduced a new national flag known as the 'king's flag' or the 'flag of Great Britain', which combined the Saint George's Cross with that of Saint Andrew, and was flown above the castles and fortifications in England and was also used by the Royal Navy. However, it does not seem to have been carried by any of the regiments at this time.

Each company of foot and troop of horse within a regiment would have carried its own colour. During Tudor times, an infantry ensign usually had horizontal stripes or waves in two or three colours, which was also popular in the Low Countries. According to one anonymous author known only as I. T. who published his *A B C of Arms* in 1616, the design was to be:

> At the discretion of the captain, either his own colours, belonging to his house; if he be a gentleman of coat-armour [i.e. have a coat of arms], or what other his invention shall best like of. But always having a red cross therein, being the badge of an Englishman: St Georges cross being peculiarly appropriated to that Nation.

Writing in 1622 Francis Markham confirms that officers used their coats of arms as a basis for their colours, but warns that he puts 'his honour to hazard' if the colours are captured. He goes on to suggest that:

> All colours belonging unto private captains ought to be mixed equally of two several colours, that is to say (according to the rule of Heraldry) of colour and metal, and not colour on colour, as green and red or blue and black, or such like, nor yet metal on metal, as white and yellow or orange tawney and white, for colours so born show bastardy, peasantry or dishonour. Now in the corner which is next to the upper point of the staff, he shall carry a fair large square or canton, containing a fixed part of the colours a plain red cross in a white field (which is the ensign of our Kingdom of England). If the colours do belong to a colonel they shall then be all of one entire colour or one metal, only the red cross or ensign of the Kingdom shall be in his due place as aforesaid. If they belong to a colonel general, to the Lord Marshal of the field or any such superior officer, they shall be all of one entire colour or metal and the red cross or ensign shall be in a very little square or canton, as in a twelfth part of the field or less if it please them. But if they belong to the General of the Field then they shall be of one entire colour or metal, without any red cross at all, as was before said.[58]

---

58   Markham, *Five Decades of Epistles of Warre*, pp.74–75; TNA SP63/242, Extract of Sir Thomas Phillips letter to the Lord Deputy, 10 January 1626, f.97.

Certainly, the colours of the London Trained Bands in 1588 and 1599 show no regimental uniformity whatsoever, with each ensign having a variety of different colours and each had a 'metal' colour and another colour, in horizontal stripes.[59]

When Jacques Callot was commission to produce several illustrations of the Île de Rhé Expedition, in the margins he recorded some of the captured English ensigns having horizontal stripes, eight of which have Saint George's Crosses, some larger than the others, and four were without a cross. Certainly, a Frenchman, Fontenay Mareuil, records in a letter dated 22 December 1627 that, all the captured colours had a three foot square white St George's Cross in the top corner nearest the colour pole. However, Callot's drawings of the St George's Cross are not as large as Mareuil suggests.[60]

Even when English and Scottish soldiers served abroad, they insisted that they serve under the colours of their own nation. Cecil wrote 'it grieves me to see English colours carried against English colours and that His Majesty should lose his subjects blood both ways'[61] When Christian IV tried to impose Danish colours on MacKay's Regiment which was first formed to serve with the King of Denmark's army, Christian IV 'would have the officers to carry the Danes cross [white cross on red] which the officers refused'. The dispute was brought before Charles I and on 21 February 1627 Secretary Conway wrote to Anstruther informing him that the Council of War had decided that all Scottish regiments in the Danish service were to have the St Andrew's Cross 'in the principal place next the staff [and] may carry what other arms the king of Denmark should be pleased to direct in their colours'.[62]

However, it was not just the Scottish regiments that objected, on 3 December 1627 John Chamberlain wrote to Sir Jacob Astley that:

> The king [of Denmark] offered us Danes [sic] colours which we refused and the king was very angry with us and told us he would make us English colours, but we sent His Majesty word again that we could not receive any without order from our general. Now they begin again to speak of giving us colours and it is like that towards the spring they will urge it hard upon us if we have not other order from our general.[63]

Unfortunately, the outcome of this dispute is not known.

There is only one payment for ensigns in the archives, dated 10 July 1625, which is a warrant for '100 colours at £5 apiece and 200 drums at 40 shillings

---

59 Lt Col J. Leslie, 'Survey or muster of Armed and Trayned Companies in London, 1588 and 1599' in *Journal of Army Historical Research* vol. 4 no. 118, 1925.

60 TNA AO 1/299/1135, Roll of Sir William St Leger colonel and sergeant-major general to land forces on the expedition to Cadiz, 4 July to 31 October 1625; SP 16/521/75, Secretary Conway to Sir John Oglem 8 June 1625, E351/282, Army Paymasters, Sir William St Leger, Colonel Major-General to land forces on the expedition to Cadiz, July-October 1625; SP 16/66 f.19; Jean Rey *Histoire du drapeau, des coleurs et des insignes de la morarchie Francaise* la papier du Richelieu vol. 2 p.677 Letter to A M de Bouillon-Malherde *Oeuvres de Malherde Poesie et prose* p.199.

61 Quoted in Dalton, *Life and Times of General Sir Edward Cecil*, p.7.

62 Monro, *Monro His Expedition part* 1 p.2; TNA SP 75/8, Conway to Buckingham, 20 October 1625, f.14.

63 TNA SP 75/8, John Chamberlain to Sir Jacob Astley, 3 December 1627, f.412.

apiece'. These drums and colours were to be made by John Bone and James Reynolds who curiously were also to supply the army with shirts and shoes. Unfortunately, there is no description of these ensigns, but they were usually six foot square and made of taffeta. Since they came from a central source there may have been more uniformity in the regiment than in Tudor times. Certainly by 1640 there appears to have been such uniformity in the colours carried by the Earl of Stafford's Army in Ireland and that of Charles' during the Bishop's War, for example the Earl of Essex's Regiment in 1640 carried orange and white colours and in Stafford's Army William St Leger carried Watchet (sky blue) and white ensigns. Unfortunately, it is not known whether they chose similar colours for their ensigns in the 1620s.[64]

However, by June 1627 the colours which had been issued two years previously had worn out, which prompted Sir John Burgh to write to Buckingham that 'the colonels and captains desire to have new colours, the charge will be about £300'. However, there is no reference of any new colours being issued in 1627, although there are no signs of wear and tear on Callot's Île de Rhé illustrations, but this might just be artistic licence on his part.

The design of the colours carried by the Earl of Morton's Scottish and Sir Piers Crosby's Irish Regiments are also not known, but Morton's almost certainly bore the St Andrew's Cross, either in the canton or over the whole colour, although ensigns carried by Scottish regiments during the English Civil War were not necessarily in the national colours of blue and white. We do know that Crosby's Regiment was paid £200 in May 1627 'to furnish colours, drums and halberds', but whether Crosby's Regiment had the Irish flag, a red cross on green, on the colours it is not so certain, since they were nationalist colours. They probably would not have had the religious slogans and images that Irish Catholic Regiments bore during the Civil War period.[65]

Neither do we known the design of the cornets carried by Montjoy's or Cunningham's troops of horse, but these would be about two feet square with a fringe around three sides and unlike the infantry colours often had elaborate designs of figures and animals, while others had just a plain motto. The troop's trumpeters also had a banner on their trumpets, which usually bore the troop commander's coat of arms, or if he did not have armorial bearings then they were a copy of the troop's cornet. Drums also appear to have born the captain's coat of arms. Among Buckingham's accounts for the campaign of 1627 is a payment of £7 to George Porter for 'gilding the banners' and a further £7 was paid to 'His Majesty's trumpeters for seven banners'.[66] However, these may have been for Buckingham's own use rather than for his cavalry, because generals did have their own standards. Mansfeldt's personal standard is known to have had laurel leaves around the edges and was divided into eight segments by laurel leaves. Within each segment was

---

64 Stuart Peachey and Les Prince, *ECW Flags and colours, 1: English Foot* (Leigh on Sea, Partizan Press, 1991) pp.27, 89–90; Stephen Ede-Borrett, *Flags of the English Civil Wars* (Leeds, Raider Books, 1987) pp.31–35.
65 *CSP Ireland* Charge of recruiting apparelling, victualling and arming 3,000 men out of Ireland, May 1627, p.240.
66 BL Add ms 12,528 Sir Sackville Crowe's Book of accompts containing receipts and disbursements on behalf of the Duke of Buckingham, 1622–1628.

the initials EM, for Ernest Mansfeldt under a crown and in the centre was a wreath within which was the motto 'Force Mest Troup'. Unfortunately, the colours are not known.

When Buckingham's army was routed in October 1627 many colours fell into the hands of the French. Early French reports put the number of colours captured at about 20, which an English naval captain appears to confirm when he writes 'In the retreat we lost all the colours of two regiments save only one which was snatched by a young gentleman… and brought off by his swimming in a dike with it'.[67]

However, according to a letter from Richelieu to M de Guise dated 14 November 1627, 44 colours had been captured, while another letter states 45. An English source also states that '44 colours taken from us with one horseman's pennon'. The Italian Ambassador, Alessandro Antelminalli also writes of 'the loss of forty-four colours, which remain as trophies in the hands of the French is deeply mortifying'. At least some of the colours must have belonged to Colonels Bingley's, Morton's, Courtney's and Rich's Regiments of Foot which are known to have been badly cut up on 29 October 1627.[68]

The colours and three captured artillery pieces were paraded through the streets of Paris on 22 December 1627, and were hung in Notre Dame Cathedral along with other trophies that French Armies had taken over the centuries. However, at the beginning of the French Revolution all these trophies were removed from the cathedral and stored at Les Invalides, but in March 1814 with the imminent surrender of Paris to Napoleon's enemies, all the standards and ensigns which had been captured up to then were burnt to stop them falling into the allies' hands and the debris swept into the River Seine.

Repeated attempts to find the descriptions of these captured colours in French libraries or what happened to the other colours once the regiments were disbanded in 1628 has proved fruitless, apart from one. When Sir John Burgh was killed during the Île de Rhé Expedition, his body was brought back to England for a state funeral and when his body was buried in Westminster Abbey his colour was laid to rest with him, the colour pole being broken in two.

Artillery

When Cecil sailed for Spain in 1625, he had '10 drakes or field pieces' with him and in 1627 Buckingham's army was equipped with 22 pieces and was supplied with a further nine pieces of artillery the following year.[69]

---

67 BL Add ms 26,051, Journal of the Voyage of Rease 1627.
68 Pierre Grillon (ed.), *Les Papiers de Richelieu* vols. 2 (Paris, 1977) pp.645, 65; BL Sloane 826, Papers of concerning the proceedings in both Houses, 1621–1629 f.31; HMC 11 Report Appendix 1 p.133; Thomas Birch, *The Court and Times of Charles the First* (London, Henry Colburn, 1848) pp.286, 291.
69 TNA SP 16/473/37, Certificate of the officers of the Ordinance to the Commissioners of Saltpeter and Gunpowder, 9 December 1640.

ARMS AND ARMOUR

Artillery at this time was composed of many types of artillery pieces, but the usual field pieces were the demi-cannon, the culverin, the demi-culverin and the drake. Even the names of these pieces did not always correspond to a particular type. On 28 August 1628 two long drakes of brass, four saker drakes and three drakes of iron were delivered to Colonel Peblis, the Master of the Ordnance of Lindsey's Army. The demi-cannon was the heaviest artillery piece with a bore of between 6 and 6¾ inches, whereas the saker had a bore of about 3¾ inches.

Unlike the infantry and cavalry, there were few guides for the gun crews. They could use William Bourne's *The Arte of Shooting in Great Ordnaunce* which was published in London in 1587 or Thomas Smith's *The Art of Gunnerie* published in 1600. However, it was not until 1626 that John Smith published his *An Accidene or The Path way to Experiences*, which although too late for Cecil's Expedition may have been used by Buckingham's gunners. Smith's treatise was followed by Robert Norton's *The Gunner* in 1628. These treatises gave the would be gunner advice on the various artillery pieces and how to load them, although no doubt the more experienced gun crews would have taught the others about gunnery practice.

However, neither Smith nor Norton mentioned the 'drake' which was imported into England in 1625. The drake appears to have first been used by Prince Maurice at the siege of Bergen op-Zoom in 1622 'some [fired] a hundred [and] others seven score musket bullets'. However, the name seems to have been referring to an artillery piece that was not of the standard specifications, i.e. a culverin drake, which had a shorter barrel than the standard culverin. What dimensions the drakes used by Buckingham are unknown, but 2,400 shot for the six pound drakes and 2,400 pieces of shot for the three pound drakes were issued.

Ranges varied according to the artillery piece. A culverin had a point-blank range of 400–420 paces and a maximum range of 2,000–2,100 paces, whereas a demi-culverin's range was between 380 and 1,800 paces. The largest piece that Buckingham took with him to the Île de Rhé was a demi-cannon whose range was between 350–370 and 1,700–1,800 paces. Unfortunately, the range of the drake is not recorded, but since each piece was cast separately, each barrel might have its own quirks.[70]

The gunner would have to know mathematics so that he could calculate the range and evaluation of the shot. However, since the cannon would recoil it would have to be reset each time the piece was fired. During the 1730s, Benjamin Robins found that at 800 yards an artillery shot could veer 100 yards to the right or left of its target and due to wind resistance sometimes it would fall 200 yards short of its' target. However, at 50 yards it usually found its mark.[71]

At first only one mortar appears to have been sent out with Buckingham, however later an additional four brass mortars were sent to France.

---

70 Robert Ward, *Animadverions of War*.
71 A. R. Hall, *Ballistics in the Seventeenth Century* (Cambridge: Cambridge University Press, 1952) p.53.

Unfortunately, neither Smith nor Norton refers to mortars, but Robert Ward in his *Animadverions of War* published in 1639 records:

> The mortar-pieces are very necessary and useful, as well for the assailants, as the defendants; for being duly used they annoy the enemy much in his forts or trenches. First by shooting Granados either single or double, iron bullets, stone or leaden shot… [they] are ordained and fitted to shoot in an oblique or crooked line to convey their fire-balls or other shot over the walls, hills or any obstacle… [because] they shoot right upwards, but most commonly they are mounted from 80 degrees to 70, 60, 50 or 45 according as the distance is in farness from the piece, or height of the hills, walls or houses.

In 1625 an inventory of the ordinance in the Tower of London records found only 16 demi-cannons, 21 culverins, and 33 demi-culverins in the armoury, so in April and May 1625 contracts were issued to produce new pieces of artillery. Richard Pitt was paid £102 5s 9d for 'new casting and repairing' three demi culverins and three sakers. Once the cannons were cast a touch hole was drilled into the barrel and the Royal Coat of Arms was engraved on them.[72]

Matthew Banks 'His Majesty's carpenter for the Office of the Ordnance' was employed making the gun carriages for six demi-cannon, six culverin and three demi-culverin 'being all of very strong and dry stuff and of extraordinary workmanship' for which he received £75. He also received £10 for four block carriages 'made all of choice timber at 50 shillings a piece' and 'for letting in of the iron work into the field carriages and block carriages, £5'.

In 1627 Matthew Banks was again ordered to make a field carriage for a culverin and another for a demi-culverin, with two spare carriages for demi-culverin and two spare carriages for a drake. On 6 November 1627 he was again ordered to make various items, including two field carriages for culverins at 46s 8d apiece, for which he received £68 12s 8d. In August 1628, a further payment was made to cast three brass culverins for £66 and two brass mortars for £36 4s 8d. An order was also placed for 1,000 demi-culverin iron round shot, 1,500 saker shot and 2,000 minion shot, but by this time La Rochelle had surrendered so they were not needed by the army.[73]

The wheels for the artillery pieces were made by a wheelwright and then finally a painter would be employed to paint the carriages in a 'fair lead colour'. Unfortunately, what this colour was is not recorded, but it has been suggested that it was either red or grey, both colours being used by the Royal Artillery in later centuries.

To fire an artillery piece a linstock was needed, which according to Smith was 'a short staff of wood more than half a yard long with a cock at the one end to hold fast his match, and a sharp pike in the other to stick it fast upon the deck or platform upright'. However, on 16 July 1625 John Harrington,

---

72 On 30 June 1628 John Brown was also paid for 'engraving his Highness' arms very large and fair upon 108 pieces of cast iron ordnance': TNA E407/13, Ordnance Quarter Book, January–December 1625.
73 TNA WO 49/59 Debenture Book, 1627, ff.2, 190, 202, 203.

gunmaker, received £4 10s for '10 snaphaunce [or flintlock] linstocks', although since the artillery pieces continued to be fired with match presumably these flintlock linstocks were not a success. On 12 April 1627 George Hambleton leather seller of London received £4 16s for 48 gunpowder bags. Canvas was also needed to make cartridges to ram down the barrel of the artillery piece.[74]

For the Cadiz expedition 50 horses were impressed to pull the cannon, however only 44 arrived at Plymouth on 15 August 1625, of these 21 horses were lost at sea, eight at Cadiz itself and four on the return trip. This left just 16 horses, five of whom had been left behind at Plymouth when the fleet sailed because they were unserviceable. Five were landed in Ireland and seven at Plymouth where they were sold. During this time 40,000cwt of hay was bought to feed these horse at 25s per cwt. Even if all the horse had survived there would not have been enough to pull all of Cecil's artillery pieces, since a single demi-culverin required seven horses and a culverin eight.

Among the items purchased for these horses were slings to ship the horses at Plymouth. Fifty saddles were also bought for 30s each 'thereby upon occasion to make them a troop of horse for land service'.[75]

The 1628 inquiry into the campaigns found that the artillery pieces were poorly cast and they should be melted down and re-cast. It also found that the Office of the Ordnance could only 'furnish six pieces [of iron cannon] completely mounted upon any occasion'.

---

74 TNA WO 49/55, Debenture Book, 1625; TNA E407/13, Ordnance Quarter Book, January–December 1625.
75 TNA E351/285, Army paymasters and Treasurers of the Forces, Viscount Valentia, Master of the Ordnance on the Expedition to Cadiz, 13 August 1625–30 June 1627.

THE FIRST BRITISH ARMY 1624-1628 (REVISED EDITION)

Blow off your coal

5

Provisions

Throughout history commanders have known that, to quote Napoleon, 'an army marches on its stomach'. Therefore, it was essential to keep an army well supplied. However, this was easier said than done, because in 1625 in theory a soldier's basic daily ration was two pounds of bread, one pound of meat or cheese and eight pints of beer, or to put it another way for a regiment of a 1,000 soldiers, they need 2,000 pounds of bread, 1,000 pounds of meat or cheese and 8,000 pints of beer per day. The soldiers' rations were to be divided into a 28-day month or a seven-day circle, 'for the first three days… bread and cheese, and for the latter four days bread and beef, and the drink for the whole time [was to be] beer or beverage [cider]'.[1]

On 27 April 1627, the Privy Council ordered that the soldiers were to receive:

> Wheat (after the proportion of one pound [450g] of bisket to a man by the day) three-quarters of a peck to a man by the week, butter one-quarter of a pound with half a pound of cheese to a man by the day or with fish half so much butter and cheese the same quantities in each kind… for four days in the week and the other three days in the week the same quantities likewise of flesh… viz beef, two pounds to a man by the day or one pound of pork or bacon with a pint of pease.

Pease was measured in a liquid measure rather than a weight.

In 1627 the recruits being transported to Denmark were not so lucky, because they were ordered just to have a four pints of beer, three cakes of biscuits weighting 1.5lbs, ¼lb of butter and 6oz of cheese per day.[2]

The concept of calories and vitamins in food during the seventeenth century was unknown, but modern nutritional charts for wheat and rye, which was used to make bread, gives about 2,466 calories or roughly 3.5–5oz of protein. Calculating the nutrition of meat is more problematic since it depends on the part of the animal the soldier received and whether

---

1 John Glanville, *The Voyage to Cadiz* p.52.
2 Lyle *APC 1627*, A letter to the Mayor of Hull, 28 February 1627, p.105, Minutes of the Privy Council, 27 April 1627, pp.248–249; TNA SP16/71/33, Duke of Buckingham to Nicholas, 18 July 1627.

it included bone or fat. Modern nutritional charts record that in 3.5oz of beef there is between in 0.7–1oz of protein and 110–198 calories. Therefore, an average ration of beef in 1625 was 1,122.5 calories and 4.2oz of protein. Beer also causes a problem because there was 'strong beer' which was used to get drunk with and 'small beer', which was the standard drink in the early seventeenth century, so it contained about 1,720 calories or 0.3oz of protein. The soldier's cheese ration would give him 1,886 calories and 4.8oz of protein, or a total of 5,408.5 calories for 'meat' days and 6,072 calories for 'cheese' days and seven to 7.5oz respectively of protein, which is far higher than Perje's estimates that a soldier needed 3,800 calories and 3.5 to 5.6oz of protein per day when on the march. So far from being on starvation rations the soldier in the early seventeenth century had a comparatively good diet compared to a British Napoleonic soldier, whose rations gave him just 2,466 calories and 1.7oz of protein per day. Furthermore, a sergeant was luckier since he received three times the daily ration of a private soldier, rising to 12 portions for a colonel.[3]

According to Francis Markham, it was the corporal's duty:

> To receive from the clerk, sergeant or other officer all the proviant-victuals which do appertain to his squadron and to see the same weighed and truly shared among them with all indifference and equality, without any respect of person or hope of lucre and commodity.[4]

The meat or cheese would be cut into 1lb portions and weighed on a pair of scales to make sure that the soldier got his correct measure, but no allowance was made for any bone or fat within the meat.

Certainly, in the latter part of the sixteenth century these rations appear to have been usually issued in the evening once the soldiers had set up their camp after a day's march. Robert Monro also remarks when the soldiers had done their duty for the day and having meat to eat, they 'could invite his comrades to supper and make merry', until he was called to duty once more. Sir Charles Morgan believe that the soldiers liked playing dice and smoking, but only rarely is tobacco mentioned as being issued, so presumably they would have to buy it as well as any other provisions from the sutlers who accompanied the army.[5]

The weather could also play its part in the logistics of an army, because not only would wagons get stuck in the mud if it had been raining, which prevented them joining their regiments in the evening to distribute the provisions. Furthermore, at the siege of Breda in January 1625 it was reported that not only was there a scarcity of wine, beer and bread among Spinola's

---

3   G. Perjes, 'Army provisioning, logistics and strategy in the Second Half of the 17th Century' in *Acta Historia Academiae Scientiaruim Hungaricae* vol. 16 (1970) pp.12–13; Edward J. Coss, *All for the King's Shilling* (Norman, University of Oklahoma Press, 2010) p.99.
4   Markham, *Five Decades of Warre*, pp.67, 103 Turner, *Pallas Armata*, p.201.
5   William Gerrard, *The Art of Warre*, p.13; *Monro His Expedition*, part II p.18.

soldiers who were besieging the town, but also 'the soldiers have been fain to eat their bread half-baked for want of dry wood'.[6]

On the other hand, Robert Ward, recommended that in, 'hot weather the sutlers belonging to every company may march with their wagons between companies whereby they may the sooner relieve [the soldiers] with meat and drink, which otherwise they cannot come unto but once a day'.[7]

One account records that the soldiers were given 21 days rations on their voyage to France in 1627, but this is likely to have been given to the regiment who then distributed the provisions to the men, either way the soldiers quickly consumed their rations and so began to go hungry.

Once they had received their rations the soldiers would keep their provisions in a knapsack. Illustrations of the soldiers' knapsacks suggest they were large enough only to keep a day or two's rations in, although several days rations could be issued if the soldiers needed to move quickly as in the landing at Cadiz in 1625.

These knapsacks were issued by the Office of the Ordnance. Among those contracted to supply these on 16 July 1625 was Thomas Horsell, a London merchant, who was to receive £536 for 5,000 leather knapsacks 'at several rates' and on 23 April 1627 William Legate and George Hambledon each received £46 17s 6d for supplying 750 knapsacks of black leather at 15d a piece.

Unfortunately, despite being used by soldiers throughout Britain and Europe, only two are known to exist. These were recovered from a shipwreck off the coast of the Netherlands, and both are tubular leather bags with one end closed in what appears to be a knot, while the other end is fastened with cord to pull the bag shut. Two leather straps are attached to the bag so that the soldier could carry it over his shoulder. The second knapsack also appears to have two straps, but only fragments remain. Robert Monro records that when the English and Scottish troops, who were eating 'hard biscuit and beer', discovered that the newly recruited Danish soldiers had been issued with 'dry beef and bacon', they cut the cords of the Danish soldiers' knapsacks and ran off with them. It was only when Munro, and other officers, were threatened with punishment by the King of Denmark that they put a stop to this practice.[8]

However, in all the warrants for the army there is no mention of plates or bowls, knives, spoons or drinking vessels being issued to the soldiers, therefore presumably they would have to supply these items themselves. Neither are there any references to any water bottles being issued to the soldiers even though leather and pottery ones are known to have existed at this time. However, in an engagement at Oldenburg in 1626 Robert Munro records that a barrel of beer was brought up to the McKay's Regiment and the men used their hats and helmets as a drinking vessel, although the soldiers

---

6   TNA SP 101/28, German newsletters, 1620–1625.
7   Robert Ward, *Aminadversions of Warre* (1639) book 2, p.23.
8   TNA WO 49/55, Debenture book, 1625; WO 49/58, Debenture Book, 1627 f.143; *Monro His Expedition* p.61.

were probably improvising what was to hand in the heat of battle rather than using them daily.

It is unlikely that the soldiers would carry pottery items since they are easily breakable and pewter plates were expensive and so are more likely to be used by the officers rather than their men. Paintings of camp life do appear to show soldiers with tankards which look like pewter, but these may have been the property of the sutlers who sold the drink to them. Wooden bowls and spoons have been found on the *Mary Rose* some of which have marks on, it is presumed that they were made by a member of the crew to mark his property.

It was the responsibility of the Commissary General to find the provisions for the army, which at this time was Sir Allen Apsley, who was also responsible for the navy. He, or his deputy, Henry Holt, would send out warrants to the local inhabitants to purchase grain, et cetera. After Apsley's death 100 creditors petitioned for payment for victuals and so forth, which they had supplied for the Cadiz and Île de Rhé Expeditions. These included Michael Bremen, a butcher of Chichester who had supplied beef to the army in 1628 and petitioned the Privy Council on 16 March 1635 for the £143 which he was still owed, and as late as October 1639, 34 contractors still had not been paid.[9]

The contractors would bring their provisions to one of the various government store houses which were set aside for the navy until the stores were needed. These houses were situated on Tower Hill in London, at Rochester, at Dover and at Portsmouth. Tower Hill also accommodated the navy's bake houses, garners, Cowper houses, pond yards and the slaughter and cutting houses. The cattle would be brought to Well Close in London and then slaughtered at East Smithfield. Once the cattle had been slaughtered their meat would then be cut up and salted to preserve it. During the Napoleonic Wars it was reckoned that 2lb of salt was enough to preserve eight pounds of meat and so often had to be soaked in water overnight to wash out most of the salt to make it edible.

At Dover there were also bake houses, mills and granaries, while at Portsmouth there were bakeries and brew houses. Among the suppliers in Hampshire was Apsley's deputy, Henry Holt, who claimed that he could supply 80 tonnes of beer per week from his breweries at Portsmouth, Southampton and Newport as well as bake 30,000 biskets. However, Portsmouth beer does not appear to have been very good because on 18 July 1627 Buckingham ordered the soldiers to be issued with 'London beer' rather than the poor-quality local brew, because 'the soldier is better satisfied with his beer (if it be good) than with his victuals'.[10]

In 1640 a recipe for bread for the soldiers was two parts rye and one part wheat or barley, while the French soldiers appear to have bread baked from two parts wheat and one part rye. No yeast is mentioned so the flour was probably mixed with water, left to ferment for a few days, and then baked.

---

9   TNA SP 16/28 f61; SP 16/43 ff.30, 135; SP 16/316 f.68.
10 TNA AO1/1798/370, Sir Allen Apsley's Account, 1 January 1627–31 December 1627; TNA SP16/71/33, Duke of Buckingham to Nicholas, 18 July 1627.

Officers would probably get better quality bread than their men, however since loaves were supplied by the dozen and weight, unscrupulous bakers might 'half bake' their bread so that the water would not entirely evaporate and thus the loaf would be smaller but meet the required weight. There is even one account by Father Pietro Drexel, who accompanied the Catholic League's army during the 1620 campaign in Bohemia, that he bought a loaf of bread from a soldier which was 'black and a wet mass', which he could push his finger through like 'dough'. On the other hand, the bread might be too mouldy to eat.[11] A solution to this was to issue 'biskets', which were probably much like the hard tack biscuits in later centuries, being supplied by the bag rather than the dozen and could last months if not years.[12]

According to John Glanville's account of the Cadiz Expedition, the provisions that were not issued to the soldiers were ordered to be:

> Carefully stowed and sent in bags or vessels fit for the purpose with an inscription to be set upon every such bag or vessel, making mention of the quantity and quality of every parcel of meat and drink therein contained and for how many soldiers, and what company or regiment the same was assigned.[13]

Provisions would only to be issued when a warrant was presented to the commissary, which had been signed by an officer of the designated regiment or company. The commissary would then make a tally of all provisions received or issued.

This was procedure was fine in theory, but during the Cadiz Expedition some soldiers did not receive their rations while others devoured several days' rations in one day leaving them hungry, unless they could find additional food. On 26 October 1625 Glanville found 'many soldiers complaining of sickness and others of faintness, affirming that they had not tasted any meat nor drink' for four days.

Even when the soldiers did receive their rations it was not always edible. According to Captain George Took, who served in the Cadiz Expedition:

> Our cheese was such, that, though amongst soldiers and sea stomachs, it is not yet digested, much of the butter might have been mistaken for cattle soap and … our pork that they failed much in quantity…Concerning the rice… either little was laid in or much was embezzled.[14]

Neither did the soldiers receive their full ration entitlement as Sir Michael Geere, who also served on the Cadiz Expedition records, 'our victuals or flesh [are] cut at half the king's allowance and that so stinks that I presume

---

11　Father Drexel's diary in Sigmund Riezler Kriegstagebucher, *Aus Dem Ligitischen Hauptquarter, 1620* (Munchen: G Frans'chen, 1908) p.161.
12　TNA SP 28/135. Part 6, Nicholas Cowling accounts, September 1643–April 1645; C. H. Firth, *Cromwell's Army* p.211.
13　Glanville, *The Voyage to Cadiz*, p.53.
14　Captain George Tooke *The history of Cales passion: or as some will by-name it, The miss-taking of Cales presented in vindication of the sufferers to forevvarne the future* (London: W Hunt, 1652), pp.7–8.

[it] hath been the cause of death and sickness which is amongst us. No dog of a Paris Garden I think would eat it'.[15]

Took also criticised the 'purveyors, butchers and pursers' for the lack of meat on the expedition, while Walter Yonge, who was back in Devon, believed Sir Allen Apsley and Sir James Bagg should be punished because they were 'unfit to be employed in the same [position] again' for letting these abuses happen.[16]

Unfortunately for the soldiers nothing appears to have been learnt from the failure of the provisions during the Cadiz Expedition, because when the fleet sailed in June 1627 it had been planned that the sailors would have three months rations and the soldiers six. However, Sir Allen Apsley only gathered enough provisions to last them until mid July, probably believing that the island would quickly surrender and that La Rochelle would supply the army with its needs. Unfortunately, not only did this not happen, but the Île de Rhé was not a very fertile island, being known for its production of salt and wine rather than crops, which would quickly become exhausted as the siege wore on. In fact 120 soldiers are said to have died from 'the bloody flux' after gorging themselves on grapes used to make the island's wine and at least once during the siege the army had its daily supply of rations cut, which prompted Sir Alan Apsley, who had accompanied Buckingham on the voyage, to write to a Mr Alcock on 16 July, stating that there was no bread or beer for the soldiers and so they had to make do with wheat and wine, but they had no means of grinding the wheat into flour. With victuals running low Buckingham sent Sir William Beecher back to England to plead for more provisions.[17]

On 13 July 1627 arrangements were ordered to be made to victual 4,500 soldiers for 70 days, but at this time Buckingham's army was much stronger than this figure. It was suggested that they use the rations left over from the Cadiz Expedition, but when these were examined, they were found to be rotten.[18]

The soldiers' diet contrasted with that of Buckingham's, whose food bill included £26 8 shillings for poultry, £4 for bacon and £7 for quails, and £23 16 shillings on herbs to flavour the meat, not to mention £125 9 shillings on confectionery washed down with £232 worth of wine and £13 3 shilling of lemon juice. He also took £7 15 shillings worth of oranges and lemons and paid an additional £47 for fruit.[19]

An undated document by Sir Sackville Crow, believed to be written about August 1627 reported that he along with Sir William Russell and Burlamacchi had collected three weeks provisions for 12,000 men at Weymouth, and a similar amount at Bristol and the promise of the same amount at Plymouth. In addition, there were victuals for two months for 12,000 at London. All of

---

15 TNA SP 16/11/111, Sir Michael Geere to his son, 11 December 1625.
16 John Glanville, *Voyage to Cadiz in 1625* (London: Camden Society, 1883) pp.52–53, 60, 69–70; G. Roberts (ed.), *Diary of Walter Yonge* (London: Camden Society, 1848) p.89.
17 TNA SP 16/289 f 39; TNA SP 16/72/8, Nicholas's statement of the period of which the forces in the Île de Rhé were victualled 4 July 1627.
18 *CSPD 1627–1628*, pp.257, 264; SP 16/73/100, Sir Allen Apsley to Nicholas, 14 August 1627.
19 BL Add ms 12,528, Sir Sackville Crowe's Book of accompts.

which was waiting to be transported to Portsmouth, but there does not seem to have been enough wagons to transport it to Buckingham's army, which badly needed these supplies.[20]

The total rations for the Île de Rhé Expedition and that of La Rochelle amounted to 312,245cwt (hundredweight) of salted beef and 20,213cwt of salted pork. In addition, 59, 615cwt of cheese and 617,303¼cwt of bisket, plus 60,998cwt of butter and 4,693 bushels of pease. There was also 204,789cwt of dry fish and 31,500 'Canada' fish, plus 99 bushels of salt, 3,900 bushels and one peck of wheat, 1,799 bushels and one peck of beans and 792 bushels and three pecks of oatmeal and 114 tonnes of water, 4,430cwt of bacon, 3,419 bushels and one peck of rye, 4,424 bushels of barley and 63 bushels of meal, with 920 bushels of oats and 97 oxen and 220 weathers. 6,300cwt of hay was also sent to feed the livestock. The cattle were probably to provide fresh meat for the officers and possibly the wounded soldiers. Rice, oatmeal and sugar had also been provided for the sick and wounded for the soldiers of the Cadiz Expedition.[21]

For the soldiers going to the relief of La Rochelle in June 1628, the Privy Council ordered the Lord Admiral to cut their provisions to 2lb of bisket, a 4lb of beef, 0.4oz of butter and 0.8oz of cheese, although this still cost six pence per man, but this order was superseded on 2 July 1628 because Apsley was ordered to receive £9,800 for victualling 4,000 men for 84 days, with rations of bisket, beer and flesh for two days per week and bisket, beer, butter and cheese for the other five days. This would cost 12d per man per day.[22]

The cavalry was luckier because they could forage for food more easily. They could ride ahead of the infantry and demand food from the inhabitants of a town. However, they needed to provide feed for both their horses and themselves. In June 1627 Francis Woodhouse, quartermaster of the army received £100 for oats and hay for 150 horses belonging to the officers and the cavalry. A further £326 7s was received for two months' supply of oats, hay and straw to feed the horses and this was even before they had left England. In addition, a Monsieur Barbeau, a servant to Monsieur Sabies was also paid £50 for oats and hay for his horses at Portsmouth.[23]

Wagons were needed to transport the provisions for the soldiers, and horses, to and from the supply stores as well as with the army. If a regiment was marching through friendly country, then they could order supplies to be collected at a magazine in a town on their route of march, but if they were marching through hostile territory then they needed to take supplies with them or demand provisions from the local populace. According to Edward

---

20   TNA SP 16/66/24, Sir Henry Glenham to Marmaduke Moore, 4 June 1627.
21   TNA AO 1/1798/372, Sir John Bagg, Vice Admiral of the South of Cornwall, Victuals, Munitions et cetera, sent to the Île de Rhé for the fleet and army there, 1 March 1627–30 September 1631, 366; TNA SP 28/135, The account book of Sir William Waller's Commissary General during the Civil War records the issuing of fresh meat to the officers and wounded soldiers.
22   TNA SP 16/106/4, Order of Council that the Lord Admiral should give order for providing victuals for 7,000 land soldiers for three months 1 June 1628, E403/2747, Order Book, 1628, f.31.
23   TNA E351/205, Army expeditions, Sir R. Graham, buying horses and provisions for an expedition to the Île de Rhé, 1627.

Cooke's *Prospective Glass,* which was published in 1628, an army should try to preserve its own provisions, but 'seek by all means to intercept your enemy's victuals and lay siege to those places from whence their chief relief of victuals doth come'.[24]

Writing in 1633 Thomas Raymond, who was serving with an English regiment in the Low Countries recalls:

> Wherefore not daring to trust to the country for victuals, we had abundance of ammunition wagons laden with bread and we had mills and ovens carried in wagons, whereof we made some use, but they answered not expectation, they not being able with a whole weeks cooking to furnish the army on days bread. And a greater number of them would not only greater increase the charge and hugely in accommodate the march of the army.[25]

Fortunately for Thomas Raymond and his comrades the bread on the wagons was not needed and they were able to feast themselves on the bread so that it would not go to waste. This was a typical life of a soldier having barely anything to eat one week and plenty the next.

When an army could not rely on the local bakers then brick ovens were built. Perje estimates that it would take about 500 bricks to build an oven, which was big enough to bake between 599.85–1,125kg of bread per day. Therefore, an army of 10,000 men would need between eight to 15 ovens or 4,000 to 7,500 bricks, which would have to be transported along with the firewood to heat the ovens, as well as the ingredients to make the bread, and the bread itself once baked. Where possibly these ovens would continue to bake the bread for several days while the army was stationary.

By 1640 the Dutch Army were using portable copper ovens weighing about 694–793lb, so two or three could be transported on a wagon. They could bake about 1,322–2,480lb of bread per day, and it was suggested that the English Army also use them. Unfortunately, the design of these ovens is not known, and they do not appear to have caught on and it would not be until the Crimean War (1854–1856) that Alexis Soyer invented the portable army stove which was used up to the twenty first century.[26]

How many wagons Buckingham's army took with them is not recorded, but in 1625 the Spanish general, Ambrogio di Filippo Spinola (1569–1630), ordered the Estates of Brabant to supply 500 wagons to carry 'provisions of meal and oats' for his army, which was estimated at 10,000 strong. When the town of Breda fell to the Spanish in 1625 Spinola demanded a further 4,000 wagons to carry away the corn which the Spanish had found in the town. How many wagons he already had is not recorded, but in 1622 Duke Ernest of Saxe-Weimar's army of 3,000 foot and 1,000 horse had 500 wagons or a ratio of eight soldiers to one wagon. On the other hand, in 1626 Mansfeldt had 800 wagons for his 18,000 men and 36 pieces of ordinance or a ratio of four wagons to nine soldiers. One source suggests that the regimental wagons

---

24 Edward Cooke, *The Prospective Glass* (1628) pp.1, 3.
25 John Cruso, *The Art of Warre,* pp.143–145; Raymond, *Autobiography of Thomas Raymond,* p.41.
26 G. Perjes, 'Army provisioning, logistics', pp.8–9; TNA SP 16/414 f.107.

were to be distinguished by flags identical to those carried by the regiment, but there is no evidence to suggest that these flags were ever made.[27]

In addition to the wagons for provisions an army would need to transport large quantities of ammunition needed to keep their troops supplied, plus the senior officers might also take their coaches with them on campaign so adding to the trail of the army which further extended an army's baggage train. Therefore, to capture the enemy's baggage train was of strategic importance, so soldiers were detached to guard it. These wagons also needed horses to pull them. In April 1643, the Royalist Oxford Army contained 50 carts which were pulled by 367 horses, which in turn needed to be fed.[28]

---

27 TNA SP 77/18, List of Spanish forces in the Netherlands, 1625, f12–13, Substance of Spinola's letter to the State of Brabant and their answer, 16 June 1625, f.169; SP101/28, Germany newsletters, 1620–1625 f.152; SP 101/29, Germany newsletters, 1626–1631; BL Harl ms 7364A, Book of tactics in Charles I's time.
28 TNA WO 55/457, Royalist Ordnance Papers, 1643, f.62.

THE FIRST BRITISH ARMY 1624-1628 (REVISED EDITION)

Cock your match

6

Quartering

Part of the soldiers' welfare was the need for shelter, particularly during inclement weather. There were two types of quartering; military encampments when an army was quartered in the countryside, which could last just a day or two to several months if an army were besieging a town. Alternatively, soldiers could be billeted upon the local population, which could also last a day or two if the army was on the march or months if not years if they formed part of a garrison of a town or city. Either way it was the responsibility of the Quartermaster General of the army and regimental quartermasters, who would ride ahead of the army and allocate an area or town where a regiment was to quarter. Whenever possible, towns or villages were preferred, but when these were few and far between then the quartermasters might decide, quarter the cavalry in the towns and establish an encampment in the open fields for the infantry. The Spanish army preferred to demolish the houses within a settlement since the timber would make more shelters for soldiers than if they quartered in the house itself.[1]

Military Encampments

Among the many military manuals of the day, there were diagrams and instructions of how to form these encampments, with each company or troop being gathered together within the space designated for the regiment. The camp would be set out like a small town with lanes between the rows of soldiers' huts. Henry Hexham says that a general should choose the ground where his army is to encamp carefully, so that it is not too near any woods or hills where the enemy might be able to observe the soldiers' movements or launch an attack. He also recommends that a ditch be dug around the camp, four feet deep by three feet broad, to protect it from attack. However, the military engineer Samuel Marolois, suggested that the ditch was to be six to

---

1 Henry Hexham, *The Art of War* (1638), p.21; Geoffrey Parker, *Army of Flanders and the Spanish Road* (London: Routledge, 1997) p.166.

eight feet broad and five to six feet deep 'and the parapet of the same breath and height'.²

Hexham also suggests that, if possible, the camp should be by a river, not only for defensive purposes but also that any waste may be thrown into it. If there was not a river nearby, then it was recommended that the remains of the cattle and any other waste, was to be buried outside the camp.

However, the single unit in these encampments was the hut, where the soldiers would live. None of the accounts of Cecil's or Buckingham's armies mention the issue of tents for the soldiers, and even in the Civil War they are rarely mentioned, apart from sheltering equipment belonging to the artillery train. In fact, the soldiers seem to have preferred huts to tents. In 1641, the diarist John Evelyn served as a soldier in the Low Countries and recalls that huts, 'during the excessive heats was a great convenience… [whereas] the sun piercing the canvas of the tent, it was during the day, insufferable, and at night not seldom infested with mists and fog, which ascended from the river'. Samuel Marolois, also stated that huts were 'warmer [at night and] cooler against the sun and more durable'.³

Unfortunately, no instructions survive on how to build one of these huts, but they were made of any materials at hand, for example wood covered with turf or straw. Although William Gerrard, writing in 1591, records that:

> Having marched all day and coming at night to the place where they must camp, one sets out the driest and warmest plot of ground he can get in the quarter, which is appointed to his band for lodging… whilst another makes provisions with one of their boys, in some adjoining village (if time and safety from the enemy doth permit) for long straw both to cover their cabin and make their bed…another with a hatchet… doth cut down forked bows and long poles to frame and rear up their cabin withal and provide timber and fire wood if it be winter or if need requires.⁴

Hexham also states that the soldiers made 'their huts of forks, lathes, withes and straw or for if they were staying just one night, sticks and boughs or as such things as they can get'. Some illustrations of encampments show these huts made from wooden planks over a frame and open at both ends, while others show a more study construction with walls, doors and even a window. One contemporary source records, that:

> In each hut or cabin you shall lodge two common soldiers and so you shall have 50 huts for each company, divided into ranks, each rank consists of 25, every hut is to be six foot broad and seven foot long, each hut shall be distant one from another rank [by] two feet. The ranks shall be divided by a street or lane of 10 foot broad.⁵

---

2 Marolois, *The Art of Fortification* (1638) p.32.
3 William Bray, (ed.), *The Diary of John Evelyn* (Frederick Warne & Co, 1818) p.26; Marolois *The Art of fortifications* p.38.
4 Gerrard, *The Art of Warre*, (1591) p.13.
5 BL Harl no. 6,344 *A Short military treatise concerning all things needful in an army* ff.160–169.

Other sources suggest that the hut should be eight feet by eight feet, including Robert Ward, who writing in 1639, records that the Swedish armies had three men per hut because they drew up six files deep, whereas two soldiers seem to have been normal. Although when a soldier brought his wife or 'whore' with him, then they would share a hut together along with any children they might have.

Whatever the size of the hut and the number of occupants it was the sergeants' duty to make sure that:

> The soldiers build their huts even and keep exactly the measurements so the streets may be even and that if it be possible that the huts be built all of one height and after one and the same fashion… All the doors of the huts must be open to the lane that is between the ranks of the huts.[6]

Thomas Raymond, who was a 'gentleman of the pike' in Colonel Packenham's Regiment in the Low Counties in 1633, does not describe his accommodation but records that, he 'only wanted a good bed and sheets', to make himself comfortable. He managed to purchase 'an old tent' to make a bed with which he stuffed with straw and 'supported with four crutches two foot from the ground' and managed to sleep 'excellently well'.[7]

The standing orders for Sir Horace Vere's army, dated about 1614, record that the:

> Quartermaster ought to go through every captain's quarters to see if the soldiers have built their cabins high and wide enough and have made their beds above the ground. The captains ought every three weeks to get fresh straw to give their soldiers… and to air their cabins and to keep their quarters swept every day very clean to avoid sickness.[8]

The soldiers' arms were to be left outside these huts.

Francis Markham states that the corporals should make sure that the men of his squadrons should 'live together like lovers and companions', that is helping each other rather than any sexual connotations.[9]

All the manuals agree that it was up to the provost and the corporals to make sure that the camp should be kept clean, especially when an army quartered in the same place for a long time, for example when besieging a fortress or town.

In front of the company was quartered the captain, ensign, clerk and drummers, whose tents or huts were 20 feet from the soldiers' accommodation. The lieutenant, sergeants and the surgeons were quartered behind the company, with a 10 foot road separating them from the company. Officers also appear to have preferred huts, but senior officers would have to

---

6   Robert Ward, *Animadversions of Warre* (1639) book 2, pp.32, 38; BL Harl 7364.
7   Raymond, Autobiography p.38.
8   TNA SP 9/202/1/18 Compendium of the Discipline of the Art of War under Sir Horace Vere, 1603–1625.
9   Markham, *Five Decades of War,* p.67.

live in tents, which they were expected to purchase themselves. According to Thomas Raymond:

> My colonel had a kitchen tent, a servants' tent, a lodging tent for himself and another, the largest and fairest tent of all, wherein he eat, gave audience to his officers and entertained. Which also was our church having there every Sunday morning a sermon preached by the chaplain to the regiment.[10]

There was also to be a place behind the company, 20 feet wide and 24 feet long 'to wash and dress meat'. The regulations stated that a whole company should occupy an area of 300 feet long by 24 feet wide.[11]

Horses were to be also housed in huts, which were to be:

> Seven foot long and six broad and lodge 25 horses in one rank, behind that you shall leave a void place of 10 foot broad and then lodge 25 horses more in such manner as all the horsetails may be towards that void place and their heads towards their masters' huts. Between which and the horses' heads you shall leave a space of five foot for the horsemen to put their forage and provender.

The horseman was also making his hut seven feet long and five broad and leave two feet between every two huts so that for 50 horses and their master's there must be allowed a place of 50 foot long and 48 broad. A road was also to divide these huts into troops. The treatise continues:

> It is to be noted that the horses' dung may be cast upon the place where the wagons stand but no other kind of dung [or] filth because horses dung does not infect the air nor is not unwholesome as other dung and it is the provost marshal's part by means of his sergeants to look unto this and that the quarters be kept clean.[12]

The men and horses of the artillery and wagon train were also to be lodged in a similar fashion.

However, when on the march a soldier might not be able to erect a hut and like Thomas Raymond:

> Had nothing to keep me from the cold wet ground but a little bundle of wet dried flax, which by chance I lighted on, and so with my boots full of water and wrapped up in my wet cloak, I lay as round as a hedgehog, and at the peep of day looked like a drowned rat.[13]

Another officer found a small piggery, which he used as a shelter one night, but not before he had eaten the previous occupant.

---

10 Life of Alexander Leslie, p.72.
11 BL Harl no. 6,344 *A Short military treatise concerning all things needful in an army* ff.160–169.
12 *Ibid.*
13 Peter H. Wilson, *The Thirty Years' War*, a sourcebook (Houndsmill: Palsgrave-MacMillan, 2010) p.163; Raymond, Autobiography, pp.38, 39, 40.

QUARTERING

Sutlers scene from Jaques Callot's *Siege of Breda*. (from Philippson's *Geschichte des Dreissigjahrigen Krieg*)

There were strict rules on how the soldiers and camp followers should conduct themselves in camp. The quartermaster:

Shall suffer no soldier, his wife or child or any other in their behalf to draw any beer, sell any goods or sutler within the regiment for to this end there are sutlers ordained and allowed and their cabins appointed behind the regiment.

Secondly, he shall not permit any places for the dressing of meat or dicing tables or any holes for what purpose so ever to be dug neither before nor behind nor near the quarter being dangerous in the night time both for men and horse to pass such ways, especially upon any occasions of alarm which may breed great inconvenience and disorder, for behind the sutlers cabins there is a firing place appointed to dress their victuals aside from all other passages.

Thirdly he shall not suffer any building of cabins, placing of tents or shops, selling of goods upon wagons or cars [carts] or anything else within the foresaid alarm place or ways before, behind, within or about his regiment but command all such to repair to the market place.

Fourthly considering that butchers are permitted to lodge here in the cabins behind the regiment and that their proceeds much filth and stink from the beasts that are killed causing great sickness and infection, the quartermaster shall cause the marshal to take care that a pit may be digged at a convenient depth and largeness in the ground behind the quarter which shall be covered with strong spurs and planks laid over in the midst of which planks shall be a door to shut and open of one foot and a half square. These pits shall be made at the charge of the butcher or butchers, that shall quarter with the regiment. If necessary, also that every colonel or great officer that keeps a table should have such made near their lodging that a whole quarter might be kept sweet and wholesome as well as their own lodging which otherwise in small time must needs grow very nasty and noisome.[14]

The sutlers' tents and huts were also organised in a street of two rows of stalls. These streets are often recreated by contemporary artists, such as Palemdes Palamedesz or Philip Wouverman, who wanted to create a picturesque scene of camp life. These artists usually depict sutlers' stalls with laurel wreaths, a tankard or a flag hanging outside the tent as a symbol that a soldier could buy food or beer there, and probably other things as well.[15]

Billeting Upon the Population

The preferred option was to quarter the soldiers on the inhabitants of a town, especially during the winter when armies usually ceased campaigning due to the inclement weather. If a soldier remained in a garrison for a long time, then they might not just settle down and start a family, but also rent or even buy a house to live in. In 1625 at least three soldiers in the Portsmouth garrison owned their own homes, for example James Jeliffe, who appears to have been a sergeant, lived in a house with three rooms, a garden, barn and a field and even had a maid. Nicholas Turner, who was a musketeer also had a three roomed house modestly furnished and 45 shillings of apparel, which was equivalent to one suit of clothing. And Thomas Granger left his house to his son Richard.[16]

However, for the majority of soldiers their only option was to be billeted on the local population, either for a night or longer. In May 1625, the Deputy Lieutenants of Devon were ordered to expect 10,000 soldiers and to provide 'lodging and diet by the week'. This billeting money was to be three shillings per week for a soldier or lesser officer, five shillings for an ensign and seven shillings for a lieutenant. The money would either be given to the soldiers (who were more likely to spend it on other things) or to the householder directly, although it was often in arrears. Captains and more senior officers had to come to their own arrangements with the householder, although they were more likely to get the better quarters. The quartermasters would chalk on the door of a house the regiment or company and the number of men that the property was to accommodate. A company was to be quartered in the same town or village, if possible, to make training easier and they were to set watches to keep order. Unfortunately, for the local population they had no say in the matter and with most houses having only a couple of rooms, and probably only one bed, this would bring misery on the householders.[17]

The Justices of the Peace were also to make a list of all the soldiers' arms and apparel within their county and were to make sure that the markets

---

14 BL Harl 7364.
15 Beelden, Van een Strizd een strijd: oorlog en kunst vóór de Vrede van Munster, 1621–1648 (Zwolle: Waanders, 1998) p.251.
16 HRO 1626/B/057, Will of William Hare, senior, of Longparish, Hampshire, 1626.
17 TNA SP 16/34 f.134.

were well stocked, and the provisions were sold at reasonable prices to avoid profiteering by the local inhabitants. The Justices were also to issue a proclamation imposing martial law on their county and that no officer or soldier stray from the town they were quartered in without a pass.[18]

There were also strict rules issued to the householders and the soldiers billeted upon them. If a soldier was not present at the set time for dinner and supper then he would 'lose their meal and the host shall not be bound to give him anything until the next' meal, unless he was on duty. A curfew was also to be imposed upon the soldiers from eight o'clock in the evening in the winter and an hour later in the summer, and the householder would have to report any soldier to an officer if he went absent during this time. The rules also stated that the soldiers were not to 'strike his host, hostess, child or servant'. However, many soldiers probably took over the houses where they were billeted, forcing the householders to feed them when and where they wanted. On 28 January 1626 Sir William St Leger wrote 'the men live at more ease than if they were at home'.[19]

At first, the inhabitants of the Isle of Wight who were billeting soldiers found it was not as bad as they had thought, as Oglander records:

> Sir Alexander Brett's and Sir Henry Spry's two regiments came to us about the 6th of May 1627 and stayed with us until the 21st and 24th of June following. We at first thought our island could not have victualled them, but we found no want of provisions and seeing we were well paid for their board could have been content on the same conditions to have kept them longer.[20]

But when a new regiment arrived on the island, the inhabitants' experience was totally different, as Oglander's entry for 14 August 1628 records, 'our island being miserably oppressed with the [Morton's] Scotch Regiment'. On 3 September it was with joy that he wrote 'we were freed from our Egyptian Thraldom or like Spain from their Moors for since the Danish [Viking] slavery never were this island so oppressed'.[21] The inhabitants of the Isle of Wight petitioned the Privy Council complaining that a Scottish soldier had murdered 'one Stevens', while a Highlander had raped a woman at Northwood, and they had committed, 'diverse robberies' and had killed 'divers of His Majesty's deer'.[22]

Although, in theory, the householders were paid for this inconvenience, the quartering of soldiers usually fell upon those who could least afford it. A petition from 26 inhabitants of Odiham in Hampshire complained that 67 soldiers of Sir Sheffield Clapham's company were quartered upon the poorer people of the town at a cost of £13 8s per week, whereas the wealthier parishioners had left the town to avoid the soldiers and so 'the tradesmen of

---

18 *Ibid.*
19 TNA SP 16/19 f66, SP 16/2 f.122; BL Egerton 2087 ff.24, 45–46.
20 W. H. Long (ed.), Oglander Memoirs, extracts from the Ms of Sir John Oglander (London: Reeves and Turner, 1888) pp.29–30.
21 Long, *Oglander Memoirs*, pp.36–38.
22 TNA SP 16/88/54 Misdemeanours committed in the Isle of Wight by some of the Scottish Regiment there billeted, nd [1628].

our town [are] very much decayed'. Other Hampshire towns also complained that the soldiers were 'being billeted in the houses of such poor and indigent persons'. In one incident to ease the county's burden the Hampshire authorities moved some soldiers to Farnham, in Surrey, without the lord-lieutenant or deputy-lieutenants of the county of Surrey's permission.

In Gloucestershire, the soldiers were not satisfied with their quarters, 'being billeted in the houses of such poor and indigent persons as are not able to provide for him according to the entertainment allowed by His Majesty'. Therefore, the deputy lieutenants were ordered to move them 'to such householders as may be of competent ability to provide for them'. Fortunately, for at least some of the inhabitants, half of the 600 men of Colonel Spry's Regiment who were quartered in the county were sent to Herefordshire.[23]

By 1628, the southern counties of England had billeted soldiers for three. The accounts of Guildford, in Surrey, record that 125 soldiers were quartered on the town for 14 days in January 1628, followed by a further 98 soldiers for 26 weeks. Each of the inhabitants had between one and three soldiers billeted upon them at a time, at a cost of £499 4s 8d, but only £24 10s was paid during this time.[24]

In April, the Mayor of Sandwich refused to quarter some soldiers of Crosby's Irish Regiment pretending that he had not received any notice of their arrival and later the Privy Council wrote to the lord lieutenant of Surrey that the soldiers of Bartie's Regiment had been left 'destitute of all means' after being evicted by the population of Kingston and other places in Surrey, who refuse to house or feed them. The inhabitants of Bath also refused to quarter the soldiers of Colonel Grenville's Regiment any longer, because they had not been paid for the quartering the men up to then.[25]

When Charles called his third parliament in April 1628, the Commons presented a petition to the king concerning the effects the billeting of soldiers was having on the local population. It claimed that:

> The service of Almighty God is hereby greatly hindered, the people in many places not daring to repair to the church, least in the meantime the soldiers should rifle their houses.
>
> The ancient and good government of the country is hereby neglected and condemned.
>
> Your officers of justice in performance of their duties have been resisted and endangered.
>
> The rents and revenues of your gentry [are] greatly and generally diminished [because their tenants]; farmers to secure themselves from the soldiers insolence by clamour and solicitation, of their fearful and injured wives and children enforced to give up their wonted dwellings and retire themselves into places of more secure habitation…

---

23 HMC 12 Report Appendix XI Records of the Corporation of Gloucester 13 February 1628 Privy Council to Lord Lieutenant of Gloucestershire pp.483, 485.
24 SHC LM 1330/73, billeting accounts for Guildford.
25 Lyle, *APC 1628*, pp.434, 389, 424.

> Tradesmen and artificers almost discouraged, and being enforced to leave their trades and to employ their time in preserving themselves and their families from cruelty of the soldiers.
>
> Markets unfrequented as our ways are grown so dangerous that the people dare not go to and fro upon their usual occasions.
>
> Frequent robberies, assaults, batteries, burglaries, rapes… and barbarous cruelties and other most abominable vices and outrages are generally complained of from all parts where these companies [of soldiers] have had their abode, few of which insolencies have been so much as questioned and fewer, according to their demerit, punished.[26]

It also claimed that 'many of those companies… do openly profess themselves Papists [Catholics] and therefore to be suspected if occasion serve, they will rather adhere to a foreign enemy of that Religion, than to your Majesty'.[27]

In May 1628, the House of Commons proposed the Petition of Right which included the clause:

> Whereas of late great companies of soldiers and mariners have been dispersed into divers counties of the realm and the inhabitants against their will having been compelled to receive them into their houses and there to suffer them to sojourn against the laws and customs of this realm and to the great grievance and vexation of the people.[28]

The 'laws and customs' referred to in this clause were those established by *Magna Carta*. After three weeks debate between the Commons and the Lords the petition was reluctantly ratified by the King on 7 June.

Meanwhile during the debate over the petition, the billeting of soldiers continued, and on 2 June 1628 the Mayor of Canterbury complained that:

> The soldiers threaten to take the meat from the butchers' shops and others take away men's wares and break open the doors of the billeters, yea some of the officers would have killed their billeters in their own houses. The inhabitants… stand in fear of their lives.[29]

Unfortunately for the inhabitants of towns and villages where the soldiers were billeted, the King could not understand their hardship of not being able to afford to feed their own families, let alone lodge and feed the soldiers with no financial help forthcoming. He had promised that they would be paid and therefore anyone who refused to billet the troops must be 'out of some diffidence of His Majesty's royal promise, or by the example and encouragement of some persons ill affected to His Majesty's service or out of sinister and false apprehensions'. Therefore, the King was willing to 'pass by

---

26 Cobbett's, *Parliamentary History of England* vol. 2 (London, 1807), p.283
27 Ibid.
28 Petition of Right clause XI.
29 CSPD 1628–1629 p.145.

their contempt without censure or punishment' and that they would soon be paid when Parliament voted him the subsidies needed.[30]

The Subsidy Bill completed its passage through the House of Lords on 17 June 1628, which granted the King five subsidies, each valued at four shillings in the pound, but the money would come in slowly and whether the inhabitants ever received their full entitlement of quartering money is doubtful.

The ratification of the Petition of Right did not bring about an abrupt end to the quartering of soldiers. This would not come until the army was disbanded in October 1628. Even then the inhabitants would still be forced to quarter soldiers being raised for foreign powers, although billeting would not be on such a vast scale until the outbreak of the Bishops' Wars in 1639 and 1640 and then the Civil Wars, when both king and parliament would forget the 'laws and customs' of the people of Britain.

---

30   HMC Report 12 Appendix XI Records of the Corporation of Gloucester, pp.485–486.

QUARTERING

Try your match

7

Discipline and Pay

According to the Imperialist general Raimondo Montecuccoli, 'for war you need three things, 1, money; 2, money; 3, money', but unfortunately for Charles he did not have enough of it to fight one war, let alone three. Therefore, the easiest option was not to pay the merchants or soldiers the money that was due to them. It was estimated that the pay of a company of foot in 1625 was £14 19 shillings for a week, and that it cost three shillings and six pence per week for a healthy soldier and four shillings and six pence for a sick one. In 1627 the wage of a common soldier was eight pence a day or two shillings and six pence per week, three shillings 6 pence for a corporal and 12 pence a day or five shillings for a drummers and a sergeant. A lieutenant could expect to receive four shillings and an ensign two shillings and six pence a day.[1] However when it came to the captain, his pay could vary, a captain in the field or in the Jersey garrison could expect to receive eight shillings a day, whereas a captain on the Scilly Isles and at Dover Castle received 10 shillings a day. On the other hand, a colonel could expect £1 per day, plus his salary as a captain of a company or a troop and if he was also a general then he could also receive £4 per day or £10 if he commanded an army.[2] However there were deductions taken from the soldiers' wages to pay for his clothing and other items.

By April 1626, the soldiers of Cecil's Army were owed £11,000 in arrears. There was some attempt to pay the soldiers, but their wages came in slowly, £2,910 was ordered to be paid on 6 April 1626 to those at Plymouth and a further £900 on 20 May 1626 with an additional £900 being ordered to be paid on 20 June 1626, although by now the arrears had increased further. However, Charles could not cut his losses and disband the army due to the fear of invasion and he knew that if the army were disbanded then he would have to pay the soldiers all their arrears. Therefore to cut down costs, on 19 January 1627 the Privy Council ordered that all officers and lesser officers who were not with their companies would not receive their pay, but there also

---

1   Lyle, *APC 1625–1626*, the list of a week's pay for 100 foot, 15 June 1627, p.276.
2   Lyle, *APC 1627*, pp.63, 382, 439, Minute of the Privy Council re garrison of Jersey, 16 February 1627, establishment of fort at Landguard Pointe, 1627, and an estimate of his Majesty's weekly charge for 200 men in Jersey, 23 July 1627.

seems to have been an ad hoc arrangement over paying the soldiers because although Captain John Mason was officially responsible for distributing the pay to the soldiers, in May 1627 he complained that he had paid the captains in Dorset five months pay for their companies, but they had also received 'one thousand marks' from another source while other regiments had not received a penny. On 27 May 1627 he warned Buckingham that:

> The pay must be without fail [delivered] otherwise mutiny and disbandment will follow which was hardly prevented in Sir John Burgh's Regiment at Winchester by reason of eight weeks pay arrears and to the poor billets of that town, chiefly caused by the default of Berkshire loans not supplied.[3]

In April 1628, several troopers of Montjoy's troop complained that they still had not received their pay. Many officers, sergeants, corporals and drummers were petitioning the Privy Council in March 1629 for their arrears some of which dated back to the Cadiz expedition. The lesser officers claimed that they were being 'driven to such extreme necessity' being unable to afford accommodation or buy food and clothes.[4]

An undated order written either in 1627 or 1628 records that the senior officers who had taken part in the Cadiz expedition were to receive just eight of their 19 months arrears of pay; while the company officers and lesser officers would receive just four months pay. This pay was to come from the loans where each regiment was quartered. It was not until May 1636 that Sir William St Leger finally received his arrears of pay for the Cadiz expedition.[5]

Charles had not been granted the subsidies he needed to wage war when he dissolved his first parliament in August 1625, so he turned to demanding loans from his subjects. In November 1625 letters were sent out to all Justices of the Peace to introduce the loans systems, but with the money he needed not forthcoming Charles was forced to summon his second Parliament in February 1626. This parliament did agree that Charles should have four subsidies, but only if the king agreed to their demands to impeach Buckingham on the charge of treason. Charles could not possibly agree to this, and he dissolved Parliament once more without it granting him his badly needed subsidies. Charles looked to the loans system once more as well as Tonnage and Poundage, which was a tax upon wine, imports, and exports.

Warrants were again sent out to the Justices, often the demand for the loan would be sent out using a printed form with the name of the individual and the amount due being written in the appropriate spaces. On 10 July 1626 William Ireland of Saint Margaret's, Westminster received one of these forms stating that he was charged with loaning Charles £10, 'which we do promise in the name of Us, our heirs and successors to repay to you or your assigns within eighteen months'. William Ireland paid his £10 on 18 July 1626.

A booklet was even printed for the Justices called *Instructions which His Majesty Commissioners on the Loan of Money to His Majesty* on how they

---

3 TNA, SP16/64/7, Captain John Mason to Buckingham, 27 May 1627.
4 TNA SP 16/138/77 Petition of the Sergeants, Corporals and Drummers, 14 March 1629.
5 SP 16/61/99 Particular account of money due to the officers that went to Cadiz, April 1627.

were to collect this money. Among its 17 points was that the money should be paid to the JPs within 14 days of them informing those who had been chosen to lend money and if possible, it should be in full. Moreover, the booklet called for confidentiality when it came to how much an individual had to pay, which led to suspicion that those in rural areas were paying more than those in the towns. The Forced Loan managed to raise over £240,000, but not before 76 of the leading gentry were imprisoned without charge for not paying this loan. However, in December 1627 it was estimated that £600,000 was needed to pay Buckingham's Army and re-equip the fleet so that it could relieve La Rochelle. The exchequer accounts show that £184,820 0s 10d was received and a further £82,244 15s 1d was paid on billeting, coat and conduct money, a total of £267,064 15s 11d. Some counties were better than others at collecting the Forced Loan, Berkshire, Cambridgeshire, Hampshire, Sussex and Suffolk paid the most, whereas Northampton, Lincolnshire and Gloucestershire paid the least since the gentry failed to cooperate with the collectors and were sometimes openly hostile.[6]

In 1628 the Petition of Right would supposedly guarantee that 'no person should be compelled to make any loan to the king against his will', and that no one should be imprisoned without 'the lawful judgement of his peers'.

Unfortunately for Charles the money received from the Forced Loan and other money raising schemes came nowhere near meeting the amount he needed, which according to Cust was £1,000,000 per year just for fighting Spain and France, but he also needed subsidies for the Danish war effort as well as paying for Morgan's men. The sale of crown land brought in another £120,000, although it was worth £145,000, so Charles had no choice but to recall Parliament. In 1628 Charles' third Parliament finally voted him five subsidies worth £350,000 but he dissolved it the following year and he would not call another Parliament again for 11 years in which time he continued to rely on non-Parliamentary taxes and imprison those who refused to pay them.[7]

Meanwhile this lack of pay was given as an excuse for the bad discipline of the soldiers. On 9 January 1625, the Earl of Lincoln, who was a colonel in Mansfeldt's army, complained to Secretary Conway that, 'I have received... miserable complaints from my captains of want of money for the payment of their companies which occasions some to run away others to steal and some to faint such for want of means'.[8] The inhabitants on the roads to Dover in Kent would be the first to feel the soldiers' indiscipline. Luckily, they would only have to endure the soldiers for a night or two, but when about 8,000 men unexpectedly arrived at Dover on 25 December 1624, they found that no food or lodging had been provided for them and so they rioted. The following day the lieutenant of Dover Castle, Sir John Hippisley, complained, 'the soldiers

---

6 *Instructions which His Majesty Commissioners on the Loan of Money to His Majesty* (EEBO); John Bruce *Letters of the Verney family* (Camden Society, 1858); Kevin Sharpe *The Personal Rule of Charles I* (London: Yale University Press, 1992) p.20; Cust, Richard *The Forced Loan and English Politics*, (Oxford: Clarendon, 1987), pp.3, 92.
7 Cust, *The Forced Loan*, p.75.
8 TNA SP 14/181/25 Sir Thomas Dutton to Lord Chamberlain Pembroke, 7 January 1625.

commit great outrages, pulling down houses and taking away cattle et cetera and when some were taken and imprisoned, the prison was broken open and they were rescued'. On 27 December, another resident of Dover complained that the soldiers were all 'rascals and gaol birds' who commit all, 'disorders of rapine… killing sheep in abundance and all other things they can catch, threatening the breaking into our houses and sting fire to the town upon any want they may have'.[9]

Another protested that although they had been given money to buy food, they steal 'all they can lay [their] hands on. They also stole 140 sheep from one man in a night and people dare not open their doors'. Since the soldiers claimed they did this because their pay was in arrears the townsfolk themselves had a collection to pay off the soldiers and to ease matters and between 2,000 to 3,000 soldiers were sent to stay at Folkstone and Sandwich. This was not enough, and the abuses continued. On 31 December Hippisley wrote that the 'country [around Dover] is utterly wasted for 10 or 12 miles because the soldiers, plunder all about' and that the soldiers had 'ravished a lady and her two daughters'. Hippisley also claimed that the officers were 'careless about the employment', but what he did not know was that the soldiers had even attacked them and so they were afraid of their own men.[10]

On the same day, the Privy Council wrote to the mayor of Dover suggesting that the Kent trained bands should be mustered to 'reduce all to order by punishing the offenders and maintain the public peace', but added they were not to shed 'blood without great cause'.[11]

Finally, on 2 January 1625 Hippisley was able to report that the commission for martial law had arrived at Dover, but he added that imposing martial law was useless unless the soldiers were paid, although on 7 January 1625 he was finally able to report that the town was now much quieter because a soldier had been hung for burglary which 'struck terror into the rest'. After this there was no more serious outbreaks of violence. It had been estimated that Mansfeldt's force would stay just three days at Dover, but it was not until the end of the January that they finally embarked for the Low Countries.[12]

With Mansfeldt's force gone, the inhabitants of the towns around Dover could try to pick up their lives once more. However, for the inhabitants of Devon their ordeal was about to begin. On 23 June 1625, the Privy Council wrote to the Mayor of Plymouth, that the soldiers:

> Have so ill behaved themselves in their passage to Plymouth and at Plymouth and in places thereabout… that the country thereabouts is much grieved and injured

---

9   TNA SP14/177/33 Francis Wilsford to Nicholas, 27 December 1624.
10 TNA SP 14/177/18, 23, 44, 47, 48, Sir John Hippisley to Buckingham, 24 December 1624, Sir Richard Bingley to Buckingham 26 December 1624, Richard Marsh to Nicholas, 30 December 1624, Sir John Hippisley to Secretary Conway, 31 December 1624, Sir John Hippisley to the Mayor of Dover and Justices of Kent, 31 December 1624.
11 Lyle *APC 1623–1625* pp.409, 411 A letter to the Earl of Mongomerie, Lord Lieutenant of Kent, 26 December 1624, A letter to the Mayor of Dover, Sir Nicholas Tufton, Sir Edward Hales and others, 31 December 1624.
12 TNA SP 14/177 ff.10, 26.

by them... [For the officers] do very remissly perform their duties there in the ordering and governing of the soldiers committed to their charge.[13]

The Mayor of Plymouth was to report any unruly companies to the Privy Council, but the soldiers' behaviour would drastically affect the lives of the people of the town, who feared to leave their houses. Some soldiers were handed over to some 'Netherlanders for their better disciplining', but unfortunately Captain George Took of Essex's Regiment does not elaborate upon this.[14]

On 1 June 1628 £3,000 of the £18,000 lent by the city of London for the payment of the artillery was used to pay the navy who were preparing once more to sail to La Rochelle. However, this was not enough and on 17 August 1628 rioting broke out in Portsmouth. About 300 sailors are said to have surrounded Buckingham's coach demanding their pay, one mariner even pulled the Duke from his coach. One sailor was arrested and on 22 August hung on the gibbet between Portsmouth and Southsea Castle.[15]

However, it would be wrong to suggest that all soldiers were ill disciplined. In December 1624, while the outrages were being committed in and around Dover, the inhabitants of Winsborough in Kent found the soldiers quartered upon them 'demeaned themselves very honestly and civilly', and early in 1627 the inhabitants of the Isle of Wight were also surprised how well behaved the soldiers were and it was said that in 1628 100 soldiers sent to Taunton were 'billeted there with good content for two months'.[16]

Sometimes soldiers were blamed for incidents they did not commit, as at Banbury when a soldier was committed by the constable for an affray on market day. The ensign of the soldier's company, Henry Raynde, hearing this assaulted the constable, George Phillips and in the fray that followed a soldier appears to have threatened to burn down the town. On the following Sunday, 2 March 1628, a fire broke out and the Justices blamed the soldiers quartered in the town. The fire burned all night and the following day destroying, '103 houses, 20 kiln houses and other out houses' or about a third of the town, causing an estimated £20,000 of damage. The Justices complained to the House of Lords and several soldiers were summoned to appear before the Lords. The soldiers' guilt appeared to be confirmed when Reynde failed to appear forcing a warrant to be issued for his arrest.

However the Lords heard that although it was true that the soldiers had beaten the town constable, believing only their officers could dispense justice to them, they had helped put out the fire and it was later found that the fire had begun in a malt house, by the 'negligence of a maid' and so the fire

---

13 Lyle *APC 1625,* A letter to the Mayor of Plymouth, Sir John Ogle and other commissioners at Plymouth, 23 June 1625, pp.99–100.
14 Took, *Cales passion or the mistaking of Cales* p.3.
15 Lyle *APC 1627–1628* Minute of the Privy Council, 1 June 1628, p.481; Oglander *His Observations* pp.33–34.
16 TNA SP 16/181/11 Sir John Hippisley to Buckingham, 7 January 1625, *Journal of the House of Commons* vol. 1 pp.885–886.

was put down to 'God's providence' and the charges against the soldiers was dropped.[17]

Nevertheless, for failing to appear before the Lords and for 'ignominious speeches', against Lord Saye and Seale, Henry Reynde was ordered never to bear arms again either in England or the Low Countries being 'unworthy to be a soldier'. Once he was apprehended, he was to stand in the pillory both at Cheapside in London and in Banbury 'with papers on his head, showing his offence' and then imprisoned.[18]

Reynde was not the only ensign to be arrested for indiscipline, Ralph Bowyer (or Bowers), whose company was billeted in Bromsgrove, Worcestershire was arrested in May 1628 for saying that 'this country is base; and, for their baseness, I trust in God to see this country overrun by the enemies before it be long'. He also wounded a Robert Biddle.[19]

Sometimes the soldiers themselves were provoked by the local inhabitants, in one incident some soldiers of Crosby's Irish regiment were celebrating Saint Patrick's Day at Witham in Essex, by wearing red crosses in their hats. Unfortunately, 'an untoward boy' tied a red cross to a dog's tail and forehead, and one report also said to a whipping post, which outraged the Irishmen. In the rioting that followed 'many townsmen and soldiers were dangerously wounded' and between 30 and 40 were reported killed on both sides. Unfortunately, the parish registers do not survive for this period so there is no way to confirm the number of dead. However, after this the soldiers were disarmed by their major.[20]

Neither was violence on civilians always an act of indiscipline, but was also used as a tool to collect taxes or the Forced Loan. On 21 May 1628, the House of Commons heard that the Mayor of Chichester was accused of ordering soldiers to ransack the houses of those who refused to pay their taxes. While in Surrey some constables ordered the soldiers to pull down or set fire to the houses of non-taxpayers, and others gave the local population a choice either they paid their taxes or have soldiers billeted upon them. This prompted the people of Surrey to present a petition on 26 March 1628 of the 'great complaint made of the insolencies of diverse soldiers in Surrey'.[21]

There were *Laws of War* which were published which were meant to prevent these disorders. They were read out to each regiment regularly so that the soldiers would learn them, but the officers and the regimental provosts were usually unwilling, or unable, to impose discipline. Unfortunately, Mansfeldt's *Laws of War* do not appear to have survived, but all the armies he commanded were notorious for being allowed to plunder the countryside in lieu of their pay.

---

17   Relf, Frances Helen (ed) *Notes of the Debates in the House of Lords* (Offices of the Royal Historical Society, 1929) pp.66, 67, 72–78; Beesley, Alfred *The History of Banbury* (London, Nicholas and Son, 1841) pp.277–280; *Journal of the House of Lords* vol. 3 pp.698–700, 707–708, 835–836, 849–852.
18   *Journal of the House of Lords* vol. 3 pp.849–852.
19   *Journal of the House of Lords* vol. 3 pp.829.
20   G WE 'Communication, Saint Patrick's Day 1628 in Witham, Essex' *Past and Present* vol..61 pp.139–148; Journal of the House of Lords vol. 3 26 March 1628 pp.707–708.
21   *Journal of the House of Commons* vol. 1 pp.876–877, 902.

On the 3 October 1625, Cecil published his articles of war, but these were only 12 points long and included the usual clauses of a religious age including that God be 'duly served twice every day' by the soldiers and that they 'shall be discharged every watch, with the singing of a psalm and prayer usual at sea'. There was to be no 'swearing, blaspheming, drunkenness, dicing, carding, cheating, picking and stealing and the like disorders'. All companies were to 'live orderly and peaceably'. There were several articles which were commonsense while on board the ships, such as no candles were to be used without lanterns and smoking below deck was forbidden. Furthermore, no soldier was to, 'give offence to his officer, nor strike his equal or inferior'.[22]

However, an experienced soldier like Cecil must have known that these laws were totally inadequate and when Henry, Viscount Falkland, published his *Lawes and Orders of Warre* on 11 June 1625 for his army in Ireland it listed 46 offences, 18 of which carried the death penalty. It may be that Cecil published some further laws which have not survived, but if he did then they were superseded on the 3 December 1626 by a stricter 32 point military discipline code, which was accompanied by a royal warrant:

> Soldiers and mariners after the publication of this commission commit any robberies, felony, mutiny or other outrage or misdemeanour, or which shall withdraw themselves from their places of service or charge… or shall be found wither the said county or towns and county of Southampton… to be punished with death… for an example of terror to others and to keep the rest in due awe and obedience. To which purpose you will and pleasure is that you cause to erect a gallows or gibbet in such places within the said county of Southampton as you shall think fit and thereupon to cause the same offenders to be executed in open view that others may take warning.[23]

This in turn was followed by a 60 point list on 17 September 1627, 41 of these points prescribed the death sentence for various crimes, followed by 12 points which had the sentence 'according to the quality of the crime' or 'according to discretion'. Although it should be noted that during the English Civil War the Earl of Essex's Army had 95 articles and the King's Army 100.

In all there appears to have been at least five codes of conducts issued to the soldiers at this time, although only one states that:

> The provost must have a horse allowed him and some soldiers to attend him and all the rest commanded to obey him and assist him otherwise the service will suffer for he but one man and must correct many and therefore cannot be beloved and he must be riding from one garrison to another to see that their soldiers do no outrage nor scath about in the country. And these orders we hold in all counties where true discipline is esteemed.[24]

---

22 BL Lansdown ms 844, Miscellaneous articles, 1558–1726, f.309.
23 TNA SP 16/13/42, Copy of Martial laws ordained by his Majesty for the government and good ordering of troops in the Kingdom (1625).
24 SP 16/13/42, Copy of Martial laws ordained by his Majesty (1625).

Both Surrey and Middlesex are also known to have employed provost marshals and in 1627 the Deputy Lieutenant of the counties where the regiments were quartered were to place 'strong and diligent watches… upon all the passages and usual ways', and any deserters apprehended. The ringleaders were to be imprisoned until they could be tried by two Justices of the Peace and the rest returned to their regiments. Any vagabonds pretending to be soldiers were to be imprisoned.[25]

In July 1626 Martial law was imposed on Surrey and Hampshire, followed by Kent, Sussex and Dorset in August, and Berkshire in November. Unfortunately, none of the court martial papers for Mansfeldt's, Cecil's or Buckingham's armies have survived to record how effective the *Laws of War* were. However, on 31 August 1626 the Privy Council wrote to the mayor of Southampton ordering the soldiers in the local gaol be tried by common law rather than martial law. Certainly, there are references to soldiers in a county's Quarter Session and Assize records plus individual accounts of people such as Sir Walter Erle of Dorset who reported in Parliament in 1628 that:

> There came 20 [soldiers] in a troop to take sheep. They disturbed markets and fairs, rob men on the highway, ravish women, breaking [into] houses in the night and enforcing men to ransom themselves, killing men that have assisted the constables that have come to keep the peace.[26]

Sometimes the local inhabitants would also be involved. On Boxing Day 1626 11 soldiers and Giles Fisher a tapster, surrounded Mr Brereton's house near Corscombe in Dorset demanding money, one even fired his musket into the house. It was not until Mrs Brereton threw £4 10 shillings from a window that Fisher and the soldiers went away.[27]

According to the *Laws of War* the death sentence was passed, in theory, for those caught robbing, stealing, mutiny, striking an officer, discharging a firearm at night without cause or before an ambush, abandoning the Colours to be captured and spreading false rumours which might discourage the soldiers. A common soldier could expect death by hanging, but an officer might be shot or beheaded. Although there were some soldiers who are known to have met their deaths due to these laws, such as Rowland Hopkins and Thomas Browne were hung for manslaughter at Winchester in September 1627 and Thomas Reeve is recorded as being 'executed', for killing a bailiff at Andover. Most soldiers appear to have been 'reprieved and enlarged' [i.e. set free], such as John Wilson and Michael Holland who were Reeves' accomplices and 10 of the 11 soldiers who had besieged Mr Brereton's house were also freed, with the tapster and the remaining soldier being put to death. According to Captain George Took, the soldiers, 'Cast

---

25 SHC 6729/11/11 Order from the Privy Council to the Deputy Lieutenants of Surrey, 16 September 1626.
26 Gardiner, S R *History of England from the Accession of James I to the Outbreak of the Civil War, 1603–1642*, (London; Longmans & Co., 1883) vol. 6 p.253. Unfortunately, the Quarter Session papers for Surrey have not survived for this date.
27 Goodwin, Tim, *Dorset in the Civil War* (Tiverton, Dorset Books, 1996), p.5.

THE FIRST BRITISH ARMY 1624-1628 (REVISED EDITION)

Print by Jaques Callot in his series the "Miseries of War." showing soldiers being hung. (from Philippson's *Geschichte des Dreissigjahrigen Krieg*)

lots for the gallows; by which punishment [and] with a severe proclamation backing it, there ensued a temper, their fury being reduced into the former channel of obedience'.[28]

This casting of lots could either be in the form of throwing a dice and the one with the lowest score is executed or drawing tickets from a hat and the soldier who drew the piece of paper with a gallows on it was hanged.

However, it must have been extremely difficult to impose discipline on the soldiers when the government interfered with the process. On 15 June 1627, the Privy Council wrote to the Attorney General to pardon five soldiers in Dorchester gaol who had been condemned for burglary and other felonies because they were 'able and experienced in the wars'. Also, Cornelius Dermot, a sergeant in Sir Piers Crosby's Regiment who had distinguished himself at the Île de Rhé and Rochelle and lost a hand in this service, but had stolen goods worth £20 in Ireland, was pardoned on 18 March 1629 since 'he was now repentant'.[29]

Among the accounts for Hampshire in 1626 is one for a 'gibbet at the court of Guard, two shillings 6 pence' and eight shillings 5 pence for 'the setting up of a horse' at Southampton. To 'ride the wooden horse' probably came under heading in the *Laws* 'shall be punished according to the quality of the offence' or 'punished according to discretion', whereby a soldier would sit on a wooden 'horse' with his hands tied behind his back and sometimes muskets were tied to his feet for a certain length of time.[30]

---

28   Took *Cales* p.3.
29   PC2/39, Privy Council, vol. 5, Dec 1628–May 1630 f.689; BL Kings ms 265 f.82; Lyle *APC 1627* p.350; TNA SP 16/13/42 Copy of Martial laws.
30   SP 16/13 ff.77–78, Articles of War, 3 December 1626; Hampshire Record Office 44M69/G5/48/4, Account of acts of Commission of Martial Law at Winchester, 10 January 1627, ff.109–111, 118–119; TNA AO1/300/1140, N Prescods Treasurer of Loan at Southampton; SHC 6729/11/11, Copy of an order from the Privy Council to the Justices of the Peace for Surrey, 16 September 1626.

DISCIPLINE AND PAY

Print by Jaques Callot in his series the "Miseries of War," showing a soldier being punished on the Strappedo. (from Philippson's *Geschichte des Dreissigjahrigen Krieg*)

Another punishment which came under this heading was the strappado. A soldier would have his hands tied behind his back and hoisted up by the hands or thumbs so that he could only stand on tip toes or he might even be raised off the ground and then dropped at least once which would often result in his arms being dislocated. It was believed that the very sight of a strappado, 'will do good in a wicked mind'. In his *Miseries of War* Jacques Callot shows a soldier undergoing this punishment having been raised high above his comrades. The caption reads, 'It is not without cause that great captains have well advisedly invented these punishments for idlers, blasphemers, traitors to duty, quarrelers and liars, whose actions, blinded by vice, make those of others slack and irregular'.[31]

According to James Turner during the 1620s Gustav II Adolph introduced 'running the gauntlet' whereby a soldier would have to walk between two ranks of his own company or regiment while being hit by his comrades. This practice was known during the sixteenth centuries except then the offender was stabbed by his fellow soldiers using their pikes. This punishment would continue in armies into the nineteenth century.[32]

Imprisonment was also used if a soldier blasphemed for the first time but could also be used if a soldier was 'seen without his sword' or if he left his garrison or lodging without a pass, striking a fellow soldier or insolence to an officer, not keeping his arms clean or selling or pawning them. For a second offence 'blasphemers' and 'common swearers' could expect to have their tongue bored with a red-hot iron. How often this was carried out is not known. One unusual punishment was a soldier from the ship *Anthony* who, on 11 November 1625, was 'ducked at the main yard arm of the *Anne Royal* for being mutinous against the seamen'.[33]

---

31 Daniel, Howard, (ed) *Jaques Callot's Etchings*, (Dover Publications, 1974) sketch no. 274.
32 Turner *Pallas Armata* pp.348–349.
33 Dyfnallt Owen *Calendar of the Cecil Papers* entry for 11 November 1625.

However not all soldiers were court martialled, a Robert Sharpe, who is described as 'being a very mutinous fellow, did knock Sergeant Penrose down to the ground with the butt end of his musket'. No reason is given for this attack, but possibly over their arrears of pay because Sharpe, who seems to have been the ringleader of some mutinous soldiers, went on to demanded money from his lieutenant. After several abusive words Sharpe was escorted back to his lodgings by Penrose, who appears to have recovered enough from his assault. However, at his lodging Sharpe once more became violent and so Penrose was forced to run him through with his sword.[34]

Another major problem for the authorities was the soldiers who strayed from their Colours and would roam the countryside whether out of boredom or in search of plunder. These soldiers were not treated as deserters and there are countless orders circulated for them to be apprehended and returned to their regiments.

However, for the civilians, particularly in the southern counties, it would not be until the army was disbanded that they could finally feel safe in their homes.

Print by Jaques Callot in his series the "Miseries of War," showing a soldier being executed by firing squad. (from Philippson's *Geschichte des Dreissigjahrigen Krieg*)

---

34 SP 16/110/75, Certificate of Thomas Piggott and others setting forth the circumstances under which Sergeant Penrose occasioned the death of Robert Sharpe, 24 July 1628.

DISCIPLINE AND PAY

Guard, blow and open your pan

8

Tactics

As stated in Chapter 1 an officer was appointed dependent on his status in society rather than any military experience he might have, and so when it came to tactics, he had little or no prior knowledge to draw on. He may have served in one of the European armies or he could appoint an experienced sergeant, who had served in these wars. Alternatively, he could read one of the military manuals which had been published. These included Jacob de Gheyn's *The Exercise of Arms,* published in 1607 in several languages and reprinted in 1619. In 1616 a writer known only by the initials 'I.T.' who described himself as 'a friend to the friend of a soldier', published his *The A.B.C of Arms,* which was followed in 1622 by Francis Markham's *Five Decades of Epistles.* The following year an anonymous tract, *The Military Discipline,* was published and then in 1624 Ernest von Mansfeldt published *Directions of Warre,* written especially for his army, which had been raised in England.

These titles were followed by those of Gervase Markham, Francis' brother, who was a prolific writer who not only published books on military matters including *The Soldiers' Accidence* in 1625 and *The Soldiers' Grammar* in two parts in 1626 and 1627, but also books on farming, horsemanship and even *The English Housewife,* first published in 1615, which gave the housewife 'the inward and outward virtues which ought to be in a complete woman'. In 1629 James Achesone published *The Military Garden, or Instructions for all Young Soldiers,* which appears to have been the only Scottish drill manual at the time.[1]

However, all of these manuals showed how a pikeman, musketeer or harquebusier should handle their respective weapons, and their drill movements, rather than actual battlefield tactics.

It is believed that during the early part of the seventeenth century there were at least three schools of thought; the traditional Spanish school with its outdated large formations of infantry which were unwieldy in battle; the Dutch school introduced by Maurice of Nassau who formed his infantry into smaller bodies 10 men deep and finally the Swedish school, whose king,

---

1 James Archesone, *The Military Garden or Instructions for all Young Soldiers and such who are disposed to learn and have knowledge of Military Discipline* (Edinburgh: John Wreittoun, 1629).

Gustav II Adolph, introduced a more mobile form of warfare and whose tactics are seen as the birth of modern warfare. However, by studying the drill manuals of the time it is clear that many of his innovations, such as the cavalry charge or the 'Swedish volley' were already standard practice in both the Spanish and Dutch schools. Even Michael Roberts, one of the champions of Gustav II Adolph's reforms had to admit there was no Swedish drill manual.

For those who preferred the Spanish tactics they might read Johan von Wallhausen's *Kriegskunst zu Fuss,* which was published in 1615. During the same year, a French edition under the title of *L'Art Militaire d'Armes,* also appeared. A companion volume was also published for the cavalry. However, for those who followed the Dutch school, such as Mansfeldt, Cecil and Buckingham, they may have obtained a copy of Captain John Bingham's *The Tackicks of Aelian* which was published in 1616. Although Aelian had written his tactics in *c.* 100AD, they were translated into Dutch from the original Greek by Prince Maurice in 1594, who found the phalanx of pikes and formations of slingers and archers could easily be adapted and put to use with his infantry. However, these smaller Dutch formations, of about 500 men, eight or 10 men deep, had only proved successful in a few battles, such as Nieuport in 1600. When the Dutch in Frederick's army at the battle of White Mountain insisted that the Bohemian infantry form up in this fashion they were quickly overwhelmed by the larger Spanish formations, which each comprised 1,000 or more men.

The pikemen formed in the centre of these Spanish tertios or 'battles' and were surrounded by musketeers. Sometimes detachments of musketeers were also attached to the corners of this formation. They were often supported by cavalry and artillery. Despite being considered large and unwieldy, they were popular among the Catholic armies in Europe, apart from the French, who used the Dutch system, and some Protestant Armies, particularly at the beginning of the Thirty Years' War.[2] The Spanish formations would dominate the battlefield until about 1636.

However, this is not to say that the Dutch formations did not adapt after 1600, because in 1614 plans of Dutch orders of battle at Rees, Dornick and Emerick show the infantry divided into three; the vanguard, main battle and rearward, with two formations side by side drawn and up about 100 feet apart. About 300 feet to their rear are two further formations of foot, and then 600 feet at their rear is another body of troops, so forming a diamond-shaped formation. Each formation shows the musketeers forming up behind a body of pikes. Some plans also show these two bodies divided into two, 30 feet apart, (see diagram):

---

2   The word tertio is not to be confused with the *Tertios* which was used in the Spanish Army to describe some of their infantry units.

Deployment of more than one regiment according to Gervase Markham, 1627

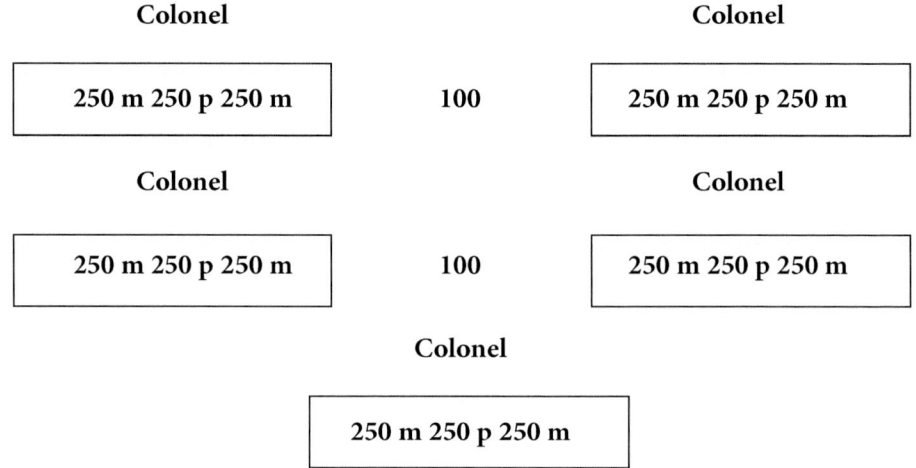

Dutch Formation for Several Regiments, *circa* 1614

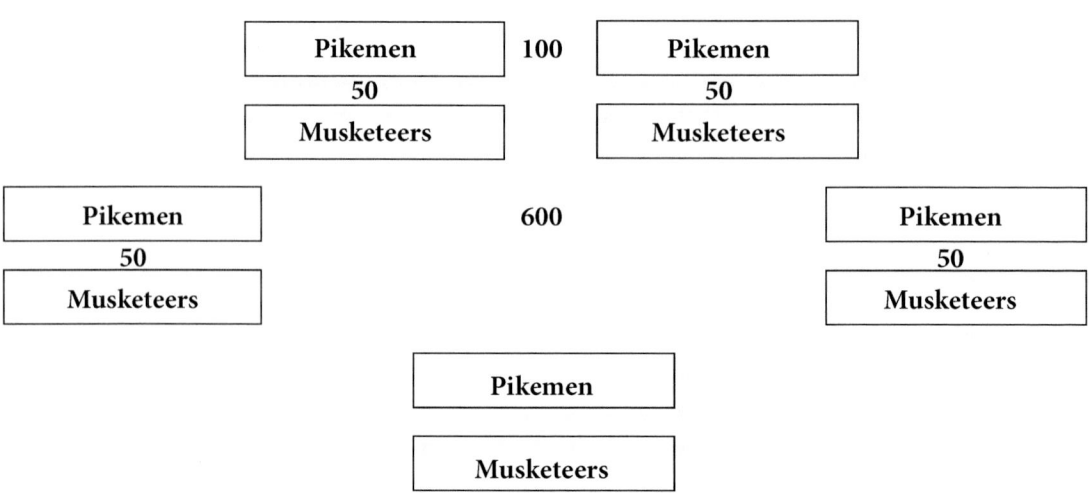

The rear regiment would often be discarded, resulting a triangular-shaped formation. These formation were adopted by the Danes under Christian IV, and Gustav II Adolph appears also to have deployed his brigades in a similar fashion, although using one formation instead of two. However, contemporary plans seem to suggest that this formation was short lived, at least in battle.[3]

In the second part of his *Soldiers' Grammar*, published in 1627, Gervase Markham records a similar formation, but now the musketeers are deployed on either side of the pike block rather than behind.[4] However, he argued that dividing the firepower of the formation weakened the impact of the musketeers' fire, but also their manoeuvring opened gaps in the formation which could easily be exploited by the enemy.

Then there were the various elaborate formations, suggested by writers such as Gervase Markham in his *Soldiers' Grammar*, who recommended drawing up a company in diamond or triangular shaped formations, which would almost certainly have caused confusion if they were adopted on the battlefield. Therefore, most officers probably followed the doctrine used by the Earl of Essex in 1642 when he wrote:

> Be careful in the exercising of your men and bring them to use their arms readily and expertly, and not to busy them in practicing the ceremonious forms of military discipline, only let them be well instructed in the necessary rudiments of war, that they may know to fall on with discretion and retreat with care, how to maintain their order and make good their ground.[5]

When a colonel or captain was appointed his commission usually stated that he had to drill his regiment or company every day.[6] Despite these Instructions there seems to have been a lack of training of the English regiments at this time. Thomas, Lord Cromwell, who was a colonel in Mansfeldt's army, records that his regiment had 'not a sword amongst them… and are obliged to train the men with cudgels'.[7] Unfortunately for Mansfeldt's officers when they did try and train their companies at Dover in 1624 their men beat them with these improvised weapons.

---

3   Markham, *The Soldiers' Grammar,* pp.48–51.
4   Plans in TNA SP 9/202/1/18, Compendium of the Discipline of the Art of War under Sir Horace Vere, 1603–1625; Cruso, John, *Military Instructions for the cavalry* (Cambridge, 1635), Appendixes; Roberts, Keith and Hook, Adam, Pike *and Shot Tactics, 1590–1660 (*Botley: Osprey, 2010) pp.18–19.
5   BL Harl ms 135, Sir John Smythe's answer to Captain Humphrey Barwike's book on military affairs, Barriffe, William *Military Discipline or the Young Artillery man* pp.206–210; Essex's quoted printed in Scott, C L, Turton, A and Von Arni, E G *Edgehill, the battle reinterpreted* (Barnsley: Pen and Sword, 2005) p.201.
6   SP 16/70/136, Commission appointing Stephan Country a captain, 13 August 1625.
7   TNA SP 14/181/23, Lord Cromwell to the Council of War, 25 December 1624.

THE FIRST BRITISH ARMY 1624-1628 (REVISED EDITION)

The New and latter form of imbattling a Regiment According to Gervase Markham

```
   Col's    1st Capt   3rd Capt    Major    4th Capt   5th Capt   2nd Capt   Lt Col
    D    E    D    D    D    D    D    E    D    D    D    D    D    D    D    E    D

sssss ppppppppp sssss sss ppppp ss sss ppppp ss sssss pppppppppp sss sss ppppp ss sss ppppp ss sss ppppp ss sssss ppppppppp sss
sssss ppppppppp sssss sss ppppp ss sss ppppp ss sssss pppppppppp sss sss ppppp ss sss ppppp ss sss ppppp ss sssss ppppppppp sss

           E              E                   E                        E              E

sssss ppppppppp sssss sss ppppp ss sss ppppp ss sssss pppppppppp sss sss ppppp ss sss ppppp ss sss ppppp ss sssss ppppppppp sss
sssss ppppppppp sssss sss ppppp ss sss ppppp ss sssss pppppppppp sss sss ppppp ss sss ppppp ss sss ppppp ss sssss ppppppppp sss
sssss ppppppppp sssss sss ppppp ss sss ppppp ss sssss pppppppppp sss sss ppppp ss sss ppppp ss sss ppppp ss sssss ppppppppp sss
sssss ppppppppp sssss sss ppppp ss sss ppppp ss sssss pppppppppp sss sss ppppp ss sss ppppp ss sss ppppp ss sssss ppppppppp sss
sssss ppppppppp sssss sss ppppp ss sss ppppp ss sssss pppppppppp sss sss ppppp ss sss ppppp ss sss ppppp ss sssss ppppppppp sss
sssss ppppppppp sssss sss ppppp ss sss ppppp ss sssss pppppppppp sss sss ppppp ss sss ppppp ss sss ppppp ss sssss ppppppppp sss
sssss ppppppppp sssss sss ppppp ss sss ppppp ss sssss pppppppppp sss sss ppppp ss sss ppppp ss sss ppppp ss sssss ppppppppp sss
sssss ppppppppp sssss sss ppppp ss sss ppppp ss sssss pppppppppp sss sss ppppp ss sss ppppp ss sss ppppp ss sssss ppppppppp sss

     L       L         L         L        L         L         L        L
```

Key

Col	Colonel
Lt Col	Lieutenant Colonel
Capt	Captain
L	lieutenant
E	Ensign
D	Drummer
s	shot
p	pike

This lack of training is confirmed by Garret Barry, who was serving with the Spanish army, who records that some of the deserters of Mansfeldt's army who fled to the Spaniards claimed, that some, 'filled in all the powder at once into their pieces' [muskets'] mouths even to the top… that there was scarce place to put in the bullet'. He added sarcastically, 'so skillful were they in their pieces'.[8]

Nothing appears to have improved for the Cadiz expedition, because Captain George Took states that during the summer of 1625 the soldiers spent little time drilling, because they did not have enough weapons to practice

---

8   Garret, *The Siege of Breda*, p.99.

with, that 'so little did our musketeers know [how] to fire, [or] advancing, retreating or to skirmish'.

However, lessons had been learnt because when the army returned from Spain, the deputy lieutenants of the counties where the soldiers were quartered were to make sure that the soldiers not only kept guard and placed watches, but also to see the officers drilled their men:

> Both [to] prevent disorders and maintain the accompts of their numbers that none run away and enable the better for the king's service and their own health. You are likewise to have a care that their powder and shot be frugally expended with exercising and to the end the powder spent may be with least waste we advise that they may be trained with false fires and when they shoot at they must be with a full charge and with a bullet at a mark.[9]

On 22 June 1627 Sir William Haydon, the Master of the Ordnance, was ordered to deliver to each regiment bound for the forthcoming expedition one barrel of powder for every company. This equated to two pounds for each man to fill their bandoliers twice, with some spare, and one barrel of 100lb weight of musket shot and 200lb weight of English match, making three lengths for every man cut into fathoms and 300 pounds of Dutch match (to be equally divided, and likewise one powder bag for every company).[10]

Whether this was because of the feared invasion or because Charles was due to visit the soldiers at Portsmouth is not known, but the Venetian Ambassador, Alessandro Antelminalli's letter dated 2 July 1627, records, 'the king remains at Portsmouth, or in its neighbourhood, profoundly interested in drilling the troops which he does almost daily'. On 9 May 1628, a further 20 barrels of powder were ordered to be sent to Earl of Morton's Regiment on the Isle of Wight, for 'training, exercising and enabling his soldiers for his Majesty's service'.[11]

For large regiments they would be divided into two divisions or bodies. The first division was drawn up with the colonel's company on the right and the major's on the left, with the second, sixth and fourth captains companies in the centre. The second division would have the lieutenant-colonel's company on the right and the first captain's company on the left with the third, seventh and fifth captains' companies in the centre. However, some drill books differed in the order the captains' companies were to be drawn up in the centre of each division, and none state the order the companies were to be drawn up if the regiment was only strong enough to form one division.[12]

---

9   TNA SP 16/34/ Council to the Justices of Kent, Sussex, Dorset and Hampshire, August 1626.
10 Add ms 37,817 Nicholas papers vol. iv, Correspondence of the Duke of Buckingham as Lord Warden of the Cinque Ports, 1624–1627, f.136.
11 HMC 11 Report Appendix 1 p.120.
12 Turner *Pallas, Armata*, pp.227–228; BL Harl 5,109 A volume of miscellaneous military matters f.112–113. This manuscript draws up the companies slightly different. The first body Colonels' company, 1st captain, 5th captain, 3rd captain, major. The second division Lt Colonel's company, 4th captain, 7th captain, 6th captain, 2nd captain.

THE FIRST BRITISH ARMY 1624-1628 (REVISED EDITION)

The old ancient form of imbattling a Regiment According to Gervase Markham

```
                Lt Col                              Col                        Major
   L      L       L      3 Capt   1 Capt   E    E    E    2 Capt  3 Capt  5 Capt   L       L
ssssssssssssssssssssssss ppppppppppppppppppppppppppppppppppppppppppppppp sssssssssssssssssssssss
ssssssssssssssssssssssss ppppppppppppppppppppppppppppppppppppppppppppppp sssssssssssssssssssssss
                   E                     E                   E                      E
ssssssssssssssssssssssss ppppppppppppppppppppppppppppppppppppppppppppppp sssssssssssssssssssssss
ssssssssssssssssssssssss ppppppppppppppppppppppppppppppppppppppppppppppp sssssssssssssssssssssss
ssssssssssssssssssssssss ppppppppppppppppppppppppppppppppppppppppppppppp sssssssssssssssssssssss
ssssssssssssssssssssssss ppppppppppppppppppppppppppppppppppppppppppppppp sssssssssssssssssssssss
ssssssssssssssssssssssss ppppppppppppppppppppppppppppppppppppppppppppppp sssssssssssssssssssssss
ssssssssssssssssssssssss ppppppppppppppppppppppppppppppppppppppppppppppp sssssssssssssssssssssss
ssssssssssssssssssssssss ppppppppppppppppppppppppppppppppppppppppppppppp sssssssssssssssssssssss
ssssssssssssssssssssssss ppppppppppppppppppppppppppppppppppppppppppppppp sssssssssssssssssssssss
   1 S    3S      5S      D        L      L    L    6S     D       4S      2S      D
```

Key

Col	colonel
Lt Col	lieutenant colonel
Capt	captain
L	lieutenant
E	ensign
S	sergeant
D	drummer
s	shot
p	pike

An infantry regiment was to be drawn up in 10 (later eight) ranks deep with each soldier having his own place of importance in the following order:

File	Order of Importance
1	1 (leader)
2	5
3	9
4	8
5	4 (middle man)
6	3 (middle man or half-file leader)
7	7
8	10
9	6
10	2 (Bringer up)

Interestingly, James Achesone's *The Military Garden* refers to these as strings and string leaders rather than file leaders. When the order was given to, 'double your files', the 10 ranks would become five.

The seniority of the rank was as follows:

| Ranks | 10, 9, 8, 7, 6, 5, 4, 3, 2, 1 |
| Order of Importance | 2, 6, 10, 7, 3, 4, 8, 9, 5, 1 |

Therefore, a body of either musketeers or pikemen would consist of 100 men.

There were three distances between the ranks and files; 'open order', where each soldier stood six feet apart; 'order', when each soldier was three feet apart, which was used when manoeuvring the troops, and finally 'close order', which was 'when the files join shoulder to shoulder; and the ranks come up to the swords point'. This last distance was used when 'you are forced to break through an enemy, or when you receive a charge of horse, or of foot, that purposeth to break through you'.[13]

There was no reference to marching in step, but in 1614 Sir Horace Vere stated that the pikemen should advance at a slow or 'soft pace for not breaking of the ranks', while Buckingham instructed in 1627 that he, 'will not have that any pace or paces should be taught to the musketeers but their own paces which God hath given them'. Presumably, this also applied to the pikemen.[14]

To make fighting easier one English volunteer in the king of Bohemia's army recalls, 'our general gave command to draw out part of every company to be of the forlorn hope to give onset… [we] threw away our knapsacks, with all our provisions, both linen and woollen because it would be troublesome to us'.[15]

Although contemporary manuals refer to volleys as just one rank firing at a time, it is generally believed that Gustav II Adolph, introduced the practice of more than one rank firing a volley. However, Gervase Markham in his 1625 drill book also mentions this type of firing, describing it as an 'ancient and vulgar manner of discipline', because:

> In teaching to give volleys… (which is that the whole volley shall be given of all the shot in one instant as well of them behind as before) [in front] is utterly to be condemned for either the hindmost must venture to shoot their fellows before through the heads or else will overshoot and so spend their shot improfitably. Besides the volley being once given, the enemy comes on without impeachment or annoyance.[16]

He does not mention this type of volley in the first part of his *Soldiers' Grammar*, but in the second part refers to it when repelling enemy cavalry, insomuch that it is better to fire single volleys when surrounded by the pikes:

---

13 HRO 44M69/G5/48/1, Instructions for exercising foot companies, 1627; TNA SP 16/13, *Directions for Musters* (1638); Bingham, John *The Art of Embattaling Armies* p.16.
14 TNA SP9/202/1 f.188; HRO 44M69/G5/48/1, Instructions for exercising foot companies, 1627.
15 *A Relation of the passage of our English Companies from time to time* EEBO.
16 Markham *The Soldiers' Exercise* p.9.

so that the volley is given entirely and without impeachment or trouble to one… another. Whereas to shoot over one anothers shoulder or by making the first man kneel, the second stoop, the third bend his body, the fourth lean forward and the fifth to stand upright and so deliver their volley both rude and disorderly, bringing great danger to the soldier and placing them in such a lame and uncomely posture.[17]

In his *Directions of Warre* Count Mansfeldt does not mention this type of volley, nor was it mentioned in the *Instructions for Musters and Armes*, which was first published by the Privy Council in 1623, and again in 1626, which records that the musketeers would form up 10 ranks deep, and:

In advancing towards an enemy when they do not skirmish loose and disbanded, they must give fire by ranks after this manner. Two ranks must always make ready [load] together and advance 10 paces forwards before the body. At which distance a sergeant (or when the body is great) some other officer must stand, to whom the musketeers are to come up before they present and give fire. First the first rank and whilst the first rank gives fire the second rank keeps their muskets close to their belts and their pans guarded and as soon as the first are fallen away the second present and give fire and fall after them. Now as soon as the two first ranks do move to their places in the front, the two ranks next it must unshoulder their muskets and make ready: so as they may advance forwards 10 paces (as before) as soon as even the first ranks are fallen away and are to do in all points as the former. So, all the other ranks through the whole division must do the same by twos on after the other.[18]

Monro, whose regiment served with the Danish and then the Swedish armies, advises that musketeers should be taught to fire in single, double and three ranks, although he says that firing by three ranks is, 'ordinary in battle before the enemy join or against horsemen'. Therefore, Gustav II Adolph might not have invented the 'Swedish volley', but he may have reintroduced it, if it ever went out of fashion.[19]

Whether a volley was delivered by a single rank or more, Sir John Smythe records that, 'a few volleys well given with pieces leisurely and therefore well charged shot, within 20, 30 or 40 paces do work more effective than a great number of volleys out of pieces charged in haste or a great distance'.[20]

When it came to hand-to-hand fighting Johan Jacob von Wallhausen suggested that the musketeers not only use their muskets but also their bandoliers, musket rests, helmets and fists and there are many references during the English Civil War of musketeers 'falling on' with 'clubbed muskets'.[21]

---

17 Gervase Markham, *The Second Part of the Soldiers Grammar* (London: A Matthews, 1927) p.23.
18 *Instructions for Musters and Armes, 1623* (EEBO).
19 Monro, Monro *His expedition,* p.190.
20 BL Harl ms 135, Sir John Smythe's answer to Captain Humphrey Barwike's book on military affairs.
21 Bingham, *Tacticks of Aelian*, p.84, Johann Wallhausen, *Kriegkunst zu Fuss*.

TACTICS

Illustration from Johann von Wallhausen showing the ways musketeers could fight. (from Philippson's *Geschichte des Dreissigjahrigen Krieg*)

Unfortunately, when it comes to the pike in battle the drill books are unusually silent. Sir Thomas Kellie records that they stand, 'as an idle spectator, serving as a prop to the enemy's shot and never being able to offend them which is a pitiful condition'. Robert Munro also records that at Oldenburg in 1626 the pikemen of his regiment stood for two hours under constant enemy fire suffering more casualties during this time than the musketeers who were engaged in the fighting. In 1642 Donald Lipton went so far as to produce a tract, *Warlike Treaties of the Pike*, in which he gave examples of how useless the weapon was in battle.[22]

However, James Turner, writing between 1670 and 1671 strongly disagreed with the pike being useless, and he used a whole chapter to discredit Lupton's treaties. The Imperialist general, Raimondo Montecuccoli, agreed with Turner arguing that, 'The pike is the only weapon, it has no equals, for resisting cavalry attacks. Without it a force is completely disjointed, like arms and legs lacking a torso. It is a wall the stability of which permits an army to stands its ground it holds the foe at bay'.[23]

Unfortunately, apart from Gervase Markham none of the English drill manuals records how the infantry were to receive cavalry, although he does record in the second part of his *Souldiers' Grammar* an elaborate way of receiving enemy cavalry in a wedge shaped formation, whereby one

---

22 Kellie, Sir Thomas Pallas, *Armata or the Art of Instructions for the Learned: and all generous spirits who effect the Profession of Arms* (Edinburgh, 1627) p.107, Lipton's tract is printed in Firth *Cromwell's Army* pp.387–390; Monro, *Monro His Expedition* pt I p.18.
23 Turner, *Pallas Armata,* pp.178–186; Thomas Barker, *Military Intellectual in Battle* (University of New York Press, 1975) p.88.

of the sides of the square is open thus letting the cavalry obligingly enter the formation to be attacked on three sides. A Dutch manual written by Simon Stevin, the quartermaster general of the Dutch army at the turn of the seventeenth century, refers to a 'Pyckschasen' or 'Pike Redoubt' which he defines as, 'An array of pikemen, so placed that, while the points of the pikes are extending everywhere towards the enemy, [the] musketeers are protected, can constantly fire at the enemy and also leave and enter the pike redoubt without bringing it in [to] disorder'. Stevin illustrated several versions of the redoubt, but all are shown as a hollow square of alternate blocks of musketeers and pikemen and was to be trained to manoeuvre if necessary.[24]

One of Jacob de Gheyn's illustrations, which was published in 1607, shows a pikeman in a position 'charge for horse and draw your swords', which shows a pikeman using his foot to steady the end of the pike while he holds the point at the height of a horse's chest. This position is repeated in later military manuals, which also suggest that the ends of the pike be reinforced with metal langets at least three feet long to prevent a cavalrymen cutting off the pike's head.

Despite suggestions that the pikemen stood idly by in battle there are many accounts of pike bodies coming to 'push of pike' to show the fierceness of a battle. One English participant at the siege of Bergen op Zoom in July 1622 boasted that, 'we were one-quarter of an hour at push of pike'.[25] Pikes were between 16 and 18 feet long, which were cumbersome to march with and there are various secondary accounts of soldiers cutting them down to make them more manageable, but when it came to the push of pike this would mean there would be weaknesses in the pike block, because according to John Bingham's *Tacticks of Aelian*, 'Short pikes against long have a great disadvantage. With the long pike a man is able to strike and kill his enemy, before himself can be touched, or come into danger of a shorter [one], the pike keeping the enemy out so far as the length [of] it'.[26]

Although the drill manuals mention that pikes should be at their 'close order' with at least the first four ranks having their pikes at the 'charge', i.e. horizontal at chest height and the remaining ranks at the 'port' or at a 45 degree angle few record the mechanics of the push of pike, but there appears to be several versions. The *Instructions for the management of the pike and musket* states that the pikemen, 'Without shaking or clattering their pikes. And that they push not till they come upon push of pike with the enemy. And then that they push together, both forwarded and retreating; and upon every push that they make a little stand to recover their strength again'.[27]

Whereas Sir John Smythe, writing in 1595, suggests that rather than standing still the pikemen make a 'pushing and foining [i.e. thrusting]' as they press forward without standing and 'give a puissant thrust' into the faces

---

24 W. H. Schukking, *The Principal Works of Simon Stevin*, pp.527, 493.
25 TNA SP 101/45 f.184.
26 Bingham, *Tacticks of Aelian*, p.83.
27 BL Kings mss 265 f.12.

of the enemy and so force the front ranks onto the rear ones.[28] Sir Thomas Kellie, writing in 1627, also states that, 'when battles [i.e. formations] come to push of pike, good commanders sayeth that your pikemen must not push by advancing and retiring their arm as commonly is done, but only go jointly on together without moving their arms'.[29]

Which one of these push of pikes was best is not known, but it would take a brave man to withstand the sight of hundreds of pikes coming towards him, which is why the file leader was considered the most important soldier in a file, while the 'bringer up' was the second in that he would try to prevent the rest of his file from breaking.

While at 'push of pike' the musketeers would also fire into the ranks of the pikemen which further helped to disorder the block. Once the pikemen could no longer use their pikes because the enemy had come within the length of them, then according to Smythe they often resorted to their swords, although at the landing of Buckingham's Army at the Île de Rhé one account records, they, 'fell to it with swords and push of pike, until they were breathless on both sides. The French finding our pikes to be longer than theirs, threw away their pikes and went to it with stones and so did our men, but ours beat them out and made them fly'.[30]

When it came to the cavalry little had changed in the way of tactics. In 1616 John Bingham published his *Tacticks of Aelian*, but this was based on mounted warfare of the second century, although he did include a chapter on modern warfare. Gervase Markham's *The Soldiers Grammar*, published in 1626 also had a section on cavalry, but this extended to just five pages, plus various illustrations of formations, which like the infantry tactics included some unusual formations such as the rhombus and the wedge, which also probably never materialised on the battlefield, although they were repeated in later works on tactics, such as Robert Ward's in his *Animadversions of Warre* which was published in 1639.[31]

It would not be until 1632 that John Cruso publish his *Military Instructions for the Cavalry*, which was the first modern book on cavalry tactics written in English. Cruso was not a soldier, and it is likely that he wrote his *Military Instructions* while he was still a student at Cambridge. Nevertheless, it is clear that he had read works by Jacob Wallhausen, Giorgio Basta and Ludovico Melzo who were some of the leading tacticians of the day. The latter first published his work on cavalry in 1611 and by 1626 it had been republished 16 times and translated into German and French in addition to the original Italian.[32] Therefore these works must have been imported into England and may also have been used by Cunningham and Montjoy to train their cavalry.

---

28 Smythe *Observations* pp.23–27.
29 Kellie, *Pallas Armata* p.25.
30 Smythe *Observations* pp.23–27; Gareston, Mr *A Continuing journal of all the Proceedings of the Duke of Buckingham on the Isle of Ree* 1627 p.7.
31 Ward *Animadversions of Warre* pp.309–313.
32 Wallhausen, Johann *Art Militare a Cheval* which had been translated into French from its original German in 1616, Giorgio Basta *Il Governo della cavalleria Leggera* (1612) and Ludovico Melzo *Regole Militari del Cavalier Melzo sopra il governon e servitio della cavalleria* (1611).

According to the *Instructions for Musters and Arms,* and Gervase Markham the cavalry should be drawn up in six ranks, 'and to be doubled to three deep on occasion', which also pre-dates Gustav II Adolph's reforms. However, John Bingham suggests a five rank formation, although the majority of European sources stated that cuirassiers should be drawn up in six ranks and harquebusiers in eight. These ranks had a similar superiority to the infantry, with the file leader being the most important and the bringer up the second.

Formations for a troop of Horse

```
            Capt                              Capt
             T
                                       T               T
          H H H H
          H H H H                H H H H H H H H H H H H H H H H
          H H H H                H H H H H H H H H H H H H H H C
          H H H H                C H H H H H H H H H H H H H H H
          H H H H                H H H H H H H H H H H H H H H H
          H H H H                H H H H H H H H H H H H H H H C
                                 H H H H H H H H H H H H H H H H
            Ct                   C H H H H H H H H H H H H H H H

          H H H H
          H H H H                              Lt
          H H H H
          H H H H
          H H H H

            C

          H H H H          **Troop in column of march    Troop drawn
          H H H H          up for service or exercise**
          H H H H
          H H H H
          H H H H
          H H H H

            C

          H H H H
          H H H H          Key
          H H H H
          H H H H
          H H H H          Capt      Captain
                           Lt        Lieutenant
            T              Ct        Cornet
          H H H H          C         Corporal
          H H H H          T         Trumpeter
                           H         Horse
            Lt
```

There was also disagreement on the distances between each trooper, which according to John Cruso, 'Some make close order to be two paces, open order four paces, and so for double, triple, and quadruple distances proportionable. Others make by two kinds of distances: close order, which is three foot, and open order which is six foot'.[33]

Since the troopers were often supplied by the nobility, they may not have been familiar with riding, therefore Johan von Wallhausen in his *Art Militaire a Cheval* suggests they be taught how to mount a dummy horse, at first without arms and then with. Once they had mastered this, they would then progress to the horse itself. After this part of the training was complete, they would then learn how to ride and manoeuvre in formation."

In June 1627 Cunningham was paid £200 to train his troop while it was quartered at Southampton, which presumably was mainly for their gunpowder weaponry so that they could become proficient in firing their carbines and pistols. Like the musket, the carbine or pistol was laborious to load, consisting of 24 moves to load and fire the piece. However, in battle this was reduced to just three movements, 'make ready', 'present' and 'fire', which according to Gervase Markham, 'the first is done standing or marching, the second in the charge and the last in the face of the enemy'.[34]

The cavalry of the Herefordshire Trained Bands, which were formed of lancers and harquebusiers, appears to have also relied on a letter from Barnabas Scudamore giving his brother John advice on training cavalry, by suggesting:

> Divide your troops into two divisions and skirmish one against the other with false powder or fire. Everyman to discharge his pistols to the air in the meeting when you have done, order the troop and after, draw them out by files to give fire against a mark whereby to make the men and horses ready.[35]

This is similar to that suggested by John Bingham and John Cruso for harquebusiers, which:

> Must be exercised to give fire by ranks. The first rank, having given fire, is to wheel off to the left (unless the ground will not permit it, but that it must be to the right) making ready and falling into the rear, the second rank immediately gives fire upon the wheeling away of the first, and so the rest successively… But others do utterly reject it, as too much exposed to danger. In their firing by ranks, the first rank advanceth some 30 paces before the body, first on the gallop, then in career (as some direct) and so to give fire; the second doth the same and so the rest.[36]

This manoeuvre is usually referred to as the caracole, although neither Bastra and Wallhausen do not refer to it by this name. Instead Wallhausen refers to the caracole manoeuvre as when a body of cavalry splits in two allowing the

---

33   Cruso, John *Military instructions for the cavalry* (Cambridge, 1635) p.45.
34   Markham *Soldiers Accidence* p.54.
35   Add ms 11,050, letter from Barnabas Scudamore to John Scudamore, nd c.1626, f.176.
36   Cruso, *Military instructions for the cavalry,* pp.97–98.

enemy cavalry to charge through the newly made gap in the line, while being raked by pistol shot. However, it was a dangerous manoeuvre to perform because the cavalry unit might be caught while it was deploying and put to flight. Contemporary accounts do refer to the caracole when firearms are used by cavalry.

Certainly, it would be useless for a cavalryman to fire a pistol at a heavily armed cuirassiers at anything less than point-blank range since their armour was, in theory, pistol proof, although according to the Imperial general Raimondo Montecuccoli, at the battle of Nordlingen in 1634 the Imperialist cavalry did little harm against the enemy using this method. At greater distances, the harquebusiers could use their carbines to soften up an enemy formation, which would allow the cuirassiers to force their way into the enemy's line.[37]

During the sixteenth century the lancers would have led the way and forced open any gaps which had appeared, but by the beginning of the seventeenth century lancers were in decline and only appeared on the battlefield as a general's lifeguard or in a county's trained bands.

None of the manuals mentions the use of swords, but Lord Herbert of Cherbery records that when he was being taught how to fight on horseback he was given 'a reasonable stiff riding rod… about the length of a sword and so rid one against the other' fighting and manoeuvring. Certainly, this is how Johan Jacob von Walhausen suggested that they should be taught, and shows cuirassiers fighting with swords and pistols, both firing and using them as clubs, and even wrestling on horseback and once dismounted.[38]

It has been said that Gustav II Adolph introduced the cavalry charge into Western Europe, by ordering his cavalry to charge and fire their pistols only after the melee had begun. However, Robert Munro records that at the battle of Breitenfeld in 1631 the Swedish cavalry waited for the Imperialists to fire first, then after a volley of musketry, 'our horsemen discharged their pistols and then charged through them with swords'.[39]

Moreover, Munro goes on to say that the Imperialist cavalry during the battle were, 'fierce and furious; while ours were stout and slow', which suggests that the Imperialist cavalry moved faster during the battle than their Swedish counterparts.[40]

Gervase Markham records in 1625 that the cavalry, on the word *Passe a Cariere*, the trooper should thrust his:

> Horse violently forward both with his legs and body and giving liberty to his bridle. As soon as the horse is started into his gallop he shall give him the even stroke of his spurs, one or twice together and make the horse run to the height of his full speed… which will not be above six score or eight score yards.[41]

---

37 Barker, *Military Intellectual in Battle*, p.110.
38 Cherbury, *The Expedition to the Isle of Rhè* pp.84–85; J Wallhausen Art Militaire a Cheval.
39 Monro, *Monro His Expedition* part II p.65.
40 Monro, *Monro His Expedition* part II p.69.
41 Markham, *Soldiers Accidence*, p.52.

TACTICS

Sir John Smythe suggested that a squadron:

> Should charge but 30 paces, that is 20 galloping… and 10 for their full career to give them the greatest blow and shock… [any further] they shall find themselves in so great a distance greatly disordered and confounded in their ranks and their horses out of breath and thereby the force of their blow and shock greatly weakened when the come to encounter with the squadron of the enemy.

He also suggests that about 10 paces from the enemy the cavalrymen should give a 'terrible shout' and that if they themselves are charged, then to advance to meet the enemy, which would result in a general melee.[42] This passage is given as an example for lancers, but Smythe makes it clear that it should be adopted by all types of cavalry. By charging at the last moment, it would also stop any cavalrymen hanging back in the charge.

Melzo, Bastra and Walhausen, suggest that the cavalry should charge at the trot rather than the gallop in order to keep the formation tight, although even this had its problems, because if a horse was killed or stumbled then it would bring down the other horses behind it, killing or wounding their riders.

However, in the early seventeenth century soldiers were more likely to take part in skirmishes, or form part of a garrison or besiege a town than take part in a large battle. Newssheets were often printed showing how many towns a general had captured, therefore for a general, knowing how to defend or attack a town was just as important as knowing the latest formations in the field.

The engineers of the day had long since developed the art of fortifications with earthworks protecting a town's stone walls, which were no match for the artillery of the period. These fortifications also made maximum use of firepower, where an attacking force could be caught in the garrison's crossfire. The drawback of these fortifications was that they were expensive and needed not only an experienced engineer but manpower to construct them, although this usually fell upon the inhabitants of the town and its surrounding area. They also took a long time to complete, especially if they completely encircled a large town or city and were lined with bricks.

On the other hand, smaller towns like St Martins, might just have a star fort which was positioned in a strategic position outside the town. This fort was described as:

> A perfect quadrangular form, every point flanking one another, according to the precise rules of fortifications. The cavalier or keep itself is made of free stone and hath without a counterscarfe of earth, the rampiers whereof are about 18 foot high. Under this there lies three half moons, the points whereof are directed towards another counterscarfe, which being the uttermost of all, is also made of stone and serves for a circumuallation unto the rest of the works, being about an English mile in compass. The ground upon which it is seated is sandy, only to the

---

42   Smythe, *Observations*.

north side it stands on a hard rock, which being environed with the sea, make the fort unapproachable of that side.[43]

The garrison was estimated to be about 1,500, commanded by General Jean Caylar d'Anduze de Saint-Bonnet, Marquis de Toiras, who in 1630 would become a Marshal of France.

The commander of the besieging force had two opinions. He could choose to carry the place by storm, which could be very costly to the besiegers, but once the storming party had entered the town then the garrison and many of the inhabitants would be put to the sword, particularly if it was believed that they had held out longer than was necessary. On the other hand, the besiegers were usually rewarded by being allowed to plunder the town of its riches.

Alternatively, the commander could choose to starve the garrison into submission, in which case it would depend on whether the garrison's, or the besieger's, food supply lasted the longest. The besiegers would quickly consume the supplies of the surrounding countryside if the garrison had not already collected them beforehand. A relieving force might also be able to break through the besieger's lines and re-supply the garrison, but it would be impossible to re-supply a large town or city in one or two nights.

Among the leading engineers of the day was Samuel Marolois who recommended that the place to be besieged was to be reconnoitred for any weaknesses before the besiegers began to set up their batteries. A map of the siege of St Martins Fort shows that Buckingham's army set up five batteries, which according to Marolois were to be:

> In such a manner that they may beat upon the parapet of the ramparts and bulwarks of the town, to dismount the enemy's ordnance, and for this reason you must raise your batteries high according to the height of the ramparts, so that your cannon may play freely about two foot lower then the top of the parapet, according to which and in consideration of the distance, you must raise your said batteries, taking heed that your cannon be planted upon a plain superficies and elevated some 13 degrees, when the distance is far off. You must not raise them so as when the batteries are near unto the place.

Marolois continues:

> You [must] make your batteries and platforms according to the greatness and number of your pieces. For a demi-cannon being shorter then a whole [cannon] of necessity the platform of the one must be longer and deeper then the other, and seeing a cannon being mounted upon its carriage is some 16 or 18 foot long it is evident that the batteries ought to made for recoiling at the least 10 or 12 feet longer making 28 or 30 foot. Twelve or 15 of the first foot towards the parapet must be under laid with thick and strong oaken planks and the other with hurdles,

---

43   TNA SP 16/288, Buckingham to Lord Conway, 28 July 1627.

when you have not planks enough. Upon the said batteries you [must] make a parapet 12, 16 or 20 foot thick or thereabouts with portholes for your cannon…

Sometimes you set up gabions six foot high and three foot broad filled with earth, for your ordnance to play out of leaving a little space between them to put out the mouth of your cannon, which space so soon as the cannon is discharged is presently stopped and blinded with a bundle of brush full of leaves that the enemy may not discover the portholes. But when you make your batteries upon the counterscharfe or upon the brink of a moat then the portholes as soon as your cannon is shot off, are shut with doors of thick oaken planks musket proof, that the said portholes may not be seen.[44]

A battery was usually formed by four cannons, at least 12 feet apart so that the entire battery covered a frontage of about 58 feet and was under the command of a battery or quarter gunner. A short distance away from the battery, a cellar was to be dug where the gunners could store their powder, match and shot until they were required. The powder was to be covered with animal skin to protect it from any loose sparks which might ignite it. These guns would have to not only break down the wall where the besiegers were to storm, but also silence any enemy guns which overlooked the position so that the storming party would not be caught in a crossfire.

Once the batteries had been established then trenches were dug. These were to be out of the range of musket fire and could either be placed in strategic places around the town or encompass it altogether. Buckingham was lucky in that he only needed to encompass three sides of St Martin's Fort, the fourth side being open to the sea was blockaded by the navy. On the other hand, an official newssheet tried to excuse the slowness of Buckingham's fortifications being dug by the ground being too hard, although this problem does not seem to have hindered the French garrison.

When Louis XIII's army besieged La Rochelle, it was on a much grander scale. Nine star forts were erected and connected by ramparts with smaller forts inbetween which encompassed the whole town. A large fort known as Fort Louis overlooked the harbour, which was also blocked by a bridge of boats to prevent any ships sailing into the bay and relieving La Rochelle.

Once the trenches, batteries and other works had been completed the besieging force would then start their 'approaches', which ran towards the weakest place or places of the fortification or town. The aim was to make the trenches and approaches at least 6½ feet wide and 6½ feet deep, although they could be up to 32 feet wide so that 'a wagon loaded with fagots brush or gabions may go in them'. Fire steps were also cut for the musketeers to fire at the garrison or for the engineers to see that they were still digging in the right direction, since the banks of the approaches were higher than a person's height. Although pioneers were known during the early part of the seventeenth century the trenches were usually dug by the infantry, who had to work quickly because it would not be until they had dug down at least three feet and piled the earth on the sides that they were shielded from enemy

---

44 Marolois, Samuel *The Art of Fortification translated by Henry Hexham*, (1638) p.37.

musketry. For their part, the garrison would do all they could to disrupt this work by sallying forth from the town or place. Since digging these trenches was considered to be extremely dangerous it was often suggested that these men should be paid extra. However, there is no evidence to suggest that Buckingham's men were rewarded for their efforts on the Île de Rhé.

Once the approaches were near enough to the enemy's fortification, a sap would be dug three feet wide and three to six feet deep so that an attack could be launched with scaling ladders. Needless to say, the enemy would pour all the shot they could at this point to kill as many attackers as possible. It would not be until the end of the seventeenth century that Marshal Vauban would recommend that a trench should be dug parallel with the fortification so that the attacking force could go 'over the top' in First World War style.

Unfortunately for Buckingham's soldiers when they tried to storm St Martin's Fort, they found that their ladders were too short because they had been made in England without even knowing the dimensions of the walls they were to face.

If the ground was right, then the besiegers could dig a mine towards the place being besieged. Once they reached the walls, barrels of gunpowder would be placed at the end of the gallery and the entrance was blocked up to make the explosion more powerful. However, if those in the town or fortification suspected that the besiegers were digging a mine then they would commence their own countermine and try and blow up the besiegers in their mine.

If a town was forced to surrender then the garrison might be able to march out with the 'full honours of war', that is, with colours flying, drums beating, the musketeers with lighted match and the promise they would not be attacked for a certain amount of time. However, when a Major Wilson surrendered Loven Castle to Tilly's forces he forgot to include the colours of the two companies which formed its garrison in the terms of surrender so that he was made to give up the colours, which resulted in him being forced to leave the army in disgrace.[45]

On the other hand, if it was considered that the garrison that had held out when there was no hope of relief and had refused to surrender could expect no terms but to march out of the town holding a white stick, the traditional sign for a safe passage and had to swear not to fight again for a certain period of time.

---

45 Monro, *Monro His Expedition* part I, p.12.

TACTICS

Present

9

Life and Death in the Army

According to Monro, in seven years he travelled 3,967 miles (6,385km) or about 744 miles (1,198km) per year, either on foot or on board ship. Likewise, it has been estimated that the German soldier, Peter Hagendorf, marched 13,980 miles (22,500km) all over Europe or on average 559 miles (900km) per year during his 25 years in the army. Usually, an average day's march was between nine and 12 miles (15 and 20km). It was believed that the Scots and Irish soldiers were seen as being more resilient to campaign life than their English counterparts.[1]

Sometimes life in the army could be pleasant, according to Robert Monro:

> We see that soldiers have not always so hard a life as the common opinion is, for sometimes as they have abundance so they have [a] variety of pleasures in marching softly without fear or danger, through fertile soils [i.e. land] and pleasant countries, their marches being more like a kingly progress than to wars.[2]

On the other hand, in another passage, Monro recalls 'no part of our life is exempted or freed from grief or sorrow; but on the contrary we are exposed to all kind of miseries and troubles'. Thomas Raymond confirmed this view when he recalls that 'the life of a private or common soldier is the most miserable in the world'. Monro continues, 'men of our profession ought ever to be well prepared, having death ever before their eyes, they ought to be the more familiar with God… not knowing how, when or where'. If soldiers travelled in small groups, then they were likely to be ambushed either by enemy soldiers or peasants seeking revenge on any passing soldier for having previously been plundered of their belongings.[3]

Alternatively, they may be called upon to fight in a skirmish or a battle. As in previous centuries it was believed that soldiers would fight better if they knew that they would be well treated if they were wounded. However,

---

1 *Monro His Expedition* part II p.7, unnumbered pages. When Monro records that he travelled 898 Dutch miles in three years under the King of Denmark and 779 Dutch miles in four years under the king of Sweden, one Dutch mile = 3 English miles or about 5km.
2 *Ibid.*, Part II p.89.
3 Raymond *Autobiography* p.43; *Monro His Expedition* part II p.7.

despite this, it was not until later in the century that the position of surgeon and the surgeon's mate are mentioned in any military tracts. In 1659 Richard Elton published his *The Compleat Body of the Art Military* which stated that in each company there, 'ought to be a barber surgeon, for the trimming of the soldiers' [hair] who ought likewise to have some skill in surgery'. Whereas according to Sir James Turner he should 'be skilful in curing all manner of wounds (so they be not mortal) for many brave gentlemen get their bones broken with bullets, which would not so frequently prove deadly to the patients if they were attended on by good, experienced artists'.

There were several types of medical practitioners during the early seventeenth century. The physician, who was usually well educated and highly qualified by the standards of the day, a barber surgeon (or surgeon) who had completed an apprenticeship of six years, which included three years as a surgeon's mate and passed an exam in medicine. Finally, there were apothecaries who also had completed an apprenticeship in making and dispensing medicine. Only the wealthy could afford the services of a physician or an apothecary. Officers might take their own surgeons on campaign. Walter Priest was the Duke of Buckingham's personal surgeon, although he died shortly after his return from the Île de Rhé, and despite his position, left his wife and two small children with 'small means' to live upon.[4]

Despite the Privy Council asking the Guild of Surgeons to supply surgeons for the regiments, since they were paid just two shillings and six pence a day, they attracted the worst in their profession. On 25 December 1624 Thomas, Lord Cromwell, who was a colonel in Mansfeldt's Army, complained that 'there is [a] great deficiency of surgeons and drummers. The surgeons pressed were scarcely good barbers and better should be sent speedily'.[5]

Fortunately for those less qualified regimental surgeons, if they could read, then they could refer to the various published medical books, like Christopher Wirsung's *The General Practice of Physicke* published in 1605 or John Woodall's *The Surgeon's Mate* which was published in 1617 and reprinted several times over the following years. Woodall recommended that a surgeon should have in his bag or chest, razors, trepans, head saws, dismembering saws and nippers, cauterising irons and even a hammer and chisel. His tools for the extraction of teeth would include forceps and a small file. The surgeon also had various potions which he could use, and were divided into, Aqua [water], oils, syrups and Opiates and such 'purging' medicines.[6]

Woodall supplied the regimental surgeon's chests for the navy and army, but unfortunately the regimental surgeons were unfamiliar with the potions he selected, so they petitioned the Privy Council on 11 August 1626 for £10 to purchase their own medicines, fearing that they might 'rather kill than cure' their patients.[7]

---

4   TNA SP 16/102, f.51 petition of Abigal Priest, widow of Walter Priest.
5   TNA SP 14/181/23, Lord Cromwell to the Council of War, 25 December 1624.
6   John Woodall, *The Surgeon's Mate* (London, 1617), p.128.
7   TNA SP 16/33/81, The Humble Petition of the Surgeon's now pressed into his Majesty's fleet, 11 August 1626.

The following year the regimental surgeons appear to have been given the choice of choosing their own instruments, because on 12 April 1627 James Molins, a surgeon of London, received £40 14 shillings six pence for:

> Sundry dismembering instruments, saws, syringes, catheters, cauterising buttons and sundry other such instruments. Also, for emplasters, unguents, oils, gums, electuaries, herbs of divers sorts, syrups, conserves, pills, strong waters, simples and sundry other drugs for furnishing of a surgeon's chest… for the train of artillery.[8]

An additional payment of £25 was paid on 16 May 1627 to the surgeons of each regiment to provide drugs for their regiment and £5 for linen for bandages.[9] One thing that was missing in a surgeon's chest was anaesthetic. It is often said that they did not know about bacteria, but there are accounts of salt being used, this may give us the saying, 'rubbing salt into the wound' and it might have been spread over the operating table before surgery began. Vinegar was also used, especially during time of plague when coins were dropped into a bowl of vinegar to sterilise them. It was also suggested that where the plague had killed all the occupants of a house the walls were to be lime washed. Recent experiments show that this is as good as modern antiseptic sprays at killing bacteria.

The tools of the surgeon's mate suggest that they were less qualified than the surgeons, having in their bag or chest, razors, combs, 'ear pickers', looking glass [mirror], shaving linen and basins and other shaving implements to help him shave the soldiers, rather than operate on them.[10]

On 30 June 1628, the Privy Council sent a letter to the Company of Surgeons requesting them to impress 16 surgeons for the army to be sent to relieve La Rochelle and that the company should make sure they 'are best experienced in the cure of wounds made by gunshots'. However, on 22 July Pieter Thorney, who had succeeded John Dixon as the Surgeon General to the Army, ordered the regiments to find more experienced surgeons from those already serving in the army, who were to be discharged. Despite these requests, in August 1628 John Woodall found it necessary to publish *Viaticum, A path way to the surgeon's chest* specifically written for the surgeons with the army to treat gunshot wounds. This rather suggests that those impressed were not qualified to extract musket or pistol shots.[11]

A sixteenth century writer estimated that for every soldier killed by a firearm a further four were wounded, although this ratio was dismissed by Sir John Smythe, but he did not suggest his own casualty rate. Nevertheless, unlike today a casualty would usually be treated in the sitting position, rather than lying down, which was seen as the position of a corpse. Despite the unhygienic conditions they were operated on, soldiers did survive surgery,

---

8   TNA WO 49/58, Debenture Book, 1627, f.84.
9   Lyle, *APC 1627*, minute of the Privy Council, 16 May 1627 p.275; TNA E403/2746; APC 1627 p.275, TNA E403/2746.
10   John Woodall, *The Surgeon's Chest* (London, 1628), unnumbered page.
11   Lyle, *APC 1627–1628* p.517; Lyle *APC 1628–1629* p.45.

in three recently discovered mass graves from the siege of Stralsund in 1628, the battle of Lutzen, 1632 and the battle of Wittstock, 1636, which contained 25, 47 and 125 skeletons respectively, there are signs that the soldiers had survived several wounds before their last battle.[12]

In the 'Lutzen grave' the soldiers were aged between 16 and 45, although the majority were between 21 and 35 years old, whereas the youngest in the Wittstock grave was about 18 years old, although the average age was 28. The tallest was about 5 foot 11 inches, and the average height was 5 foot 7 inches. Eleven are known to have come from Scotland and a further 33 were possibly Scottish, although the tests results were inconclusive, while the remainder came from Central Europe or Sweden.[13]

Three-quarters of the remains in the Wittstock grave had suffered from malnutrition during childhood, which had led to rickets or scurvy due to a lack of vitamin C, while a large proportion had osteoarthritis in the hips and knee joints and over half had tooth decay. Many were missing teeth as well as suffering from inflammation of the maxillary sinuses or oral mucosa. Many also had syphilis, tuberculosis or periosteal inflammation and had been infested with lice or other parasites.

20 of the 80 skeletons examined from the Wittstock grave had on average two slash marks to their skulls, some even more, made either by a pole weapon, such as a halberd, or a sword when being attacked by cavalry. One soldier who by his DNA is believed to have been a 17 to 20 year old Finn had five slashes to his head, although only one of these is believed to have caused the fatal blow.

A further 24 skeletons had lead balls among their bones either in the pelvic area or in the area of the right shoulder and a further five were shot in the knee region of the leg, while some skeletons bear signs of having been attacked from behind and several skeletons had signs of stab wounds, i.e. a chip out of the rib bone or a hole in their vertebrae probably made by sword blade.[14]

Although the Lutzen grave was smaller, the remains show similar wounds to the Wittstock burials, as well as signs of disease, like syphilis and a vitamin D deficiency. The height of these soldiers was on average between 5 foot 7 inches and 5 foot 11 inches and some had signs of having badly healed broken bones. Several skeletons had marks to their forearms probably where they had desperately tried to deflect a sword slash by holding up their arm up for protection.

During the storming of Madgeburg on 20 May 1631, Peter Hagendorf, a soldier serving with the Bavarian army was wounded and records:

> I was taken to the camp and bound up. For I had been shot once through the stomach (shot right through from the front) and a second time through both

---

[12] BL Harl ms 135 Sir John Smythe's answer to Captain Humphrey Barwike's book on military affairs, 1595, f.50.

[13] Eichoff Sabine and Schopper, Franz *1636 Ihre Letzte Schlacht* (Berlin: Theiss, 2012) p.178; Sabine Eickhoff 'Das Massengrab der Schlacht von Wittstock' in *Militargeschichte* February 2013 p.49.

[14] Eichoff and Schopper, *1636 Ihre Letzte Schlacht,* pp.152–163; Eickhoff, Sabine 'Das Massengrab der Schlacht von Wittstock' in *Militargeschichte,* February 2013, p.49.

shoulders, as that bullet was caught in my shirt. The army doctor bound my hands behind my back so he could use the gouge (forceps) on me.

Without anaesthetic or sterilising his instruments, the surgeon would try to remove the ball with a pair of 'stork bill forceps', which would make it easier to grab the ball. If he failed to remove all the debris, such as cloth or fragments of bone, from the wound then the patient would develop an infection. Even worse, the surgeon might grip a piece of flesh with the forceps while trying to remove the ball, so putting his patient through even more agony. It is little wonder then that Hagendorf, at this stage of his surgery, seems to have passed out, because he continues, 'thus I was brought back to my hut half dead'.[15]

The same procedure would be used in the case of stab wounds. The surgeon would try to remove all the debris from the wound and then according to Woodall spread 'an artificial balm' or oil over the area. One of Woodall's recipes for this balm was honey and white wine, which had antiseptic qualities. However, if the wound was severe enough only then did he suggest that it should be cauterised, even though it brought 'terror' to the patient who saw the red hot iron.[16]

Where a limb could not be saved then a surgeon would have no option but to amputate it. The surgeon would cut the flesh with a knife and then saw through the bone. Surgeons prided themselves on how quickly they could amputate a limb, but often the instruments had become blunt with use since surgeons often went from patient to patient, nor would they bother to clean the instruments or wash their hands each time and so infection was a strong possibility.

Woodall then suggested tying off the veins and arteries with ligatures to stop the bleeding, but even he had to admit this was rarely successful so the stump would also need to be cauterised. However, other surgeons might not have the luxury or even the knowledge of stemming the blood flow with ligatures if they had many casualties to deal with, therefore they would probably use the cauterising iron immediately to save time. Unfortunately, statistical information on the survival of amputation cases was only collected in the nineteenth century when surgical procedures of stemming the flow of blood had changed. Moreover, this statistical information is contradictory in that a soldier was more likely to survive an amputation during the Peninsular War (1808–1814), than the Crimean War (1853–1856).[17]

Once the surgeon had treated the soldier's wounds, the area would be bandaged, although infections, such as gangrene, were an ever-present danger with such wounds, which might result in further surgery or the death of the patient.

Unfortunately, according to Richard Wiseman, who became the surgeon to Charles II, surgeons did not always diagnose the symptoms properly.

---

15 Quoted in Helfferich, Tryntje *The Thirty Years' War, A Documentary History*, p.283. The German for tent and hut at this time are interchangeable so I have used the word hut instead.
16 Woodall, *The Surgeon's Mate*, p.10.
17 Sansom, Arthur *On the Mortality after the Operation of Amputation of the Extremities* (London, 1859), p.8.

LIFE AND DEATH IN THE ARMY

During the English Civil War he records that two surgeons had diagnosed one soldier as having gangrene in the face, but on closer examination, Wiseman discovered that the soldier had just been burnt from the flash of a pistol which had been fired too close to him. Wiseman also records that a soldier had been left to die by a surgeon because his wounds were thought to be mortal, when in fact after a little treatment the soldier made a full recovery.

On the other hand, it would be wrong to say that the leading doctors of the day were any better, because in their desperation to cure Henry, Prince of Wales, they tied a split pigeon to his head and a cockerel to his feet. Unfortunately, this 'cure' did not save him, and he died of typhoid in November 1612.

Even when not in battle accidents happened. During the Civil War, a Royalist officer complained, 'we bury more toes and fingers than we do men'.[18] However, the biggest cause of death in an army was disease. Despite efforts being made to keep a camp clean from filth, thousands of men and women living is such close confinement encouraged the spread of infection.

Although salt was used as an antiseptic it was believed that diseases were spread by bad smells, rather than germs and bacteria, which upset the 'humours' of the body. Since classical times doctors believed that the body contained four humours, blood, phlegm, yellow bile and black bile, which were representatives of air, water, fire and earth. Therefore, if a patient was ill then one of these humours had become unbalanced. Even John Woodall believed infections, such as gangrene, were not caused by bacteria or germs, but the reason was 'hidden in God' and 'a distemper of the four humours'. Some doctors went even further, arguing that taking a bath should be avoided because it might unset these humours.[19]

Whatever the doctors of the early seventeenth century believed the cause of disease and infection were, it was disease that destroyed Mansfeldt's army, even before its regiments rendezvoused at Dover. In January 1625, his army mustered about 12,000 men, but a mere three months later just 3,000 were still with the colours. Some of this loss was through desertion, but 18 soldiers died from sickness in the Dover Almshouse between the 12 January and the 8 April 1625, each being wrapped in a sheet before being buried. Among them was John Archer 'a soldier under Captain Dutton' who had 5s when he died and Thomas Watts, who had just 2s. In January 1625, a further 260 soldiers were reported as being sick and 48 soldiers are recorded as being buried in the parish church of St Mary's at Dover during this time; only one, Matthew Hudeley, is described as having been 'killed', probably by another soldier. Unfortunately, the parish clerk did not record what disease killed these soldiers, but often it spread to the local population as well.[20]

---

18 Quoted in C Hibbert's *Cavaliers and Roundhead* p.77.
19 John Woodall, *The Surgeon's Mate*.
20 Quoted in Hibbert, C *Cavaliers and Roundhead*; BL Egerton ms 2087 ff.7–14; Roberts, *Diary of Walter Yonge* p.81; Kent Archives TR2451/6/7, 46 were buried between 20 December 1624 and 29 January 1625, Hudeley being buried on 16 January and entry for 29 January records the burial of a 'German soldier'. The other two were buried on 3 March 1625.

One of the diseases usually associated with the destruction of armies is the 'burning ague', which we know today as typhus, and which was also known as 'camp fever' and 'gaol fever', since it thrived in cramped and filthy conditions. It is caused by the *Rickettsia prowazekii* bacteria and is highly contagious, being spread from person to person by lice, which feed off and infect their host. The perfect environment for the louse is the human body, but when the body gets too hot due to the fever or cold which results from the typhus bacteria, the lice try to find another human host to live upon. Meanwhile the bacteria in the lice's stomach multiples and pass into its intestines where it is excreted and remain active for many days in dried faeces. During this time, the louse will have bitten its new host, who scratches themselves. If the person breaks the skin while scratching, then the bacteria enters the bloodstream and the cycle would begin all over again.

About 10 days after being infected the symptoms begin to appear, with the person developing a high fever, muscle pains, delirium, severe headaches and nausea. Five days after the symptoms begin a rash develops and the skin has a flushed appearance, before the person slips into a coma, death soon following. However, after two weeks, if the fever breaks, then the person is likely to survive, but will need a long convalescence.

The majority of cases were be found in victims aged between 15 and 30 years of age, the age group of many of the soldiers. Typhus is less common in older humans, but it is more virulent. The mortality rate of those who died from typhus during the seventeenth century is unknown, but in outbreaks of the disease, during the twentieth century has been estimated at up to 15 percent, rising to 60 percent for older people when it goes untreated. Secondary infection is also common, like pneumonia and possibly organ failure.[21]

Typhoid is similar to typhus and spread in the same way, although caused by a different bacteria and usually effects those between the ages of 5 and 19. During the first week the infected person's temperature rises, accompanied by headaches, malaise and a cough, before becoming delirious, and spots may appear on the body. They have diarrhoea with green faeces and the liver and spleen become enlarged. After about four weeks death occurs for between 10 to 20 percent of sufferers.

Although none of the accounts specify their ailments, over half of Cecil's army appear to have died from illness, which would suggest a much higher rate of mortality during the seventeenth century. Whether this was because of the advances in medicine during the twentieth century or typhus or typhoid having become less virulent is unknown.

As the skeletal remains of the mass graves have shown, syphilis was common and the sufferer would come out in sores or lesions all over their body, especially around the genital area. In severe cases the flesh would rot away which would also affect the skeleton. One of the many 'cures' was mercury, either taken orally, injected up the urethra or mixed with other ingredients to form a cream, which was then rubbed into the sores.

---

21   R. N. Mazumder et al 'Typhus Fever; an overlooked diagnosis', in *Journal of Health, Population and Nutrition* (Dhaka, Centre for Health and Population Research, June 2009), pp.419–421.

However, these sores would usually disappear after a time without treatment thereby making the person think they had been cured. If treated with mercury, then this could cause madness and damage to the nervous system. Even if this cure did not drive the sufferer insane then a long-term symptom of syphilis was madness, or what the Victorians called, 'general paralysis of the insane'.

One disease that Buckingham's army are known to have suffered from on the Île de Rhé was the 'bloody flux' or dysentery, which was also caused by unhygienic conditions. There are several types of dysentery, which is one of the oldest gastrointestinal illnesses and is caused by an organism such as bacteria or parasitic worms and is found in food or water contaminated by human faeces.[22] Depending on the type of dysentery, can include a high fever, fatigue, headaches, nausea or vomiting and abdominal pains, which can lead to complications to the internal organs, including kidney failure. However, all types of dysentery result in severe diarrhoea, which sometimes contains blood hence the name the bloody flux, which can lead to death due to dehydration or organ failure.

There had been many outbreaks of plague since its arrival in the Middle Ages, but the outbreak of the 'pestilence' in 1625 was one of the worst in living memory, which caused the King to postpone his coronation. Bubonic plague was spread by infected fleas from human to human; black rats, (Rattus Rattus) to human; or even a piece of cloth to human, as in the case of Eyam in Derbyshire, during the outbreak of plague in 1665 to 1666, which decimated that village. Once the infection got into someone's lungs the result was pneumonic plague, and they could pass on the infection from human to human by coughing or sneezing. Those infected would develop buboes, usually on the neck, armpits and groin area and death would almost certainly follow. If a soldier contracted the plague when billeted on a family, then the whole household would be 'shut up' to await their fate.

How the plague effected the mobilisation of Cecil's army in 1625 is not known. Certainly, it was not listed as one of the reasons why the expedition to Cadiz failed. Neither do we know if the movement of soldiers aggravated the spread of the disease, which is one of the traditional reasons for outbreaks in towns and villages. However, the plague occurred even when there were no soldiers in the area so it may have just been a coincidence.

There were hospitals where the sick could be treated, one of which was at Plymouth, where many of Cecil's force returned to, but many of the surgeons in the area had fled in fear of being conscripted into the army. Therefore in 1627, John Bolger, who was the surgeon of Sir William Courtney's Regiment, is known to have spent five weeks assisting the sick and wounded at the hospital.[23] Where there were no hospitals, soldiers were treated by women who were paid to nurse them back to health, but there was the danger that they also might contract the disease.

---

22 Bacillary dysentery, Amoebic dysentery, Balantidiasis, giardiasis, cryptosporidiosis and viral dysentery.
23 TNA SP 16/139/11, petition of John Bolger to the Council, 19 March 1629; SP 16/60/73, petition of the Mayor of Plymouth to the Council, 20 April 1627.

Very few soldiers found a Christian burial in a churchyard. If they were lucky their name would be written into the parish register as a record for posterity, but more often the words 'an unknown soldier' are recorded with the date of their burial. When Ensign William Goddard of Captain Henry Theobale's Company made out his will in 1624 he left his soul to God and his 'body to be buried as pleaseth either my friends or enemies'. Among his worldly goods he left to his friends, Captain Henry Spry and Captain William Cromwell, was £10 each to purchase a sword and a further £10 to be distributed among the soldiers in his company. Although he left nothing to his captain, and Spry distributed the money to Goddard's soldiers not Theobale.

Illustration of a soldier being robbed by other soldiers. Note his knapsack in which he kept his belongings. (Author's Collection).

On the other hand, in his will Colonel Edward Harwood stated that he wanted to be buried in Holland and that his lieutenant-colonel was to make a 'superscription of the time of my service'. Despite both describing themselves as being in 'perfect health' they were both dead a few months later.[24]

However, for the majority of soldiers who died in battle or from disease it would be left to the local inhabitants to bury the dead in mass pits. The 47 dead from the battle of Lutzen were buried in a pit 19 foot by 22, and with a depth of 3.5 feet; whereas the Wittstock grave was 19 feet by 11.5 feet with and estimated depth of 5 foot 3 inches to 5 foot 7 inches and the bodies were placed in the grave in three or four layers, in two rows with their head placed on the edge of the grave, so some were in the west-east position, while others were in an east–west position. Neither was there much evidence of clothing which suggests that they were striped before burial, either by the local peasantry or their colleagues. Whether the local vicar or priest conducted the burial service is unknown.

---

24 TNA Prob11/147/204 William Goddard's will was made on 8 September 1625 and proved in October 1624; Prob11/162/327 Harwood's will was written on 14 June 1632 and proved on 11 September 1632.

LIFE AND DEATH IN THE ARMY

Give Fire

10

Mansfeld and Morgan

On 25 December 1624, 8,000 soldiers unexpectedly arrived at Dover. There were no provisions or quarters for them, and so the soldiers rioted, which caused the Lieutenant of Dover Castle, Sir John Hippisley, to complain the day after their arrival, 'the soldiers commit great outrages, pulling down houses and taking away cattle et cetera and when some were taken and imprisoned, the prison was broken open and they were rescued'.

On 27 December another resident of Dover complained that the soldiers were all 'rascals and gaol birds' who commit all, 'Disorders of rapine… killing sheep in abundance and all other things they can catch, threatening the breaking into our houses and setting fire to the town upon any want they may have'.[1] While another said that although they had been given money to buy food, they steal 'all they can lay [their] hands on… [including] 140 sheep from one man in a night, and people dare not open their doors'. Since the soldiers claimed they did this because their pay was in arrears, the townsfolk had a collection to pay off the soldiers, but this was not enough, and the abuses continued.[2]

To ease matters in Dover, about 2,000 to 3,000 soldiers were sent to Folkstone and Sandwich, but on 31 December Hippisley wrote that the 'country [around Dover] is utterly wasted for 10 or 12 miles', because the plundering caused by the soldiers. Some soldiers took the opportunity to desert. Hippisley claimed that the officers were also 'careless' in disciplining their men, but what he did not know was that the soldiers had even attacked their officers. Afraid of their men they stood idly by why the soldiers committed more crimes. It was even said that the soldiers had 'ravished a lady and her two daughters'.[3]

On 31 December 1624, the Privy Council wrote to the mayor of Dover suggesting the Kent trained bands should be mustered to 'reduce all to order

---

1 TNA SP14/177/33, Francis Wilsford to Nicholas, 27 December 1624.
2 TNA SP14/177/34, William Jones to Nicholas, 27 December 1624.
3 TNA SP 14/177/18, 23, 44, 47, 48, Sir John Hippisley to Buckingham, 24 December 1624, Sir Richard Bingley to Buckingham, 26 December 1624, Richard Marsh to Nicholas, 30 December 1624, Sir John Hippisley to Secretary Conway, 31 December 1624, Sir John Hippisley to the Mayor of Dover and Justices of Kent, 31 December 1624; *Stuart Dynastic Policy and Religious Politics, 1621–1625* p.332.

by punishing the offenders and maintain the public peace', but they added they were not to shed blood, 'without great cause'.[4]

Finally, on 2 January 1625, Hippisley was able to report that the commission for martial law had arrived at Dover and five days later he was able to report that a soldier had been hung for house breaking, which 'struck terror into the rest'; after this there does was no further disturbances.

However, it would be wrong to suggest that all soldiers were ill disciplined. In December 1624 with the outrages being committed in and around Dover, the inhabitants of Winsborough in Kent, found the soldiers quartered upon them 'demeaned themselves very honestly and civilly'.[5]

One reason given for the soldiers' ill-discipline was their arrears of pay and many officers sympathised with their men. The *Apology of Mansfeldt* records:

> Now as we cannot deny, much less excuse the excess and insolencies, which the soldiers as then committed, and did commit during that war, for it is well known, that it is impossible to restrain and hold them under discipline, if their wages be not paid to them. Neither they nor their horses can live by the air, all that they have, whether it be arms or apparel, weareth, wasteth and breaketh. If they must buy more, they must have money. And if men have not given them, they will take it where they find it, not as in part of that which is due unto them but without weighing or telling it. This gate being once opened unto them, tey enter into the large field of liberty.[6]

On 9 January 1625, the Earl of Lincoln complained to Secretary Conway that, 'I have received… miserable complaints from my captains of want of money for the payment of their companies which occasions some to run away, others to steal and some to faint such for want of means'.[7]

As early as the 3 January 1625 the Prince of Orange offered Mansfeldt aid if he would relieve Breda and it was rumoured in the town itself that he was coming to relieve them and that Spinola planned to raise the siege on hearing of his advance. Therefore, the inhabitants of Breda urged the governor to relax the food rationing.

However, ever since 1622 negotiations over the marriage between Charles, Prince of Wales and the Spanish Infanta, had been going on despite most Spanish and English finding this match unacceptable. Negotiations made slow headway, but fearing Mansfeldt's intervention at Breda, the Spanish began offering better terms for the marriage. It was even suggested that James' son in law, Frederick, and his wife Elizabeth might be restored to their positions in the Palatinate. A Spanish envoy was said to be travelling to England to discuss the matter. Therefore, on 26 January 1625 the English

---

4   Lyle, *APC 1623–1625*, Letter to the Earl of Montgomerie Lord Lieutenant of Kent, 26 December 1624, p.409, Letter to the Mayor of Dover and others, 31 December 1624, p.411.
5   TNA SP 181/25i Andrew Hughes and others of Winsborough, 7 January 1625; *Journal of the House of Commons* vol. 1, pp.885–886.
6   *Apollogie of Mansfeldt* p.23.
7   TNA SP 14/181/31, The Earl of Lincoln to Secretary Conway, 9 January 1625.

colonels with Mansfeldt's force were ordered not to relieve Breda or to 'meddle with any of the King of Spain's dominion'. In fact, if both these aims could be achieved then there would be no need for Mansfeldt's force even to sail since their objectives would have been achieved diplomatically rather than militarily. However, in September 1624 the French government agreed to partly finance Mansfeldt's army, and they were eager for him to relieve Breda, which would be a blow to the Spanish war effort. Although knowing his reputation the French refused Mansfeldt's army permission to land on their territory.

It had been estimated that the English regiments would stay in Dover for just three days, but the soldiers finally embarked at the end of January, even though technically they had no objective in sight.

Furthermore, the sickness which had begun even before Mansfeldt's soldiers had reached Dover continued to ravage his regiments inasmuch that an Irish officer, Captain Hugo, who was with the Spanish army at Breda, could write:

> What with plague, with agues, with the sea and the vomiting by reason of their long shutting up in the ships with the narrowness of the room, and many filled with the filthy savour, being almost all raw soldiers and unaccustomed to tempests and stinks, were cast into the waves either dead or half alive. There was counted by some above the number of 4,000; some cast into the sea for dead, by swimming got to the shore and are yet living in the town. Many bodies floating by the shore side unburied and more elsewhere cast up by the sea on the land breathed forth a grievous plague upon the neighbouring towns of Holland. A very sorrowful spectacle to the English soldiers.[8]

It was not just hunger and disease that Mansfeldt's force was suffering from, as Hugo continues that even Heaven seemed:

> To fight against him. For the winds did so beat the arrived navy, one against the other that many of the ships being broken, did serve for nothing then, as if all the cold of the winter had been reserved for that time all the rivers were shut up with a sudden frost, that provisions could not be carried to them.[9]

Hugo also accused Mansfeldt of deliberately keeping his men on board ship to stop them running away, although he was not a witness to these scenes himself. However, he was correct when he wrote that Mansfeldt's army 'in a short time, what by sickness, what by running away, they were so diminished, so that of 14,000, scarcely four [thousand] remained'.[10]

Unfortunately for James, Charles I, Frederick V, Mansfeldt and especially the inhabitants of Breda, the Spanish were just playing for time and this diplomacy over Charles' marriage, coupled with the snow and frost of winter, had slowed the progress of Mansfeldt's force.

---

8   Gerrat, Barry *Siege of Breda* (Scolar Press, 1975) pp.98–99.
9   *Ibid.*, p.98.
10 *Ibid.*, p.99.

On 17 March 1625 the Venetian ambassador, Alvise Contarini, wrote to the Doge, 'The English of the Count's force continue to die and desert, and he has written to England for 4,000 men to fill the gaps. It is not thought that they will grant this'. By now James I was ill and he died on 27 March 1625; it was rumoured that he had been poisoned by the Duke of Buckingham. With the government in turmoil no decisions could be made and so Mansfeldt turned to the Hague appealing for money and fresh quarters in which to billet his army. Mansfeldt managed to secure 200,000 florins, but only after the English ambassador promised security on the loan and on condition that he relieve Breda.[11]

To reinforce Mansfeldt's army commissions were sent to Scotland to raise two regiments of foot under Viscount Doncaster and Colonel Gray, but it would be some time before they arrived. Nevertheless, with his finances and an objective for his army finally secured it looked like that Mansfeldt could finally launch his campaign, but with the death of James the courts of Europe looked to see whether the new king, Charles I, would continue his father's policies. However, Mansfeldt need not have worried, and he received orders to relieve Breda, which had been besieged by the Spanish since 28 August 1624. The town is said to have had a castle and about 1,200 houses, plus a 'pest house' and a 362 feet high tower from where a person could view the surrounding countryside. The rivers Monk and Aa ran through the town, which was well fortified with earthworks. Among its garrison were English soldiers, including at least one company of Sir Charles Morgan's Regiment, including Morgan himself. The rest of his regiment was with the Prince of Orange's army. By the middle of March, it was believed that Breda had only enough provisions for another seven to eight weeks, and even then the price of food within the town had increased.[12]

The Prince of Orange's army was quartered close by, but neither it and the garrison, who made regular sallies, were not strong enough to drive off the besiegers, who had erected their own fortifications. Among the Prince's regiments were those of the Earls of Oxford, Essex, Southampton and Lord Willoughby, which had been raised in England in 1624 to aid the Dutch. It would be these four regiments which would play a leading part in attacking the besiegers.

On 12 May 1625, a force of 5,000 to 6,000 foot and eight troops of horse from the Prince of Orange's army set out to, 'attempt the quarter of Terheyden'. Each nation that composed this force had drawn lots to see who would be the first to form the vanguard, the main body or 'Battle' and the rearguard. The French and some Dutch troops were to form the vanguard, while the Scottish and Frisian troops for the rearguard leaving 23 English companies, who were divided into four battalions, formed the main body. The following day it was the turn of the English to be the vanguard and it would be them who would launch the attack on the fortified village of Terheyden. The village was only

---

11   Hinds, *CSP Venetian* 1623–1625 p.614; Alvise Contarini Ambassador to the Netherlands, to the Doge and Senate, 17 March 1625.
12   Whiteway *William Whiteway of Dorchester, his Diary* (Dorchester, Dorset Record Society, 1991) p.69.

accessible by two causeways, one of which was cut off by a little river and a redoubt. According to Vere these causeways, or dikes, were surrounded by water and only wide enough for 20 men to march shoulder to shoulder, although Sir Jacob Astley says it was only six men wide. At the end of the causeway were further redoubts which protected the besiegers from attack.

Early on the morning of 15 May, the English launched their attack, a forlorn hope of 50–60 musketeers from Count William of Nasseu's Regiment, who were armed with firelocks, leading the way. They were followed by some soldiers carrying hand grenades and 20 pikemen, commanded by Lieutenant Ernely of Captain Killigrewe's company. They were supported by 400 men divided into four groups of 50 musketeers and 50 pikemen, each group was commanded by a lieutenant, an ensign and four sergeants. Lots were drawn among the colonels and lieutenant-colonels as to who should command this force and it fell to Sir John Proude and Sir Jacob Astley. This body was followed by the rest of the English battalions, the Earl of Oxford commanding the leading battalion, while Sir John Vere's battalion brought up the rear in reserve.

They crept along the causeway, but it was only when they were near the first redoubt that the alarm was given. It was now that they rushed forward, 'the soldiers', records Vere, 'entered with a great deal of courage and assaulting the first redoubt became masters of it and with the like courage assaulted the second traverse and second redoubt'.

According to Hugo, one of the redoubts was occupied by an officer:

> With some few Italians; by their [the English's] thick throwing in of fire balls, forced them to quit the place, with loss of some of their lives; which done (for the defence of the rest of their men, who yet advanced forward) placing certain musketeers behind the redoubt, in the dry ditches, cut through the causeway, with the like success and resolution, they won the half moon.[13]

Hugo blames the success of the English on the Spanish army having neglected to complete the redoubts by erecting palisades. However, now the English faced a much more formidable earthwork, which they tried to storm, 'labouring, with hands and feet to get the top of the Rampier and plant their colours upon the fort itself'.[14]

Vere takes up the narrative on the attack on the main fortification, 'but they found it so high that without scaling ladders and a supply of fireworks… it was a vain thing to attempt it and yet could they hardly be drawn off. Hereupon an hour and a half being already spent in the service and little ammunition left'.

The Spanish army quickly rallied and seeing them advancing on his army the Prince of Orange gave the order to retreat. However, the English soldiers were reluctant to withdraw from the newly captured earthworks, but such was the fierceness of the Spanish counterattack that the pikemen came to 'push of pike' and the musketeers fired at such a close range that many were

---

13 Gerrat, *Siege of Breda*, p.116.
14 *Ibid*.

shot through the head or throat. According to Hugo, Vere was everywhere leading his men by example, who, 'with the like constancy and resolution that new men should supply the places of such as were retired, fresh men made good the places of such as were hurt, and such as were tired should be relieved by others'.[15]

Thomas Stanhope, an ensign of the Earl of Oxford's regiment who had planted his colours on the fortifications, was cut down and slowly the English were forced to withdraw. According to Hugo:

> [Those] who skirmished further off were for the most part miserably butchered by our cannon, with the loss of their hands, feet and heads, almost no shot being bestowed so in vain, but that it slew divers of them at once… every man kept his own rank… The causeway lay covered thick with dead bodies, no Colour appearing upon the ground but gory blood. Men's bowels torn out with cannons, their heads struck off, their hands and feet here and there scattered and generally their whole bodies miserably butchered; a lamentable spectacle, on all sides to behold. Some had thrown themselves headlong into the standing waters others lay miserably groaning half dead, who being carried into our quarter, died afterwards.[16]

That evening the Earl of Oxford wrote home to his wife:

> Our nation lost no honour, but many brave gentlemen their lives; among which number my Lord Ambassador's newly married niece her husband, Sir Thomas Wind, killed with a great shot, Captain John Cromwell dangerously hurt, Captain Tubb I believe will not escape, Captain Terrett shot through the body close under the left papp, but I hope he will live, my ensign Thomas Stanhope killed upon the place and diverse other officers and gentlemen of good quality hurt and killed. We fought as long as our ammunition and beat the enemy out of three houses[?]. But the Dutch failing to [missing = support us?] and our powder and shot being spent we were forced to retire. In which retreat we lost most of those named I returned not without a shot on my left arm; but so favourable a one, as drew no blood, nor hinders me to relate this tragedy.[17]

Vere also confirms that the Earl of Oxford, was hit in the arm by a spent musket ball which reminded him 'of the danger he was in'. However, despite this slight wound, coupled with sun stroke, he died in the Hague a few weeks later.

Among the casualties was Sir Thomas Wynne or Wind, who was a volunteer and was killed by a 'great shot', which wounded four soldiers standing close to him. According to Hugo the Prince of Orange forces lost between 200 and 500 dead, while the Spanish lost 12–15 men. However, Vere and Astley states that the English losses were 62–63 men killed and 110 wounded. Vere's soldiers had expected to be supported by the other nations

---

15 *Ibid.*, p.118.
16 *Ibid.*, pp.118–119.
17 HMC 5th Report p.411.

within the Dutch army, but when they failed to do so this caused ill feeling amongst the English.

What part Mansfeldt's force played in this attack is not known, but it was also driven off by the besiegers. This had been the final hope of the inhabitants, a point not lost on the besiegers who shouted to the garrison of the relieving force's defeat. The price of food increased even further in Breda, which resulted in riots over the price of bread and the soldiers began to eat horsemeat. Now without any hope of relief the town's governor, Justin of Nassau, began to negotiate for the town's surrender. On 5 June 1625, the garrison was allowed to march out of Breda with full military honours. Breda's surrender is considered as a major victory for the Spanish during the Eighty Years War.

Mansfeldt's force withdrew into Brandenburg. On 7 June Lord Cromwell estimated it was just 6,000 strong, of these the French contingent mustered about 500 strong and the Dutch 1,200. He also estimated that his own regiment was just 220 strong and that they had not had any bread for four days and 'we are weary of our life'. In another letter in June, Lord Cromwell wrote, 'We live here most miserably and I protest to God were it not for dead horses and cats our army had perished since coming to Haffin[?] which is our Leaguer. All the English that is left are 600, which are put under lieutenant-colonel Hopton'. This company of 600 men was the remnants of the six English regiments which had sailed from England at the end of January, including 200 men who had recovered from their sickness. No doubt others might rejoin the company once they had recovered, but there was no incentive to do so because as Cromwell recorded in his letter, 'they that know Mansfeldt best say that he never paid any man'.[18]

As to the French forces it was said that they deserted because they had been told that they were to recapture the Palatinate and not fight the Spanish. Moreover, they had also not received any pay for seven months. According to Hugo, Mansfeldt even had one of them 'drawn in pieces with four horses' for encouraging some of his fellow countrymen to desert.[19] Furthermore many of the officers were returning to England to assume commands in Sir Edward Cecil's army which was being raised to attack Spain.

With the surrender of Breda, Mansfeldt's army no longer had an objective, although it continued to soldier on in Germany. However, on 29 November 1625, King Christian IV of Denmark entered the war, although he was fighting as the Duke of Holstein, which formed part of the Holy Roman Empire, rather than the King of Denmark, due to the Danish government refusing to raise an army. Therefore, Christian had to raise and finance an army at his own expense, so he formed an alliance with Britain and Holland, who each agreed to pay Christian a subsidy of 144,000 florins per month and Mansfeldt's army, which was then at Bremen, was to be transferred to Danish service. Despite this it would not be until March 1626 that Christian finally ratified the Hague agreement.[20]

---

18 Quoted in Dalton, *The Life and Times of General Sir Edward Cecil* vol. 2, p.111.
19 Gerrat, *Siege of Breda*, pp.109–110.
20 Parker, Geoffrey *The Thirty Years' War* (London: Routledge, 1997) p.69.

This was offset by Albrecht von Wallenstein, who had taken command of the Imperialist army. He agreed to greatly expand it at his own expense until Emperor Ferdinand II could afford to repay him. With this much larger army at the Emperor's command it altered the balance of power in Germany, since it gave the Catholic forces a major new army, but would also be a counterbalance to the Catholic League's army commanded by Johan von Tilly. Many Protestant and Catholic rulers feared that the emperor would use this new army to impose his will on them.

On 19 December 1625 Wallenstein estimated that Mansfeldt had six regiments of foot and 75 cornets, or troops, of horse, 400 dragoons and nine guns, although these units were probably greatly understrength, as the previous month it was reported that he had just 3,000–4,000 foot and 600 horse. Nevertheless, on 7 January 1626 Sir Robert Anstruther reported that Mansfeldt was again recruiting his army near Lubeck, but added 'from which town we see, as yet, small hope of good'. The King of Denmark also sent him some regiments to strengthen his army bringing Mansfeldt's force up to about 9,000 foot and 3,000 horse, with additional recruits expected. Certainly, Wallenstein estimated that Mansfeldt's force mustered about 11,000 men on 9 March.[21]

Whether Christian IV tried to keep a too tight a rein on Mansfeldt or he disagreed with the king's strategy, Mansfeldt protested that he wanted an independent command and to do 'something considerable'. Therefore, he decided to launch an attack on the Hapsburg lands, possible hoping to take Wallenstein off guard while he was still training his men. A successful campaign would enhance Mansfeldt's reputation which had suffered after a number of defeats.

Unfortunately, the winter weather slowed his advance and also prevented some of Mansfeldt's provisions arriving by ship. However, by 24 March 1626 Mansfeldt's force had already captured Standoe, Havelberg, Zerbet and 'diverse other places upon the Elbe' and was marching towards the fortified camp of Dessau with its bridge that was held by a detachment of Wallenstein's forces.

The town of Dessau is situated near the northern border of Saxony where the River Elbe loops and is intersected by the River Marle. According to *Theatrum Europaeum* the bridge spanned two rivers in a clearing between two forests which covered both banks. It was in this clearing that the Imperialists had erected fortifications to protect the bridge, guarded by two regiments and about 30 guns, under the command of Johan von Aldringen, Wallenstein's paymaster general.

On 7 April 1626 Mansfeldt finally arrived opposite the fortified crossing, with about 12,000 men and tried to storm the bridge, but was driven back. Rather than withdrawing his army, Mansfeldt decided to lay siege to the camp and for the next three weeks Aldringen held out against superior

---

21 TNA SP 75/7/6, Anstruther to Secretary Conway, 7 January 1627; SP 75/7/47, Anstruther to Secretary Conway, 23 February 1626; SP 75/7/196, Rosencratz to Conway, August 1626; Tadra, Ferdinand *Briefe Albrecht von Waldstein to Karl von Harrach Harrach* (Vienna-Oesterreichische Akademie der Wissenschaften, vol. 41, 1849) pp.312, 332.

numbers. By this time there was no trace of the old English regiments he had commanded the previous year, or the two regiments that had been raised in Scotland. In fact, it was reported he had just four German regiments and one Dutch Regiment of foot, including one he had withdrawn from Magdeburg, plus four regiments of cavalry.[22]

It was rumoured that Wallenstein had ordered Aldringen to abandon the bridgehead, but Aldringen refused and asked Wallenstein for reinforcements so that he might defeat Mansfeldt. Certainly, Aldringen did not deny this rumour, but he had been insulted a few days after the battle by Wallenstein, who either called him a 'quill-pusher' or an 'ink-swiller', because he had complained about Wallenstein to the emperor. It is said that Aldringen would not forget this insult. Whether the rumour was true or not, on 21 April Count Heinrich von Schlick arrived at the Imperialists' camp with reinforcements followed two days later by two further Imperialist regiments. On 24 April Wallenstein arrived at the camp with a large force of cavalry.[23]

Hearing that Aldringen had received reinforcements Mansfeldt called a council of war, where Colonel Dodo von Kniphausen argued since they were now outnumbered, they should withdraw. This advice was rejected by Mansfeldt who believed a withdrawal would further diminish his reputation. Therefore, early on the morning of 25 April Mansfeldt ordered his men to advance; four of his infantry regiments stormed the woods that surrounded the fortified camp, which the Imperialists had occupied, while a fifth regiment, Christian Wilhelm's, tried to storm the fortifications. Mansfeldt's cavalry was met by two Imperialist cavalry regiments north-east of the wood.

Wallenstein threw in reinforcements and Mansfeldt's regiments in the woods were thrown back in disorder and the Dutch colonel, Anhoff, was killed. Meanwhile Christian Wilhelm's Regiment was routed by Schlick's forces. With the day doing badly for Mansfeldt, he ordered his army to prepare to retreat. While the Imperialist infantry counter-attacked across the bridge, Wallenstein's cavalry, which had used the cover of the forest, fell upon Mansfeldt's left wing. Meanwhile one of the shots from the bombardment by the Imperialist's artillery hit Mansfeldt's powder magazine which exploded adding to their confusion. Mansfeldt's cavalry broke and fled to Havelberg leaving their infantry to its fate. The majority surrendered and 3,000–4,000 men of Mansfeldt's force changed sides. Among the killed were Colonels Anholff and Ferentz and 32 colours, six guns and four mortars were captured, along with 48 officers, including Colonel Kniphausen.[24]

In his report to the emperor, Wallenstein records, 'God, the supporter at all times of Your Majesty's just cause, has today given me the good fortune

---

22 William Guthrie, *Battles of the Thirty Years' War vol. 1* (London, Greenwood Press, 2002), p.139. The regiments are recorded as Ferentz's Red Regiment, Berlin's, Kniphausen, Anhoff's (Dutch), Christian Wilhelm's (Madgebury) Regiments of Foot and Geists, Sterling, Mansfeldt's and Franz Karl of Saxe Laurenburg's Regiments of Horse.

23 According to C V Wedgewood, *The Thirty Years' War* (Methuen, 1981), p.210 he was called an 'ink-swiller', but Golo Mann, *Wallenstein* (London, Andre Deutsch, 1976), p.285 states 'quill-pusher'.

24 Guthrie, *Battles of the Thirty Years' War* vol. 1, pp.121–122.

to smite Mansfeldt upon the head'.[25] Nevertheless Mansfeldt's army did put up a stiff resistance because the battle is said to have lasted for six hours. Moreover, Wallenstein did not follow up his victory, a fact not lost on his critics, so that Mansfeldt was able to escape with the remnants of his army to Silesia, where he began to recruit once more.

However, these critics were not just in Vienna, because on 25 June Sir Robert Anstruther blamed Mansfeldt's defeat on the 'abandonment of England and France and diverse others', and that he had, 'not received any instructions from England'.[26]

On 22 July Mansfeld, along with the Duke Johan Ernest of Saxe-Weimar, who Christian had sent to reinforce Mansfeldt after Dessau, crossed into Lower Silesia and marched along the banks of the River Oder, hoping to join Bethlen Gabor, the Prince of Transylvania who had sided against the Emperor since beginning of the Thirty Years' War. Wallenstein had planned to follow him, but a rumour reached him that the Swedes were about to join the war and his army might be needed to meet this new threat. However, by the time it was discovered that the rumour was false, Mansfeldt was too far ahead to pursue.

While Mansfeldt and Saxe-Weimar were waiting for Bethlen Gabor, they set about imposing taxes and plundering the local population, breaking up any unauthorised gatherings and forced the inhabitants to swear allegiance to the Danish king. However, they received news that Bethlen Gabor was unable to rendezvous with them as planned so the two generals quarrelled over what to do next. Johan Ernst of Saxe-Weimar wanted to wait for Bethlen Gabor, whereas Mansfeldt wanted to march deeper into the heart of Germany, even as far as Bavaria itself. However, at a council of war, Mansfeldt's plan was dismissed even by some of his own colonels. Moreover, by now Wallenstein had realised his mistake over the Swedish intervention in the war and was now approaching Lower Silesia. Coupled with reports of Wallenstein's approach a messenger brought news that the Danish army had been defeated in battle at Lutter on 26 August by Tilly's forces.

It was now no longer a case of waiting for Bethlen Gabor to arrive, and the two generals had to march towards Hungary to meet him, with Wallenstein hot on their heels. Mansfeldt's health was failing, and leaving the Duke of Saxe-Weimar, he set out for Venice to convalesce. However, he died on the night of 29/30 November in the village of Ratona.

With the defeat of his main army and that of Mansfeldt's at Dessau, many feared that Christian IV would make peace with the Emperor and that the Catholic cause would be victorious. Therefore Charles, who had all but abandoned Mansfeldt's army to its fate, now decided to raise a new force to help the Danish king. Fortunately, in November 1626 in an effort to cut their defence budget the Dutch discharged four English regiments of foot from their service. Charles I decided to transfer them to the Danish service, albeit under English pay.[27]

---

25 Quoted Mann, *Wallenstein,* p.286.
26 TNA SP 75/7 f.156, Draft of answers to M Zobel, June 1626.
27 TNA SP 75/15, Burliamachi accounts for service of Denmark, May 1638, f.206.

Charles had planned to appoint Lord Willoughby to command these four regiments, but he, 'refused in regard of the indisposition of his body', and Sir Edward Cecil (now Lord Wimbledon), was still suffering from the political backlash of the Cadiz Expedition. Therefore, the command was offered to Sir Charles Morgan, much to the Earl of Essex's disgust, who returned to England along with Sir Peregrine Bertie.[28]

It was not just the officers who were reluctant to serve under Morgan's command, because a Mr Beaulieu wrote on 13 April, 'I saw a letter from Holland which saith that the Earl of Essex's company, consisting of 163 men when they came to the shipside and their colours lodged all save 40 refused to go'.[29]

Meanwhile believing Morgan's force to be about 6,000 strong in February 1627, Christian IV sent orders to Morgan, who was still in the United Provinces, that his force should sail down the River Weser and occupy several redoubts and a bridge of boats near Bremen, from where a force could have easy access to Münster, Paderborn and Westphalia. In fact, Morgan's force mustered only 2,472 men, 2,236 having deserted. On 20 April Robert Anstruther recorded that he had seen Morgan's force which mustered just '2,450 able men' and that 'their clothes worn'.[30]

As well as bringing Morgan's regiments up to strength, Sir Jacob Astley, Sir James Hamilton and Sir James Ramsey were also commissioned to raise regiments in England for Morgan's force. The recruits were to rendezvous at London, Harwich and Hull, but on 27 April Mr Beaulieu records that they, 'were billeted in our suburbs this week are all embarked to go to Stade… [having] kept great disorder and were very insolent'.[31] Stade was a leading port city in Lower Saxony, lying on the Schwinge River, near to where it intersects the River Elbe.

However, hearing Morgan's force were so few in number, on 19 April Christian wrote to Charles:

> Oh, that these troops were perfect in the full number promised! Then truly we would not be wanting to take this fit time and opportunity to repel the common enemy and to further the common safety. But seeing they are so few that they are almost unprofitable. Our endeavours are hindered that most unwillingly we are forced to delay our intentions. There is none that cannot see how march harm to the common cause and how much profit the enemy receives thereby.

Christian went on to ask that recruits be speedily sent so that the entire summer campaigning season might not be lost.

---

28 Letter to Joseph Mead dated 17 November 1626 in Birch, Thomas *Court and Times* p.171.
29 Letter to Rev. Joseph Mead dated 13 April 1627, in Birch, *Court and Times* p.216.
30 TNA SP 75/8, Anstruther to Conway, 20 April 1627, f.76.
31 Colonel Morgan's Regiment mustered 650 men, 529 having runaway, Essex's (later Swynton's) 577, present 587 runaway, Borlase, 613 present, 509 runaway and Levingshire 634 present and 551 runaway TNA SP 16/88/39, Note of the number of men requiste to be supplied to the four regiments, 1627; TNA SP 75/8 f.12 Conway to Anstruther, 21 February 1627; SP 75/8 f.75, Christian IV of Denmark to Charles I, 19 April 1627, Letter to Rev. Joseph Mead dated 13 April 1627 printed in *Court and Times* p.216.

To assist Morgan, Christian IV sent several regiments as reinforcements, but by 19 June these had been withdrawn, leaving Morgan's regiments thinly spread. True, some recruits had arrived from England, and Morgan was forced to place them in the redoubts along the Weser, which he feared would 'ruin many of them'. Morgan also received several letters informing him that three Scottish regiments, including Colonel Mackay's Regiment, were quartered at Stade, although these regiments mustered just 2,000 in all. There were also 'friendly' garrisons at Buxtehude, Gluckstadt and Kempe. At first Morgan had his headquarters at Bremen, but by the 23 July he had moved it to Weserbaden.[32] Mackay's Regiment remained with Morgan for 10 weeks when it was recalled by Christian. During this time there appears to have been some jealousy between Morgan's regiments and Mackay's because, 'The want of pay at the Weser made our soldiers a little discontent, seeing the English get due weekly pay'. On the other hand, the English officers were jealous because those in Danish pay received much more than they did.

Charles had agreed to finance Morgan's force himself, and in May he sold a jewel worth £100,000. However it was estimated that Morgan's four regiments alone would cost 39,600 Reich Dollars per month and the total expenditure up to 1627 for Mansfeldt's and Morgan's forces would be £978,320 14s 8d. Furthermore, by now Britain was now involved with not just in a war with Spain, but also with France, and Charles could not afford to fight on one front let alone three, so the soldiers' pay soon fell into arrears with even Calandrini, the force's paymaster, refusing credit to the soldiers until he had been paid the money which was already owed to him.[33]

While Mackay's Regiment was with Morgan, Robert Monro recalls, there were 'many alarms, but little service', which was fortunate because many of Morgan's officers appear to have returned to England to take part in the Île de Rhé Expedition.[34]

During the spring of 1627, a further 1,300 men were to be raised to bring Morgan's regiments up to strength. Nevertheless, on 6 June 1627 Morgan's four regiments still mustered only 4,913 officers and men which prompted a new levy of 1,400 men. However, these reinforcements were offset on 10 July 1627, by Mackay's Regiment with Colonel Hatzfeld's Regiment of Horse being ordered to leave Morgan and march to join Christian's Danish forces. On their march four companies were placed in a redoubt on the Elbe under Major Dunbar, which Robert Monro calls the 'Bysonburg skonce', although a newssheet referred to it as a 'sconce and Ship Bridge' at Arsom which guarded a bridge over the Elbe. Dunbar's men were supported by Hatzfeld's regiment and some Dutch companies while the remainder of Mackay's regiment marched to Rapine. However, according to Morgan this small force failed to keep a proper watch on the approaches to the sconce and soon the Imperialists advanced on the sconce.[35]

---

32 TNA SP 75/8, Morgan to Secretary of State, 19 June 1627, ff.176–177.
33 TNA SP 75/8, State of the Danish Army, March 1627, f.60.
34 Monro, *Monro His Expedition* Part I, pp.4–5, 7.
35 Ibid., p.11 *The Continuation of Our Weekly Newes* issue 30, 17 October 1627.

According to Monro this force consisted of about 10,000, although Morgan states that it was a just a body of cavalry and about 600 musketeers; which was commanded by the Count of Anhalt. But whatever the size of the attacking force when it came into view of the sconce two Danish regiments, including one which was considered the 'bravest regiment in the king's service' fled. One Dutch captain tore the colours of his company off its staff and thrust them into his breeches and then ran away. According to Monro three times the Imperialists tried to storm the sconce but were beaten off, with the Scots running out of powder they, 'threw sand in their enemies' eyes, knocking them down with the butt of their muskets', which made the Imperialist draw off. It was only at night after receiving orders from the king of Denmark that these four companies abandoned the sconce. However, Morgan states that Dunbar's men held out until about noon and that they abandoned the sconce after destroying all the ammunition they could not carry. This detachment also destroyed part of the bridge over the Elbe to prevent the Anholt's force crossing the river in force. Anholt tried to repair the bridge, but Morgan managed to intercept a convoy carrying 'boards and beams' intended to repair the bridge, which slowed his progress.[36]

Meanwhile Christian was being pursued and ordered Morgan to join him in Jutland, but after taking advice from the English ambassador, Robert Anstruther, and Dutch and Danish commissioners he decided to disobey Christian and withdraw to Stade. On arriving at Stade, Morgan with his 4,707 men were refused entry because the burghers did not want to impose a further burden on its inhabitants, since there were already English and Scottish soldiers quartered there. When Morgan's soldiers were finally allowed to enter Stade they were given the worst quarters, moreover they found the 'red shanks' among the Scottish soldiers, 'very disorderly and rude', which caused the English to fight back.[37]

The Imperialists were not ready to launch an all-out attack on Morgan, since they were concentrating their main effort on the main Danish army in Holstein, which was being pursued by three Catholic armies, namely Wallenstein's, Tilly's and Duke George of Lüneburg. Fortunately for Christian's army, Maximilian of Bavaria had ordered Tilly, who was recovering from a wound he had received at Pinneberg, not to support Wallenstein because of tension in the Catholic camp.

On 14 October Morgan wrote, 'the cold winter weather is come upon us, our soldiers are bare and naked and no monies to be had from Calandrini neither to officer or soldier'. This had reduced Morgan's force to just 3,764 men. However, this did stop Morgan fortifying the area, because four days later he was reported to be digging sconces at Oldenburg and Vagesaeck, and planning to raise another one at Oldenburg in preparation for the inevitable Imperialist advance.[38]

---

36   Monro, *Monro His Expedition* part I, p.11; TNA SP 75/8, Morgan to Conway 8 August 1627, f.252.
37   E. A. Beller, 'The Military Expedition of Charles Morgan' *The English Historical Review* Oct 1928, p.233; *The Continuation of Our Weekly Newes* issue 30, 17 October 1627.
38   *The Continuation of Our Weekly Newes* issue 39, 24 October 1627.

With promises of help from both Charles and the King of Denmark Morgan's band spent the winter in Stade, repelling an attack by elements of Tilly's Army, which was repulsed with heavy losses. On 13 December 1627, the Venetian ambassador in the Netherlands wrote to the Doge informing him that it was rumoured that Tilly's forces had captured a redoubt on the mouth of the Elbe, which Morgan had erected for the defence of Stade, adding if true, it 'would be most serious, as all hopes of holding out depend upon keeping the coast open for succour'.[39]

With the enemy closing in upon him, Morgan probably exaggerated when he wrote on 25 January 1628, that 'our provisions [appear] to diminish so fast…[that] cats and dogs [are] our present diet', and that money would 'content the hungry and naked soldiers'. Although he was able to purchase shoes and stockings for his men. On 12 February, the Venetian ambassador who was then in Hamburg reported that, 'General Morgan is in perfect health and makes sallies daily, burning the enemy's quarters and capturing much booty, which is taken to Stade'. However. the pressure on the town was mounting which was not helped by many of Morgan's men being armed with pikes which were almost useless for its defence.[40]

On 18 March Morgan wrote to Buckingham asking that he mediate with the King because, 'he and his troops seem to be forgotten of all the world…[with] their money long since out, their provisions grow every day shorter and they have no assurance of relief'. Morgan had hoped that Buckingham's army that had returned from France in 1627 would be sent to his relief, but no support was forthcoming. By 27 March it was reported that the garrison were, 'reduced to such straits that they are eating horse flesh' and that 'they foretell the immanent fall of that important place'.[41]

Finally, after giving up of all hope of relief and ammunition running critically low Morgan had no choice but to surrender his position on 5 May 1628. What remained of his regiments, which one report said was 2,000 men and another just 800 men, were allowed to march out of Stade, 'with the honours of war, drums beating, flags flying, carrying their arms, matches lighted, bandoliers full and balls in their mouths'. However, the town's inhabitants were not allowed to leave, and Morgan's forces were not allowed to fight for the King of Denmark again for six months unless they returned to England in the meantime. Instead, they were to march to Holland and could enter the service of the Prince of Orange if they wished. Only the seriously sick and wounded were allowed to remain at Stade, while those who could be moved were to be transported by wagon or ship.[42]

On 17 May Morgan's force quartered at Zwolle, where it mustered only 2,012 men from the garrison, divided into seven regiments, his own mustering

---

39 Hinds, *CSP Venetian 1626–1627*, from Giovanni Soranzo to the Doge, 13 December 1627 p.515.
40 Hinds, *CSP Venetian 1628–1629*, Anzolo Contarini ambassador to Rome to the Doge, 11 March 1628, with enclosing a letter dated 12 February 1628, p.618, Donald Lupton's 'Examples of the defects of the pike' printed in Firth's *Cromwell's Army* p.388; TNA SP 75/9, Morgan to Conway 25 January 1628, f.30.
41 TNA SP 16/96/40, Sir Charles Morgan to the Duke of Buckingham, 18 March 1628; Hinds, CSP Venetian 1628–1629 27 March 1628 p.37.
42 Hinds, *CSP Venetian 1628–1629*, from Giovanni Soranzo to the Doge, 22 May 1628 p.96.

498 men was the strongest, while the regiments of Colonel Hamilton and Mowbray mustered just 90 men each, many having run away. Another 800 were with the King of Denmark having been detached before the surrender of the town and so did not come under the conditions of surrender.[43] On 28 May 1628 the Emperor granted Hamburg full control of the Elbe, to win them over to the Imperialist cause and the following month Tilly laid siege to Gluckstadt and Krempe.

Morgan's force continued its journey to the Low Countries, where on 8 August the Privy Council in England ordered that the four regiments which had originally been commanded by Morgan be reduced into one regiment of 12 companies and brought up to strength of 2,000 men. Morgan was to command this new regiment which was to be shipped to Harwich where it would be issued with new clothing and meet the surrender terms of Stade.[44] However, only Morgan and some of his officers appear to have returned to England, leaving his regiment still in the Low Countries, since Charles could not afford to transport the entire regiment and by now only two months remained of their parole.

Meanwhile, Christian IV was in a desperate need for troops, despite Denmark having finally mobilised its forces, but only after its territory had been invaded. Morgan's Regiment was one force available and therefore it set out once more for the Elbe region of Germany, knowing they could not expect further reinforcements from England. Therefore, Christian promised to reinforce Morgan's Regiment with 1,200 of his own recruits, but in November he reneged on this offer, because he decided to spend the money needed to raise these recruits on the Danish fleet instead. Therefore, the Venetian ambassador to England, Alvise Contarini, wrote to the Dutch government asking them to spare five soldiers out of every English company for Morgan's Regiment, but this was promptly rejected.[45]

Marching through the frost of winter, Morgan's force arrived a day too late to save Krempe and so he decided to reinforce the garrison of Gluckstadt. He arrived at the town early in December, but at first was refused entry, since he brought with him neither provisions nor money. It was only after some merchants lent him £500 that he was able to buy food for his regiment.[46]

Morgan's soldiers were finally admitted by the governor of the town, Colonel Marguard Ranzow, and they were distributed in private houses. This caused, 'grievous discontent among the citizens' and Morgan's men continued to desert since they still had not been paid. However, Colonel

---

43 Beller, *The Military Expedition of Sir Charles Morgan*, pp.534–535.
44 TNA SP 75/8 ff.9, 149, Anstruther to Conway, February 1627, Sir Charles Morgan to Conway 12 June 1627; SP 16/526/101, Note of the forces preparing to join the expedition of Sir Charles Morgan, April 1627; SP 75/8 Conway to Anstruther 2 April 1627, f.70; TNA SP 75/8 Memorandum re answer to Danish Ambassador, April 1627, f.101. The regiments were Morgan's 498 men, Borlace's 493. Levington's 471, Swyntes, 420, Spyney, 150, Hamilton, 90 and Mowbray's 90.
45 Hinds, *CSP Venetian 1628–1629*, Alvise Contarini to the Doge, 18 November 1628, pp.399–400 and letter date 2 December 1628 pp.415–417.
46 Letter Mr Pory to James Mead 12 December 1628 in Birch, *Court and Times*, p.449.

Ranzow refused to admit Morgan himself, so that he had to find quarters in Hamburg, where he tried to get clothing and food for his men.[47]

This time it was not hunger, but plague, which affected the soldiers as it spread through the town. This seems to have been too much for Morgan's Regiment and 200 soldiers mutinied over their pay, which had to be put down by the senior commanders since many of their officers not only sympathised with them but had not been paid either. Fortunately, Morgan managed to obtain £400 to pay his soldiers and when Joseph Bere, the new paymaster, arrived with a further £10,000, this went some way towards clearing the soldiers' arrears of pay. Meanwhile, back in England, Parliament had once more been called, but rather than granting further subsidies to pay Morgan's soldiers they declared a public fast day as a show of support.[48]

Moreover, the King of Denmark who had once offered Morgan a senior position in his army, now turned against him, because he blamed Morgan for the loss of Krempe, even though Christian had done nothing to support him. Nevertheless, despite the souring of relations in March 1629 with the campaigning season beginning Christian ordered Morgan to meet him at either Holstein or Jutland, from where he planned to launch a new offensive against the Imperialist forces. Morgan left behind his English troops and set sail with five companies of Scottish and Dutch infantry planning to rendezvous with a large detachment of infantry before they marched to join Christian. Unfortunately, the Imperialists had gained intelligence of Morgan's movements and prevented him from uniting with the main Danish force.

However, it did not matter, because on 3 June Morgan received news that Christian had signed a treaty with Emperor Ferdinand at Lubeck on 29 May. With Morgan's troops no longer needed he marched to the United Provinces hoping to be employed there, but on 26 December 1629 the Venetian ambassador, Vicenzo Gussoni reported to the Doge:

> Owing to the dismissal of the extraordinary troops by the States, all the English troops under Morgan's command are disbanded. They are leaving this evening with my letters to the Savio alla Scrittura, to whom they will present themselves to receive the states commands. I ask that they may receive such treatment that many others may be induced to come, without making bargains, but merely placing their services at the disposition of the state.[49]

Nothing seems to have come of the transfer of Morgan's men to the army of the Venetian states, because Morgan continued to serve in the Dutch army and was present at the sieges of S'Hertogenbush and Maastricht, where at the latter he received two musket wounds. However, negotiations continued into 1630 to try and get English troops transferred to the Venetian service. With the peace of Lubeck, it not only brought an end to the 'Danish' phase of the Thirty Years' War, but also Charles' involvement in Germany, although

---

47   *CSP Venetian 1626–1628,* Giovanni Soranzo to the Doge, 11 December 1628, p.429.
48   Beller, *The Military Expedition of Sir Charles Morgan,* p.538.
49   Hinds, *CSP Venetian 1629–1632,* Vicenzo Gussoni to the Doge and Senate, 26 December 1629, p.257.

he would allow foreign powers to raise recruits in his kingdoms for years to come. As to Morgan for several years he feared being arrested over debts he had accumulated during this time. He became the governor of Bergen op Zoom, a position he held until his death in 1643.

Dismount your musket

11

The Spanish War

Charles' war with Spain should have been popular with the English people, since from Elizabethan times Spain was seen as their traditional enemy. In 1588 Phillip II had launched the first of three Spanish Armadas, all of which were decisively defeated not only by the English navy, but also the weather, a defeat which not only brought prestige to Queen Elizabeth, but also to England.

However, this time many of the ships were in a poor state of repair and the soldiers destined for the Spanish war were no more disciplined that those under Mansfeldt had been, although fortunately for the people of Kent their ordeal was quickly over. But for the inhabitants of Devon and Cornwall, and along the roads leading to these counties, their ordeal was about to begin. On 23 June 1625, the Privy Council warned the mayor of Plymouth, that the soldiers:

> Have so ill-behaved themselves in their passage to Plymouth and at Plymouth and in places thereabout… [which] is much grieved and injured by them… [For the officers] do very remissly perform their duties there in the ordering and governing of the soldiers committed to their charge.[1]

According to Captain George Took this disorder was partly since there was a delay in appointing officers to the companies. However, despite this at Lorick, in Cornwall, four soldiers who had mutinied 'were condemned and cast lots for the gallows; by which punishment with a severe proclamation backing it, there ensued a temper, their fury being reduced into the former channel of obedience'. The mayor of Plymouth was to report any unruly companies to the Privy Council and some soldiers were handed over to some, 'Netherlanders for their better disciplining'. It was not until 3 October 1625 that Sir Edward Cecil issued his articles of war to the army; nevertheless one officer claimed the inhabitants of Cornwall, 'feared to leave their houses in case they were robbed'.[2]

---

1 Lyle *APC 1625,* A letter to the Mayor of Plymouth, Sir John Ogle, et cetera, the commissioners at Plymouth, 23 June 1625, pp.99–100.
2 Captain George Took, *Cales,* p.3; BL Lansdown ms 844 Miscellaneous articles, 1558–1726 f.309.

Meanwhile it was no secret what the objective of the army and fleet's mission was, and as early as 21 July 1625 the Venetian ambassador in The Hague reported the rumours of the attack on Spain:

> Apparently, the fleet is to take a position in one of the islands, from which to prevent the passage of gold, at which the English aim above everything else, or else to invade the coast. Some believe in a design upon Cadiz, near their strait formerly captured by the English and Dutch together, and after they had abandoned it they recognised the importance of the positions, but the place is now well defended this does not seem very likely.[3]

But as the summer wore on the English fleet still had not sailed, and on 5 September 1625 an 'offensive and defensive treaty' was signed between the United Provinces and Britain against the King of Spain, which was ratified two days later at Southampton. The treaty called for 'a good number of vessels equipped and armed for war' to threaten Spain and its fleet.[4]

On 15 September Charles I and the Duke of Buckingham arrived at Plymouth to view the army gathering there.

> [They] went aboard many of the ships and at Roborough Down took a view of the whole army, using all diligence… in encouragement to all the men employed in this service and to testify his gracious affection towards the sea and land commanders, bestowed the honour of knighthood upon divers of the captains of his own ships and upon some other captains of land companies.[5]

On 4 October, five months after the army had first gathered at Plymouth and when most armies were thinking about retiring into winter quarters, the fleet finally set sail; 'notwithstanding there is such a crying out of leaks and dangers of the King's ships, which are old and unfit indeed for these seas, especially in winter'.

The fleet was divided into three squadrons, Cecil's, he being an admiral contained three Royal Navy or 'king's ships' and 25 other ships. He made the *Ann Royal* his flagship, whose captain was Sir Thomas Love. The second squadron was commanded by the vice admiral, the third earl of Essex, on the *Swiftsure*. His squadron also composed three king's ships and 26 other ships and finally was the Rear Admiral's (Earl of Denbigh's) squadron, with three king's ships and 25 other ships. Each squadron had two 'catches' attached to it, which was a small ship usually used for carrying people and messages between ships. The fleet had a total of 5,441 sailors and 9,983 soldiers on board, although none of the soldiers were on the Royal Navy's ships. Shortly afterwards they set sail, a storm blew in the Channel which forced the ships to return to harbour again.[6]

---

3   *CPS Venetian 1625–1626,* Zuane Pesaro, Ambassador to England to the Doge and Senate, 31 July 1625, p.130.
4   TNA SP 108/296 Treaty between Great Britain and the States General, 14 Dec 1625.
5   John Glanville, *The voyage to Cadiz in 1625,* pp.3–4.
6   *Ibid.,* pp.125–127.

THE FIRST BRITISH ARMY 1624-1628 (REVISED EDITION)

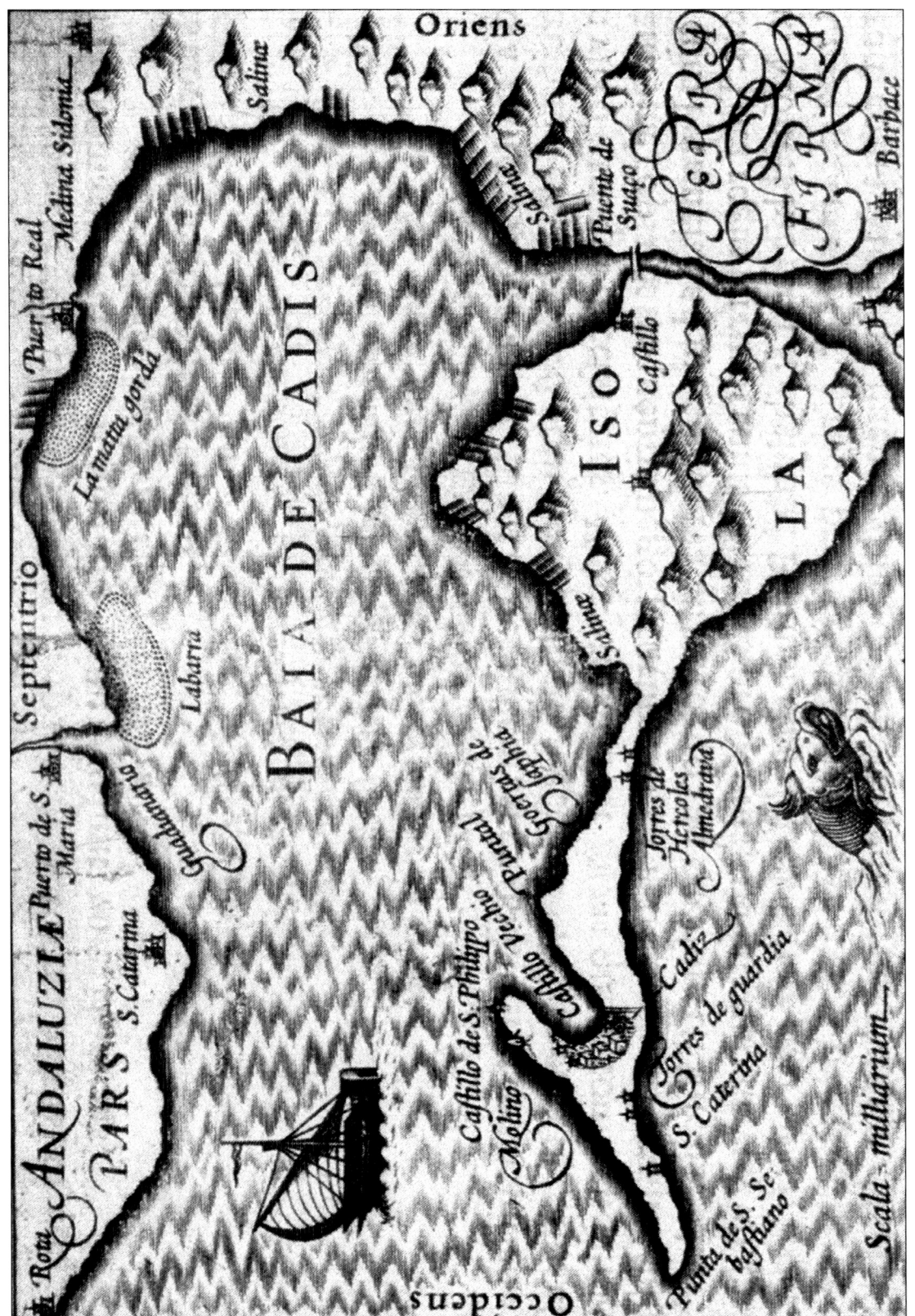

Contemporary map showing the Island of Cadiz, with the "Pillars of Hercules" on the south coast. (Author's Collection)

On 8 October with the storm abated the fleet left harbour once more destined for Spain. However, a few days later it was found that provisions were running low, which caused Cecil to order that rations were to be cut. Unfortunately, another storm blew up which lasted 38 hours and not only prevented this order being passed to all the ships but also scattered the fleet, 'our ancient seamen told us they had never been in a greater storm' recalls Glanville. Several ships were lost during the storm including the *Robert*, which sunk with the loss her crew of 37 sailors and 138 soldiers from Lord Valentia's Regiment, while other ships were badly damaged. Moreover, many of the boats which were to be used to ferry the soldiers ashore were also lost. In another ship, after the death of two soldiers and a third with sores it was suspected that plague had broken out. Fortunately for those on board it proved to be a false alarm.

During the voyage, the fleet was joined by a Dutch fleet under the command of Admiral William of Nassau. The fleet had orders to attack Spain by occupying 'some place of importance' and to hinder its commerce by capturing the gold coming from Brazil. Unfortunately, not all officers appear to have had this order and the 'place of importance' was not decided upon until 20 October when a council of war was held off the Spanish coast, whether to land at Lisbon, Malaga, Gibraltar or St Lucas, but these places were rejected in favour of St Mary Port, in the Bay of Cadiz, where the fleet could take on provisions.[7]

All this time the Spanish were watching the English fleet off the coast and so were forewarned of their coming and were able to alert the coastal towns. Even so when the English fleet sailed into the Bay of Cadiz, it managed to surprise several Spanish ships of war which were anchored off Cadiz, who believed it was the treasure fleet from the West Indies. Seeing these ships, and remembering their recent history, Essex's squadron raced for Cadiz, where Sir Francis Drake had 'singed the King of Spain's beard' in 1587 and in 1596 the second Earl of Essex had captured 12 Spanish ships laden with gold. After a brief skirmish, the Spanish ships managed to escape to Port Royal.

With the Spanish fleet out of the way it was decided to capture the island of Cadiz itself rather than St Mary Port. The town of Cadiz is at the end of an island, about 10 miles long by two miles wide in places. It is connected to the Spanish mainland by a bridge and forms part of the Bay of Cadiz. The Spanish had built a fort at Puntal, where a small piece of land juts into the harbour, and it was this fort that the fleet decided to capture. Cecil ordered the Dutch ships and the colliers from Newcastle, which were fitted with cannon, to bombard the fort. However, the colliers seem to have ignored Cecil's order because after several ships were damaged in the exchange of fire, the Dutch admiral complained to Cecil the following day that he had not received any support from the English ships. Therefore, Cecil again ordered the colliers to join in the bombardment, although he was nearly killed by friendly fire while on board the *Swiftsure*, due to the lack of training among these seamen.[8]

---

7   Ibid., p.23.
8   Dalton, *Life and times of General Sir Edward Cecil* vol.2, p.168.

It was reported that between 500 and 2,000 shots were fired at the fort by the combined Anglo-Dutch fleet before the Spanish guns within the fort were finally silenced. Therefore at about 4.00pm on 23 October, under the cover of the evening darkness a storming party, under the command of Sir John Burgh, was rowed ashore. However:

> Captain Edward Bromingham, captain of My Lord Duke's company, who coming on valiantly with his sword drawn under the fort['s] walls was slain in his boat with a musket shot and divers of his men that came in the same boat with him were killed with stones rolled over the parapet.[9]

With this repulse, Burgh decided to land his troops, including a detachment of Buckingham's Regiment, out of range of Puntal and then launch an attack on the fort, but this attack also appears to have been driven back, leaving Lieutenant Prowde wounded near the fort. A Spanish soldier then jumped over the walls and cut Prowde's throat.[10]

Early next morning with the battle for the fort still raging, Cecil was at another council of war, when according to Sir Richard Grenville, a sailor burst in and said that the Spanish had sallied out of Cadiz to attack Burgh's force. Later Cecil wrote to Buckingham, that, this force, 'so soon as they understood us they retired'.[11]

Additional troops were landed to reinforce Burgh, but when they drew up in battle formation to storm the fort, the garrison sounded a parley. The garrison, which mustered between 30–120 men were able to march out with the full honours of war, although the Spanish commander, Don Francisco Bustamante, was not allowed to take his artillery with him nor did he receive a note saying that 'he had behaved himself like a valiant and good soldier' which he had wanted. The English soldiers found that the fort was still being constructed and there were gun emplacements for up to 40 cannon, but fortunately for only eight had been installed, and two of these appear to have been dismounted during the naval bombardment. Captains Gore and Hill's companies became the fort's new garrison and provisions were landed to turn the fort into a magazine for the army. Moreover, with its capture the Anglo-Dutch fleet could anchor closer to the shore.[12]

Meanwhile, Don Fernando de Giron, the Governor of Cadiz, was at Mass when he heard that the Anglo-Dutch fleet had been sighted. He also believed it to be the West Indies fleet, but realising his mistake, ordered preparations to be made for the defence of the town. If the English had attacked Cadiz when they first arrived, the town, with its weak garrison, would probably have been forced to surrender, but Giron, who was suffering from gout, toured the town in a chair inspecting the defences and appears to have freed some galley slaves to reinforce the garrison. He also sent messages to various

---

9 John Glanville, *The Voyage to Cadiz in 1625*, p.45.
10 Dalton, *Life and times*, p.169.
11 Sir Richard Grenville, *Two Journals*, p12; TNA SP 16/9 f30 Cecil to Buckingham, 8 November 1625.
12 Glanville, *The Voyage*, pp.46–47.

THE SPANISH WAR

Spanish commanders in the area to come to his aid. The nearest of these commanders was the Duke of Medina Sidonia, the son of the commander of the Spanish Armada, who was then at St Lucar. He quickly gathered his troops together and ferried them to Cadiz and evacuated any women still in the town. Crews from some of the Spanish ships were also sent into Cadiz with a large quantity of provisions, which brought the garrison's strength up to about 4,000 soldiers, as well as the inhabitants from local towns, who arrived to help in its defence. Cadiz was quickly supplied with food and ammunition from St Mary Port.[13]

After the capture of Puntal the English began to land their forces, a task which was hampered by the loss of so many small boats during the voyage.

Another council of war was held, which discussed the soldiers' provisions, but then Sir Michael Gere, captain of *The George*, arrived and announced that large numbers of Spanish forces had been seen marching towards them. According to Cecil it was decided to march to meet this threat so that they would not be caught between the garrison of Cadiz and this force, whereas other sources state that it was decided to seize the bridge at Zuazo, which was the only route onto the island so was of strategic importance. Knowing of its strategic importance King Phillip II of Spain had ordered the bridge to be fortified before his death in 1598, and it was now guarded by Don Luis Portocamero with 2,000 men and seven guns. Therefore, Cecil decided to waste no time and set off with the majority of the English army, leaving only 600–800 soldiers who had just arrived with the Rear Admiral's squadron and so needed more time to disembark.

Unfortunately, despite their orders many soldiers had disembarked without provisions and there had been no time to issue rations before they advanced. Cecil appears to have been warned by several officers that this was the case, but when the major of Colonel Harwood's Regiment informed Cecil that three or four companies of his regiment had not eaten for several days, Cecil is allegedly have snapped back, 'That now was not the time to speak for victuals, when we were to march towards the enemy; and that want of victuals must not make men cowards'. However, Cecil had other matters to deal with in what should have been a regimental matter, although as Morgan had found that many of the officers had been appointed by patronage rather than merit so might not have had known what to do.[14]

Moreover, the officers of Buckingham's and Cecil's Regiments made certain that their men were issued with victuals, but there appears to have been little to go around since one officer of Essex's Regiment complained that 'for eighty men my land allowance was but eight little cheeses and some few biskets', while others, according to Glanville, 'had wastefully consumed at one meal that which should have served them for divers days'. Nevertheless, the Earl of Denbigh, who was left in charge of the fleet, was ordered to make

---

13   TNA SP 94/33, Anonymous Spanish account sent to the Bishop of Toledo, ff.139–141; Dalton, *Life and times*, pp.171–172.
14   Grenville, *Two Journals*, p.13.

arrangements for provisioning the army, but he appears only to have been ordered to send the victuals to the Puntal.[15]

Meanwhile, the English advanced south, away from Puntal, the bridgehead and Cadiz to the two towers on the south coast of the island, known as 'Hercules' Pillars', to await the Spanish. When the Spanish failed to appear, they continued their march towards the bridge. Essex criticised Cecil for commanding this force himself, rather than leaving it to another officer, but Essex probably wanted the glory of being the first to confront the Spanish himself.

The distance that the army marched varies according to the source. According to Cecil it was only after they had 'marched some six miles or thereabouts', when Lord Valentia, who had been riding ahead reported that the Spanish were still in front of them. Therefore, Cecil held another council of war to discuss what to do next. On the other hand, an anonymous account – probably written by Colonels Rich or Conway – states that after only:

> Half an hour's march when word was brought him [Cecil] the enemy was come out of the town and had catched some stragglers that were in a garden not far from the fort. Whereupon he sent back both Colonels Burgh and Bruce who found two of our field pieces, the horses hardly able to drawn them in great danger to be taken by the enemy.[16]

Cecil records that at the council of war they decided to continue to march towards the bridge, but that he only sent back Sir John Burgh's Regiment as it had not received any provisions since it had already disembarked when the order was given to issue the soldiers with their victuals. Although this was true, it was also a convenient excuse to cover himself for not leaving a strong enough force to guard his rear or the lines of communications.[17] Certainly both Glanville and Grenville confirm that it was only when the Spanish from the town began skirmishing with his rearguard that Cecil sent the regiments of Sir John Burgh and Sir Henry Bruce back to shadow Cadiz.[18]

According to the same anonymous account, the Spanish force in front of the army turned out to be just 100 horse, who retreated before the English advance and no opposition was encountered. Nevertheless, when they were about three miles beyond 'Hercules' Pillars' several officers, including Essex and Burgh, appear to have informed Cecil that the situation of the provisions for the soldiers was desperate, with some soldiers not having eaten for two days. In Cecil's answer to the charges against him, he denies being told this and stated that he had ordered that all soldiers should carry meat in their knapsacks. Sir William St Leger supports Cecil's opinion, when he wrote to Buckingham a few days later, 'but to speak [the] truth to your Excellency the

---

15   Took, *Cales*, p.28, Glanville, *The Voyage*, p.60.
16   TNA SP 16/10/67, Journal of the Cadiz Expedition. There were five colonels on the expedition, but the Journal mentions Horwood, Burgh and Bruce by name, leaving just these two officers. The other five regiments were commanded by more senior officers.
17   Journal of Sir Edward Cecil p.17.
18   *Two original journals of Sir Richard Grenville's* p.21; Glanville, *The Voyage*, p.59.

want of victuals was not known until we came halfway hither, the day proving very hot, our men having no water nor victuals grew very faint'. However, Cecil himself in his *Journal of the Expedition* records that 'within a mile of the bridge, the soldiers began to cry, they had neither meat, nor drink'.[19]

Therefore, Cecil ordered his men to bivouac for the night near several houses, the inhabitants having fled with their valuables. In the cellars the soldiers found barrels of wine, so Cecil gave permission for a butt (about 861 pints) of wine to be distributed to each regiment. Sentries were placed on the doors of these houses to protect the wine cellars. Unfortunately, this ration of wine was not enough for the soldiers who rushed into the houses, broke open the doors of the cellars, and drank the wine. Among the first into the cellars were the guards who were meant to protect it, and despite the efforts of their officers within a few hours the whole army fell in to a drunken stupor, in as much that no guards were set nor a password given until the early morning when the word 'Heaven bless us', was chosen. According to Sir Richard Grenville the Spanish could easily have attacked the English and cut their throats.[20]

Fortunately for the English, the Spanish did not appear and the following day, with the army in such disorder, another council of war was held, which discussed whether to continue its march or not. It appears that without accurate maps they had underestimated the distance between the bridge and the landing place, so the officers decided to retire because provisions still had not arrived from the bridgehead. The hungry soldiers withdrew the way they had come, many having thrown away their arms, either because they were too faint with hunger or hung over from the effects of the night before. Some who were still too drunk to march with the army were captured by the Spanish.[21]

Fortunately, Cecil's force managed to return to the bridgehead in safety. Even though the Venetian ambassador in The Hague in July had known that Cadiz had been fortified, it seems to have come as a surprise to Cecil when some 'slaves' informed him of the fact. Cecil rode to the town to see if it was true, only to find that:

> The bullworks were high, the town walls flanked and the ditch was 20 foot deep, cut out of the rock… There are but three ways to take any town, the first by surprise, the second by assault and the third approaches [to besiege it] and we were no ways able to attempt it by any of these means.[22]

That night the army camped in the fields, where they could be issued with their rations and early next morning, Cecil called another council of war to decide whether to besiege Cadiz itself. It was decided that town's fortifications

---

19 *Journal of the Expedition* p.18; TNA SP 16/8 f.59 Sir William St Leger to the Duke of Buckingham 29 October 1625.
20 *Two original journals of Sir Richard Grenville's* (London, 1724), pp.16–18, p.46. Although according to Took it was 'God bless us' p.30.
21 Wimbledon's answer to… charges printed in Grenville *Two original journals* p.40.
22 *Journal of Sir Edward Cecil*, p.19.

were too strong and would require a lengthy siege, which the English could ill afford. Moreover, despite the navy being in the Bay of Cadiz the Spanish ships seem to have had free movement within the bay and were able to supply the town with troops and provisions. Therefore, it was decided to evacuate the army and go in search of the Spanish treasure fleet which was expected from the West Indies any day. Cecil did toy with the idea of leaving a garrison at Puntal, but this was rejected by the council of war since it could not be easily supplied. In the end it was decided to remove the eight guns within the fort as trophies.

However, remembering some fishing boats he had seen near 'Hercules' Pillars', Sir Thomas Love, the captain of the *Ann Royal*, suggested that they try and retrieve these boats, which would prove useful for the army to re-embark. This was agreed and the army set off once more on 26 October, leaving Colonel Harwood's Regiment with Burgh's and Bruce's to continue watching Cadiz. Cecil took along with him some mariners to assist in this operation. However, they could only bring away eight of the 12 boats, so they burnt the rest. According to Glanville, amongst these boats they found one of their soldiers dead with his ears and nose cut off.[23]

This time on their return journey they were shadowed at a distance by some Spanish cavalry. Cecil decided to lay a trap for them and occupied several buildings with 300 musketeers to ambush the advancing Spaniards. However, the musketeers fired at too great a range to cause any injury in the Spanish ranks, but the sound of musketry caused panic amongst the English who thought they were being attacked by the main Spanish army.

The army was not the only ones to be sent into a panic, the fleet also saw a ship coming towards them, and fearing it was a fire ship sent by the Spanish sounded the alarm. Fortunately, they subsequently discovered that it was just an old hulk of a ship which had probably come adrift.

Meanwhile plans were drawn up to re-embark the army, but crucially they had to do so without the Spanish knowing. The regiments of Essex, Valentia and Harwood were chosen to cover the embarkation, and of these, Harwood's Regiment was to be the last to embark. They drew up on a hilly ground near Puntal hoping that the evacuation would not be discovered, but according to Sir William St. Leger, when:

> Most of the troops were shipped before the enemy did discover our retreat; but having discovered it there fell out of the town some 300 as good shot as ever I saw, and skirmished very hostile… We found the want of the use of their arms in our men; they made few or no shot to any purpose, blew up their powder, fled out of their order and would hardly be made [to] stand from a shameful flight.[24]

Geronimo de la Concepcion recalls the Spanish sent:

> Don Fernando de Giron and Don Diego Ruiz, with 1,600 infantry, sallying forth from the town [Cadiz], attacked them in the rear, killing many of them, made

---

23 Glanville, *The Voyage*, p.70.
24 TNA SP 18/8/59, Sir William St Leger to Buckingham 29 October 1625.

them abandon no small quantity of ammunition and caused them to embark more hastily than they desired. In the like manner the Duke of Medina attacked those who re-embarked at eh Isla[?] killing great numbers of them. The only loss we sustained being Don Gonzalo de Inestal, who perished in an ambuscade.[25]

During this onslaught, the regiments of Essex and Valentia disengaged the best they could, but Harwood knew that his regiment had to buy time for these two regiments to embark. Cecil sent several orders for Harwood to withdraw, but he knew he must hold out a little longer. According to another English account:

> Before Colonel Harwood's Regiment could come off, the enemy came in great number down, close up to the hills, where he had placed his musketeers (even to the giving fire to the backs of our men) he put all the musketeers of his regiment… [to] work which presently wanted and called for powder and shot sent to my Lord of Essex and Valencia for more musketeers entertaining the enemy, as well as he could with such kind of men that the other regiments before him might the better ship [leave].

After several hours Harwood ordered his regiment to retire:

> [He] commanded his lieutenant-colonel Sir Thomas Morton with the one-half of the pikes of the regiment to advance a little towards him to countenance the musketeers. Whereupon the enemy came to a stand time he took to come off and with little loss. The enemy following and skirmishing with him, until he came close under the fort where he having the advantage of an old house and placing musketeers in it and about it… [and placed] one drake, which was sent unto him by my Lord Marshal. The enemy left him and he shipped his men quietly'.[26]

Another source says several drakes were sent to support Harwood, which were hidden in an old house near Puntal. These fired on the Spanish as they drew near and slowed their advance, which gave Harwood's Regiment time to withdraw. According to Cecil:

> Notwithstanding our own men's baseness and the enemy's offence, which was well followed with excellent musketeers and reaching pieces, we got our men aboard with little loss, the enemy still pursuing, I commanded fire to be given to two drakes, which laden with shot did instantly so scatter the Spaniards that they took flight and never came on more.[27]

However, what Cecil does not say is that Sir William St Leger ordered several ships in the bay to cover the embarkation with their cannon.[28]

---

25 Geronimo de la Concepcion quoted in Charles Dalton *Life and times*, p.187.
26 TNA SP 16/522/65, Journal of the Expedition to Cadiz, 14 December 1625.
27 TNA SP 18/9/30, Cecil to Buckingham 8 November 1625.
28 Glanville, *The Voyage*, p.75.

A Spanish account sent to the Bishop of Toledo, records that the English's embarkation was not that orderly:

> On Thursday, the enemy began to embark and our [troops] charged them till within the reach of the artillery of the ships and although the weather was [un]civil we fought valiantly killing more than 350 of theirs besides many which were drowned in the fury of embarking… [our soldiers] took from the enemy many muskets and weapons and some coins which was found about the dead with the image of Queen Elizabeth.[29]

This account records that only one Spaniard was killed and seven wounded.

That night only Sir John Burgh, with 100 musketeers and some others remained on shore, trying to load the horses on board. In the morning, the musketeers also abandoned Puntal and made their way to the ships where they finally embarked. Aaccording to Glanville, Burgh had the honour of being the last one to leave, since he had been the first ashore.

With all the army on board, the fleet set sail to find the West Indies fleet with its cargo of gold as they passed Cadiz the town's artillery gave them a departing salvo. However, the Bishop of Toledo's account records that, it was:

> Reported that the enemy hath much spoil, spilling the wine and firing some houses, although it proved not so much as was thought. This day came a boat with a flag of truce and brought three of our people to change with so many of theirs, which was accordingly affected.[30]

Meanwhile, the Anglo-Dutch fleet set out to hunt down the treasure fleet and to make this task easier it spread out to wait for their prey. A week passed without incident, but on 7 November several ships saw some sails on the horizon and gave chase, only to discover they also belonged to their own fleet. For a further 20 days the English fleet waited off the South Cape of Spain, but in vain. Little did they know that several days after they had left the Bay of Cadiz, the treasure fleet arrived safely.

By now provisions were running low and many of the soldiers and crew were either dead or too sick to sail their ships, just 32 of the 250 seamen on the Earl of Essex's ship were well enough to man their stations, which had to be divided into two watches. It was decided to sink several of the smaller ships so that the crew could be used to man other ships and those who died at sea were thrown overboard. However, the ships themselves were in just as bad shape, the foreyard of the ship Sir Thomas Love was on was 'spent' in four places, while the foremast was cracked in two places and most of the sails were torn and the ropes and tackle in bad condition. While one of the masts of the ship John Glanville was on, was found to be so rotten that it split.[31]

---

29   TNA SP 94/33, Anonymous Spanish account sent to the Bishop of Toledo, ff.139–141.
30   *Ibid.*
31   HMC Cowper ms vol. 1 p.238, Sir Thomas Love to Sir John Coke; Glanville, *The voyage*, p.102, pp.107–108, 109.

Many of the ships had long since given up the search for the treasure fleet and already had slipped away. On 8 December Cecil, with just five ships left under his command, finally gave the order for them to return to England. However not all ships made it to a British port, the *Mary Constance* with 120 men on board sank, with only the officers being saved.

One of those ships which slipped away was the *Swiftsure*, which had the Earl of Essex on board. This arrived at Falmouth on 5 December, and Essex lost no time in hurrying to present his view of the campaign to the king.

Meanwhile it was rumoured in England that Cadiz had been captured, albeit with the loss of 400 men, but at Plymouth Sir John Eliot soon found out the truth:

> The miseries before us are great and great the complaints of want and illness of the victual. There is now to be buried one Captain Bolles [of St Leger's Regiment] a landsman, who died since their coming in… The soldiers are not in better case, yesterday fell down here seven in the streets. The rest are most of them weak and unless there be a present supply of clothes there is little hope to recover them in the places where they are lodged.[32]

On 11 December Glanville's ship dropped anchor in Kinsale Harbour in Ireland, one of the last ships to arrive home. With the fleet once more back home, the recriminations began as to whose fault it was that the expedition had failed so dismally. According to Joseph Mead, Cecil was accused of 'not only to have [a] want of judgement but to have been wilfully faulty, yet others think he will come off easily enough', whereas John Chamberlain put it, 'Some lay the blame on the design or council and the soldiers on the General Viscount Sitstill (as they now style him), he on the sea, but most on his Grace [Buckingham] and he on Sir Thomas Love and so from post to pillar'.[33]

Many blamed Buckingham, because as Lord High Admiral it was his duty to command the fleet not Cecil, but fortunately for the Duke he had the King's protection. There were demands for an inquiry into why the expedition failed in its main aims, and of course those officers whose reputations had been stained by the campaign pointed their fingers at Cecil; whose ship had been forced to sail to Ireland and so he could do little to defend himself.

The Secretary of State, Sir John Coke, began to gather the evidence to discover the truth. A Captain Levett listed the following reasons for the 'evil success' of the expedition:

> For delaying for so much time before they sent away the fleet.
> For sending it away in Winter when most of the ships were not able to carry forth their ordnance either by way of offence or defence.
> The sending away the fleet with so little provisions.
> For not giving the captains their orders or commissions before they went to sea by which much confusion might have been prevented.

---

32 Quoted in Charles Dalton, *Life and times*, pp.194–195.
33 BL Harl ms 390 f26 letter from Joseph Mead dated 11 March 1626, John Chamberlain wrote to Sir Dudley Carlton, 19 January 1626: *Letters of John Chamberlain* vol. 2, p.628.

> For sending unexperienced soldiers such as was neither willing nor able to do service, but, on the contrary, mutinous.
>
> That no course was taken to block up the galleys in St Mary's Port, which might have been done with six ships by which neglect they did continually carry soldiers into the town [of Cadiz].[34]

Another point can also be given as to why the expedition failed was that when the fleet sailed, they had no other plan but to 'hurt' Spain, as the Venetian ambassador put it. Also the loss of the longboats which were needed to land the soldiers was a major blow, but all sources agree that it was the soldiers getting drunk that cost the English the campaign. However, Cecil in his report to Buckingham failed to mention that he had initially ordered some of the wine to be distributed to the soldiers.

All could have been different if Cecil had been more decisive, although his hands had been tied because his commission said that he had to seek his colonels' advice. Moreover, if he had not been given two completely different objectives, the outcome of the campaign might have been very different. Despite the findings of the inquiry, Charles decided that no further actions should be taken and therefore no lessons were learnt and Cecil was rewarded with the title Viscount Wimbledon as he had been promised.[35]

However, with the unsuccessful expedition against Cadiz, fear began to grow that the Spanish might invade Britain. In January 1626 it was rumoured in England that the Spanish General Spinola was at Dunkirk, Nieuport and Ostend, with 'all his forces' and on 13 February 1626 the Venetian government wrote to their ambassador in England, 'Our ambassador in Spain informed us of the resentment of the Catholics at the attempted surprise of Cadiz by the English fleet and that he intended reprisals and was preparing a fleet'.[36]

King Charles was informed of the build up of the Spanish fleet, and to counter this new threat two fleets were to be raised, one 'to put to sea to divert or break the joining of the enemy's forces… and another greater fleet preparing for the guard of the coast and Narrow seas'.

However, just 4,628 of Cecil's army had returned to England out of about 10,000 men, which amounts to about a 55 percent casualty rate. Some companies suffered more than others, with some having just 12 or 13 men while others had between 80 and 90 men. Moreover, by April 1626 the soldiers of Cecil's army were owed £11,000 in arrears, and their pay came in slowly; £2,910 was ordered to be paid on 6 April 1626 to those at Plymouth and a further £900 on 20 May 1626 with an additional £900 was ordered to be paid on 20 June 1626.

Since Cecil's army was not strong enough to repel an invasion alone if or when it came, the country would have to rely on the militia or trained bands which were summoned in times of emergency. Each county had its own regiments of trained bands for its own defence, the number of which varied

---

34 HMC 12th Report Appendix one Coke ms vol. 1, pp.243–244.
35 Hinds, *CSP Venetian 1625–1626*, letter Marc Antonio Morosini, Venetian ambassador to Savoy to the Doge and Senate, 3 May 1626, pp.537–538.
36 *Ibid.* p.320.

from county to county, so that Somerset had five regiments, Wiltshire four and Surrey three. The London Trained Bands, which is usually considered to be the most professional force, having four regiments, known as the North, South, East and West. Each county also had at least one troop of horse. In theory the trained bands were to meet on holidays and other convenient times to practise company drill, however this had long been neglected and despite various initiatives over the years to improve the militia nothing had been done. According to Cecil, 'the trained bands being no soldiers either in discipline or anything else'.[37]

Efforts had been made to improve the trained bands during the reign of James I, which called for all the bore of the muskets and the lengths of the pikes of the trained bands to be the same size 'for the uniformity of training', but it did not say what the new regulation size should be. Moreover, the years of peace had made the trained bands lazy in their duties and from May 1620 the Privy Council ordered that each county should hold a general muster of their trained bands and that the soldiers were to be issued with 'serviceable armour and weapons', where appropriate and the vacancies that existed in the officers' position were to be filled with 'sufficient and fit persons'. Despite this order many militiaman still continued to use their old defective weapons.[38]

Nevertheless, the reforms pressed on and in 1623 the *Instructions for Musters and arms and the use thereof* was published, based on the Dutch military practice which became the official training manual for the trained bands throughout the country. Despite these orders, training was sometimes interrupted due to other reasons, for example on 28 November 1625 the Lord Lieutenant of Surrey reported that he could not muster the trained bands of Southwark and the surrounding area because of the 'great infection of the plague' and that many were 'either dead or departed from thence'. It was not until 1626 that the county of Leicestershire's accounts record, 'paid to a scrivener for writing of five book[s] of the postures for the exercising of the soldiers, 10 shillings'.[39]

As well as the new training manual, 84 sergeants from the Low Countries were to be distributed to various counties to instruct the soldiers in their drill. Several captains who had served in the Low Counties were also sent into various counties to train the trained band cavalry. The sergeants, at least, were to stay in their appointed counties for three months, although some stayed longer. In Suffolk two sergeants trained the militia in detachments of 200 men over four days between late February and early March until all those that made up the trained bands had received proper instruction in

---

37   Lyle, *APC 1626*, A list of what counties and what numbers of men within each county are to be in readiness at such certain notice and alarm given of the approach of an enemy towards them, 15 July 1626 pp.89–90; BL Harl ms 390, Letter from Joseph Mead, dated 28 January 1626, f.17; TNA SP 16/522/86, A list of the companies which already arrived in Cornwell and Devon from the Expedition to Cadiz, December 1625; BL Royal 18 A LXXVIII, Memorial by Sir Edward Cecil to Charles I, 1628.

38   Lyle, *APC 1620–1621*, A minute of letters to the Lord Lieutenants and Commissioners of musters of the several counties, 31 May 1620, pp.215–216.

39   *Ibid.*; TNA SP 16/10/45, 61, Charles, Earl of Nottingham to the King, 28 November 1625 and Sheriff and Deputy Lieutenants of Nottingham to the Council, November 1625.

their chosen weapon. On 2 October 1626, the chief constables of Suffolk were ordered to hold a muster of the trained bands within their Hundred 'upon the ninth day of October by nine of the clock in the morning at such place as you shall think most fit… with their arms completely furnished, there to exercise and train [for] two days and so every five weeks until you shall have notice to the contrary'. For these two days training a lieutenant would receive 20s and a common soldier 1s 4d. The musketeers would also receive a pound of gunpowder and two yards of match. These instructions were probably not unique to Suffolk. All the county's trained bands were ordered to muster on the same day to prevent any militiamen lending their arms to other companies. These measures appear to have brought about a marked improvement in the standard of the Militia at this time.[40]

However, it was the attitude of the militiamen themselves that made the trained bands so poor. In 1635 Thomas Palmer records that these drill sessions were:

> Sometimes so trivial, so superficial, that it improves rather our sport than our skill. The ignorant rustics wink when they give fire. The captains wink at it, and the spectators laugh at it. Such a military discipline is fitter for a May game than a [battle]field. It may be heartily wished that their martial instruction was more industrious their training more frequent and serious.[41]

With the threat of invasion, on 15 July 1626 the Privy Council drew up a list of how many men aged between 16 and 60, they expected each county to supply if the enemy fleet was sighted in the Channel. On paper this forced was estimated at between 118,400 and 120,400 men, but Berkshire, which was to supply 3,000 men, complained that it could only arm 1,000 men because, 'arms cannot be obtain'. Derbyshire could only manage 400 trained soldiers and 400 untrained men in the event of a 'sudden occasion', no doubt these counties were not alone.[42]

On the other hand, some counties were better organised, Bishop Neale of Durham examined the trained bands of his county and found them 'well experienced in the exercise of their arms', and that the arms were 'reasonably complete, but some of the common corsletts [pikemen] heretofore allowed of at musters are not (as is required) according to the modern fashion'. William, Viscount Mansfield of Nottinghamshire boasted that his trained bands were, 'completely furnished every way after the modern fashion.'[43]

---

40 TNA SP 16/522/105, Petition of Captain John Gunter; SP 16/88/48, Council to Secretary Conway, Lord Lieutenant of Hampshire, 1627; SP 16/124/24, Minute of petition to the king from Sir Thomas Longfield and others, 1628; BL Add ms 39,245 Muniments of Edmond Wodehouse.
41 Thomas Palmer, *Bristol's Military Garden* (1635), p.29.
42 TNA SP 16/13/43, Suggestions for the more effective arming and exercising of the Trained Bands, nd; Lyle *APC 1625–1626* p.37, p.89; Lindsay Boynton *The Elizabethan Militia* pp.246–248; TNA SP 16/33/66 William Viscount Wallingford to Conway, August 1626; SP 16/34/80 Bishop Neile to Council, August 1626; TNA SP 16/35/103, William, Earl of Devonshire to Council, 16 September 1626.
43 SP 16/3316/33/66, William Viscount Wallingford to Conway, August 1626; TNA SP 16/34/80, Bishop Neile to Council, August 1626.

However, serving in the trained bands had one main attraction, they were exempt from being conscripted into the army. Therefore, during Elizabeth's reign measures were taken to root out 'base, poor men', within the trained bands who had 'crept in there to hide themselves for foreign service'. Instead, they were to be filled with 'gentlemen, or farmers and the best enabled yeomen and husbandmen'. On 29 August 1605, the deputy lieutenants of Somerset were ordered to 'release those poor men already enrolled in the trained bands, who in regard of their poverty and inability are ready to sink under the burden of so great a charge'. In another undated letter the deputy lieutenants were to 'take care hereafter that poor men be put out of the trained bands and the better sort taken in their places'. Those that were to be enrolled were to be 'gentlemen, yeomen, and other householders'. A warrant from the Privy Council during the 1620s called for the men of the cavalry to be 'able and serviceable, not of the rustic sort altogether ignorant of the use of arms and no way capable of instruction, but that it be especially recommended unto the gentry to set men upon some quality upon their horses'. Officers would be chosen on their social standing within the county.[44]

In addition, on 14 February 1626 Edward Harwood recommended that those in the trained bands 'pay no extraordinary loans, nor taxes' and 'shall not be subject to arrest for debts'. Furthermore, they should have a 'place of honour assigned [to] them in their several parishes'. The militiamen were to be identified by wearing 'some order on their outward garments, the yeoman in cloth, the gentlemen in silk and the officers in silver and gold'. It is not known if Charles took these recommendations into considerations.[45]

As early as May 1624 plans had been prepared in case England was invaded. These included a survey of all fortified strategic positions within the country and that beacons were manned so that they could be lit if an enemy fleet appeared. When these beacons were lit a body of 1,000 horse and foot would rush to the scene where an invasion was expected to land, either prevent a landing altogether or delay it enough so that the trained bands in the surrounding counties could muster and march to the scene. Moreover, the local population was to be evacuated and if necessary, all the 'corn and fodder which they cannot carry away be burnt', and 'the principal places of passage and bridges… be kept with a strong guard and as occasion serve the bridges broken down'.[46]

There was also a fear that the Spanish might invade Ireland, many of the inhabitants being Catholic and with whom they had strong connections through trade, but apart from the companies from Cecil's army who had landed there, the Irish army in November 1625 mustered just 395 horse and 3,600 foot which was spread in garrisons throughout Ireland, and the 'pay of

---

44　*HMC Salisbury* vol. 12, p.181, p.478; W. P. D. Murphy, *The Earl of Hertford's Lieutenancy Papers, 1603–1621,* p.72 (Devices, 1969); TNA SP 16/88/48, Petition to the King from Sir Thomas Longfield and others, 10 August 1626.
45　TNA PRO 30/53/9/10, manuscript in the hand of the first Lord Herbert of Cherbury, nd ?1603–1630.
46　BL Kings ms 265 f.345.

all is much in arrears… the foot are in decay and the troops more ready to mutiny than to fight'.[47]

On 16 June 1626 Joseph Mead wrote, 'there is much said of the Spanish preparations against us, which the hand of God must dissipate, in whose power alone it is'. With invasion imminent on 10 July 1626 the Privy Council sent a circular to the lord lieutenants warning them of the 'great and threatening preparation made at this present [time] both in Spain and Flanders' for an invasion. They were to issue to the trained bands gunpowder, match, shot and knapsacks and for every 1,000 soldiers, 100 pioneers were to be raised who were to be provided with spades, pickaxes, shovels, hatchets, bills and other tools, each county was also to provide 'nags' to mount the musketeers on 'for the more speedy conveyance of the foot'. If an invasion did come then the trained bands should 'march instantly' with 10 days provisions to their rendezvous, which were usually the ports along the coast.[48]

At the end of July word came that an invasion was imminent and according to Joseph Mead:

> A warrant came to the city [London] to arm 4,000 men even to their knapsacks (over and besides the trained bands) and all to be householders and ready at an hour's warning to go quarter themselves in the Isle of Sheppey, but their was made a double demure one because the letters are from some of the lords and not the king, secondly by their charter they [the trained bands] are for the defence of the city, not to go further than the lord mayor goes, unless it be for the guard of the king's person.[49]

Also, about this time the new fleet was ready to sail out to meet any new threat, but Buckingham as admiral was away on a diplomatic mission, and his vice admiral had retired to his estates. Therefore, the command fell to Captain Pennington who was the senior captain. Fortunately, the invasion scare was a false alarm and the London Trained Bands and the fleet were not needed after all.

Meanwhile, Devon and Cornwall had been billeting the soldiers of Cecil's army since their return from Spain. Therefore, on 24 August 1626 instructions to relieve these counties from the burden the Lord General's, Harwood's and Rich's Regiments were ordered to be sent into Kent, while the Marshal's and Valentia's Regiments were to go to Sussex. St Leger's, Essex's and part of Bruce's Regiments had the easiest journey because they were to be quartered in Dorset, and Conway's and the remainder of Bruce's Regiment in Hampshire. By distributing the army in the coastal counties, they could also support the trained bands in the event of an invasion. It was decided

---

47 *CSP Ireland 1625–1632*, pp.49–50 The numbers breakdown as follows Ulster 150 horse and 1,400 foot, Connaught, 50 horse, 500 foot, Munster 75 horse and 750 foot and Leinster 120 horse and 250 foot.
48 Lyle, *APC 1626*, Directions to the Lord Lieutenants of several counties, 10 July 1626 and 15 July 1626, pp.71–73, pp.87–90.
49 BL Harl ms 390, Collection of letters of Joseph Mead, Jan 1626-Apr 1631 f.99.

THE SPANISH WAR

that Cecil's army should be reduced to four regiments, although this was later changed to five, and brought up to strength again.[50]

As the summer turned to autumn and then to winter there was still no sign of the Spanish invasion fleet, but then, in November, a rumour began to spread that the Scilly Isles had been invaded and fortified by Dunkirkers and that this was just the beginning of Spain's conquest either of England or Ireland. This rumour was soon proved to be incorrect and that several ships had just anchored off the Scilly Isles before sailing away again.[51]

Certainly, the Spanish wanted revenge on the British for invading their territory, but they also feared another invasion, 'Their anxieties', wrote the Venetian ambassador on 8 September 1626 from Madrid:

> about a fleet from England and other allies increase. They say that the king there at first intended to give leave to all his subjects to fit out privateers against the Spaniards, but now he is wanting them all in a formal fleet. Here they are accordingly hastening their preparation. In Seville, besides additional fortifications and munitions at Cadiz, they are equipping 21 galleons under three commanders, and they expect 15 more from Naples and Sicily. They are also making provisions at Lisbon. They wish to defend these realms and also to send to meet the [gold] fleet, although it has an escort of 21 armed galleons. They say it will bring eighteen million in gold, plate and goods, and it is thought they have orders to follow a different route from the ordinary.[52]

With no invasion materialising and bad weather keeping many trained bands men away from the summer muster of 1627, the standard of the militia began to decline once more. The following year Charles lost patience with the counties who had failed to improve the trained bands. On 10 January 1628, the Privy Council sent out directions to the lord lieutenants of various counties calling the trained bands to make 'warlike preparations' being at this time 'generally so ill provided and furnished… many for the saving of charges borrow horses and arms to show as their own'. Many did not even bother to borrow arms. The horse was to rendezvous at Hounslow Heath on 21 April at 9.00am to be reviewed by the King himself, with 'all the men, horse and arms be all fit for service'. Before then the lord lieutenants were to see that they were drilled two or three times a week. Charles appointed persons 'of trust' to review the horse of other counties. Nothing was said of the trained bands' companies of foot.[53]

However, there was no review because it 'might have been very chargeable to the country'. Instead, the trained bands were to muster in their own counties and that the deputy lieutenants erect beacons and 'a convenient number of nags [be provided] to mount shot [musketeers] upon'. The deputy

---

50 Lyle, *APC 1626*, Letter to the Deputy Lieutenants of Devon and Cornwall, 24 August 1626, pp.216–219.
51 BL Harl ms 390 Letter from Joseph Mead, dated 1 December 1626, f.167.
52 Hind, *CSP Venetian 1625–1626*, Marc Antonio Correr and Anzolo Contarini to the Doge and Senate, 16 September 1626, pp.537–538.
53 Lyle, *APC 1628*, A minute directed to the Lord Lieutenants of Several counties, 10 January 1628, pp.227–228.

lieutenants were to report back to Charles by 10 September, but like all the other instructions relating to the trained bands this one was probably also ignored. Sussex was typical of this attitude, in 1627 the Trained Bands mustered for training for three weeks but the following year it trained for just four days.[54]

However, as 1627 began with no sign of any invasion from Spain and with relations between France and Britain deteriorating, Charles and Buckingham looked at launching another expedition.

---

54   *Ibid.*, A minute concerning musters directed to the Lord Lieutenant of the various counties, pp.470–471; Fletcher, *Sussex,* p.181.

THE SPANISH WAR

Uncock your match

12

The Île de Rhé Expedition

Since 1562 France had been torn apart by its Wars of Religion, fought between the Catholics (Royalists) and the Huguenots (Protestants). The Huguenots leaders, Henri de Rohan and Benjamin de Soubise, were forced to sign the Peace of Montpellier on 19 October 1622. This peace forced the Huguenots to surrender all claims to approximately 100 fortified towns which had been captured by the Royalist army and there was to be no political gatherings. Only La Rochelle remained of the once vast Huguenot territory, the city itself was considered to be impregnable.

However, in 1625 Rohan again raised a Protestant army and encouraged the French provinces of Poitou and Languedoc to rise up, while Soubise conducted a naval campaign against the French fleet. They also appealed to England for help, but not only was it distracted with a war against Spain, but also Anglo-French relations had greatly improved, with the Louis XIII's sister, Henrietta Maria, marrying Charles I. Therefore, Charles rejected supporting the Huguenots, but lent the French king, Louis XIII, English ships. It was even rumoured that Charles might send troops to support Louis as well, much to the English Parliament's disgust, which blamed Buckingham for this support and demanded the Protestant religion in La Rochelle should be protected.

Despite several initial successes there was not a mass uprising as Rohan and Soubise had hoped, and the Huguenots were once again defeated, and La Rochelle besieged. By now the inhabitants of the city were weary of war and on 6 February 1626 the Peace of La Rochelle was signed bringing an end to the conflict. Soubise was offered a royal pardon in return for his loyalty, but he fled to England instead.

After the Peace there was a decline in Anglo-French relations, Charles hated his new French bride, although they would eventually come to love each other, and sent her court back to France. Neither did Charles introduce the religious freedom to Catholics which had been promised during his marriage negotiations. Why did Britain go to war with France in 1627? The official reasons were:

- France's sudden retracting permission to Mansfeldt's army landing in France, which cost the lives of 12,000 Englishmen.

- France's unsatisfactory conduct when Charles tried to intercede on the Protestants' behalf over the disadvantageous peace.
- Secretly making peace with Spain over the affairs in Italy and protecting their merchandise, which frustrated English efforts in her war with Spain.
- Using English ships against the Huguenots.
- Abusing 'His Majesty's patience' by France pretending to negotiate for peace and seizing ships in Bordeaux.[1]

Of course, many of Mansfeldt's men were sick even before they reached their rendezvous at Dover, so they would have died anyway. Moreover, Mansfeldt had the reputation for plundering friend and foe alike and Charles and Buckingham had stood idly by during the Huguenot uprising. They probably knew that the English ships would be used against the Huguenots, and this was one of the charges brought against Buckingham in Charles' second Parliament. Buckingham had ordered Captain Pennington's naval squadron and privateers in the English Channel to seize French ships, so the French action in this respect was just tit for tat. Finally, it was incompetence and lack of resources which caused Cecil to be defeated, rather than any French involvement. Therefore, these reasons for war were nothing more than excuses.

According to Buckingham's biographer, Roger Lockyer, the reason why Charles went to war was because of a French military build up, including 55 ships which would patrol the English Channel and threaten the English navy, and there was even the possibility of an invasion of England![2] However, Buckingham had been branded a coward for not leading the army to Spain and there were calls for his impeachment, so a war with France would not solve this and would even enhance Buckingham's reputation of being the saviour of Protestantism in France. However, Britain could not afford the two wars it was already involved in, let alone a third, but no doubt Rohan and Soubise persuaded Charles and Buckingham that if they did launch an invasion of France, then the Huguenots would rise up to support them.

Buckingham's commission appointed him 'Admiral, Captain-General and Governor the Royal Fleet with such soldiers and land forces as shall be conveyed therein'. This force was intended to fight against the king's 'brother in law's [Frederick V] and sister's enemies or the enemies of the Crown of England'. He had been criticised for leaving the expedition against Spain to Cecil, but now he began openly to boast at court 'without either fear or wit', records one eyewitness that he was going to lead a new expedition this time to France, 'whereby the worst of our ill willers… might know all'. A Florentine agent agreed with this statement when he wrote, 'by undertaking this [expedition], he [Buckingham] will make so favourable an impression on the people that he will regain their confidence'.[3]

---

1   TNA SP 16/527/53, A summary relation concerning the causes of the Île de Rhé Journey by the Duke of Buckingham, nd, 1627.
2   Roger Lockyer, *Buckingham* (London: Longman, 1981), pp.359–370.
3   Rushworth, *Historical Collections,* p.429; BL RP 8006 Anon account of Île de Rhé Expedition, although credited with a William Fleetwood when it was published. Florentine agent's account

Therefore, on 14 March 1627, the regiments of Cecil's old field army, who had remained idle in their quarters since their return from Cadiz at the end of 1625, were ordered to rendezvous 'with all expedition' for 'the defence of his Majesty's Kingdom'. The companies quartered in Sussex were to meet at Chichester and those in Hampshire at Southampton, before marching to the general rendezvous at Portsmouth. The soldiers were to march 15 miles per day to their rendezvous, so that 'they may not fail to be there the 27 of this present month'.

The Deputy Lieutenants of the counties where the regiments were quartered were to take measures to return the soldiers who had deserted to their regiments by 'strong and diligent watches… upon all the passages and usual ways'. Any band of deserters discovered the ringleader was to be imprisoned and the rest returned to their regiments. Likewise, any vagabond pretending to be a soldier was to be imprisoned.[4]

Money was short for this expedition and Buckingham is said to have cut off some diamond buttons on his clothes to help pay for it, plus there was the cargoes that Pennington had seized from French ships, which was estimated to be worth £130,000.[5]

However, there does not appear to have been a centralised system funding the war, because although Captain John Mason was officially responsible for distributing the pay to the soldiers, in May 1627 he complained that he had paid the captains in Dorset five months' pay for their companies, but he soon discovered that they had also received 1,000 marks from another source while other regiments had gone without. The soldiers had been happy to live off the inhabitants of the town where they were quartered, but now they were mobilising for war they wanted their pay, which forced Mason to warn Buckingham on 27 May 1627, that:

> The pay must be without fail [delivered] otherwise mutiny and disbandment will follow which was hardly prevented in Sir John Burgh's Regiment at Winchester by reason of eight weeks pay arrears and to the poor billets of that town, chiefly caused by the default of Berkshire loans not supplied.[6]

While his men were owed their arrears of pay, Buckingham was spending £10,207 1s 6d on equipping himself and his servants for the forthcoming campaign. These 'essentials' included £20 6s for perfume and £367 8s 6d for a silver perfuming pan. He was a wealthy aristocratic and other personal campaign expenses of generals have not survived, but these items suggest more of Buckingham the courtier, rather than Buckingham the army commander.[7]

---

quoted in Thomas Cogswell 'Newsbooks and the Duke of Buckingham' p7.
4   Lyle, *APC 1627*, Privy Council to Deputy Lieutenants of Kent, Sussex, Hampshire, Dorset, and Berkshire, 14 March 1627, pp.130–131.
5   Lockyer, *Buckingham*, p.368.
6   TNA SP 16/62/3, Captain John Mason to the Council, 1 May 1627; SP16/64/75 Captain John Mason to Buckingham, 27 May 1627.
7   Lockyer, *Buckingham*, p.373.

THE ÎLE DE RHÉ EXPEDITION

On 5 June, the king left London for Portsmouth and was followed by Buckingham on 13 June with money to give his soldiers four months' pay. Buckingham divided the main armada, which consisted of 100 ships, although only 10 belonged to the Royal Navy, into three squadrons. He would sail on the *Triumph* which bore the standard of England in the main top mast, and the admiral's red standard. The Earl of Lindsey as vice admiral would sail in the *Rainbow* bearing the king's usual colours in the fore top and a blue flag in the main top, and so was admiral of the blue squadron. While Rear Admiral Lord Harvey was in the *Repulse* bearing the king's usual colours in his mizzen mast and a white flag in the main top and was admiral of the squadron of white squadron. The Earl of Derby was admiral of a squadron with the English flag of Saint George on the main top. Derby's squadron would protect Buckingham's right wing, while Captain Pennington's squadron protected the left wing. On the main top of Pennington's ship was the Scottish flag of Saint Andrew.[8]

Strengths of Buckingham's Army, June –October 1627

Regiment	Number of Companies	27 June 1627 embarked at Portsmouth	20 July 1627	26 October 1627
Sir John Burgh	10	800	787	
Sir Charles Rich	10	800	749	617
Sir Alexander Brett	10	770	668	572
Sir Edward Conway	10	791	704	596
Sir William Courtney	10	800	758	663
Sir Peregrine Bertie	5	400	348	638
Sir Thomas Morton	10	800	738	596
Sir Henry Sprye	10	773	773	780
Sir Piers Crosby			841*	827
Sir Ralph Bingley			823*	823
3 coys of Col. Bertie			235*	
Colonel Hawley				772
	75	5934	7424	6884

On 20 June Buckingham ordered his troops to begin embarking on the ships. Three days later the king left for London after saying goodbye to Buckingham. On 24 June, the Duke boarded the *Triumph* and the following day the fleet weighed anchor. Two days later the ships' captains opened their

---

8 BL Add ms 9298.

THE FIRST BRITISH ARMY 1624-1628 (REVISED EDITION)

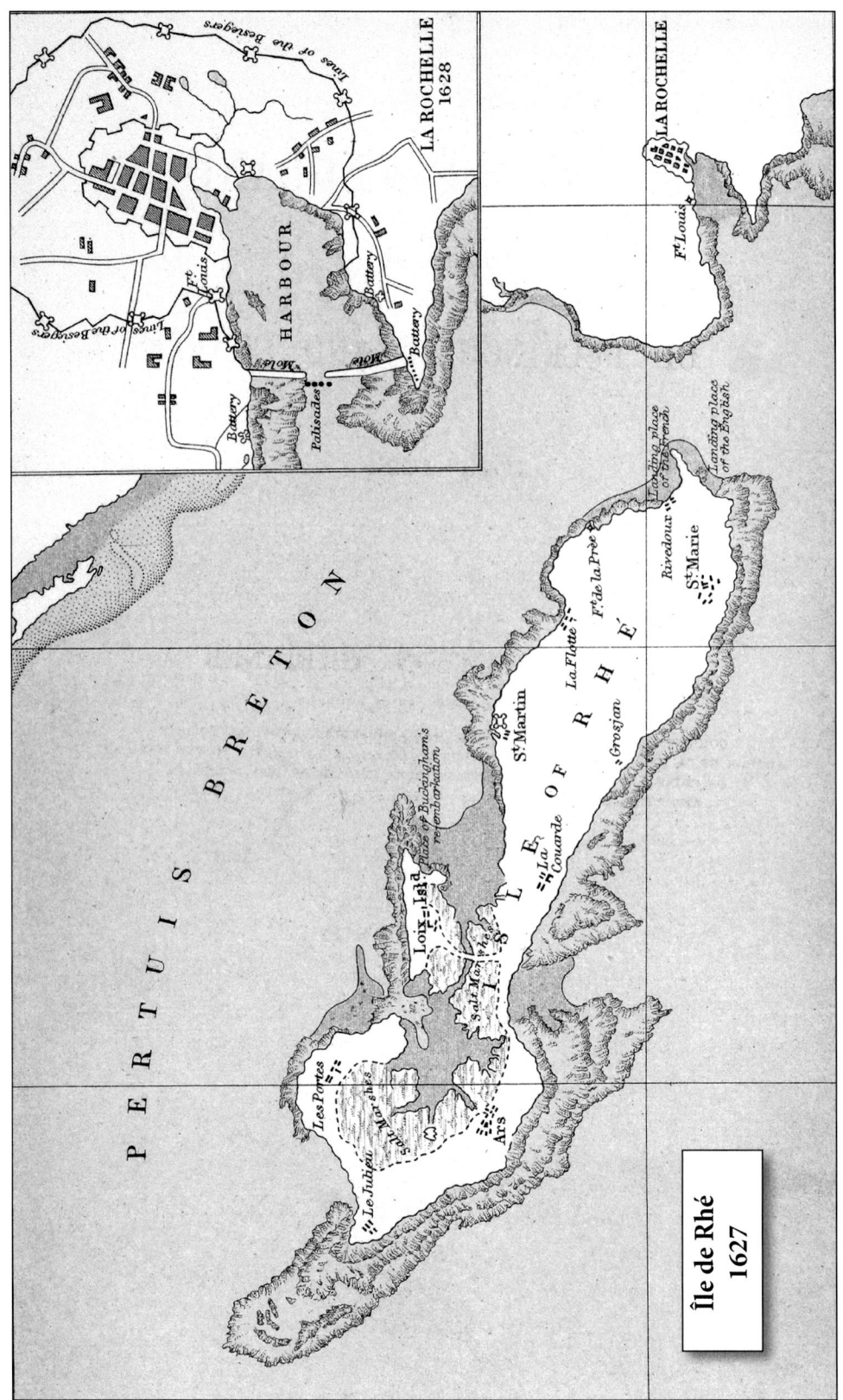

Map of the Isle de Rhe, (Gardiner's *A History of England under the Duke of Buckingham and Charles I* vol. 2)

sealed orders on their destination. However, since Buckingham had boasted at court what he had planned to do these orders were hardly secret.⁹

During the voyage, the fleet skirmished with some Spanish ships disguised as Dunkirkers which were beaten off, but in the ensuing pursuit the English ships were scattered. On 4 July Buckingham held a council of war, where it was decided to land on Île de Rhé on the west coast of France near to La Rochelle. Despite the soldiers being allowed 21 days rations when they embarked the council of war heard that they had not only eaten these rations, but needed to be re-supplied urgently.¹⁰

On 8 July, a storm scattered the English fleet and caused the *Nonsuch* to lose her fore mast. Therefore, it was not until the 10 July that the first ships of the English fleet weighed anchor off the Île de Rhé, and the remainder of the fleet arrived the following day.

The Île de Rhé is a small hook shaped island about nine miles long and 1¼ miles on the west side and 1½ miles on the east side wide. On the north coast of the island are the towns of Saint-Martins, which was the principle town on the island, la Prèe and La Flotte. These towns each had earthworks for protection, the one at St Martins being the strongest. On the north-western part was St Mary and to the east, Ars. Just off the south coast is a smaller island called the Isle de L'Oye. The island was known for its salt and wine industry, so it would be impossible for an army to live off the land for any length of time when its provisions were exhausted.

On 11 July fort at la Prée was bombarded by the *Abraham* and the *Convertine* and the following morning a council of war was held, where it was decided to send Sir William Beecher to La Rochelle with Buckingham's *Manifestation* which recorded that the king's aim was to rally, 'the [Huguenot] Churches, his interests is their good and his aim their contentment'. Certainly, the Huguenots would have welcomed British aid several years before but now they only wanted peace. Beecher returned shortly after to report that the mayor and council declined to see him because they were celebrating a fast. However, as Rushworth records, 'Rochelle who once much longed for their [English] coming…now shut their gates at their appearance'.¹¹

Despite their diplomatic language it must have dawned on Buckingham that without the help of La Rochelle there was no hope that the campaign could be a success. Neither could he return to England without accomplishing anything because he had seen how Cecil's reputation had suffered due to his failure at Cadiz.

Nevertheless by 3.00 a.m. on 12 July Buckingham had risen from his bed and at 5.00 a.m. received the sacraments. At 7.00 a.m. Buckingham asked Beecher to return to La Rochelle, this time with Soubise and Saint Blancart, who were two of the few Huguenots still willing to continue the struggle and who had accompanied the English fleet. However, it was only reluctantly

---

9   BL Add ms 26,051 Journal of the Voyage of Rease 1 May–7 November 1627. Another source says the soldiers embarked on 27 June and it was the 1 July when they opened their sealed orders.
10 Lockyer, *Buckingham*, p.380.
11 Rushworth, *Historical Recollections*, pp.467–469.

that the town admitted the three of them due to the intervention of Soubise's mother, the dowager Duchess of Rohan. Beecher read out Buckingham's *Manifestation*, but although it was well received by the inhabitants, the mayor and council decided to consult with the other Huguenot towns before they gave their support to the English. Therefore, Buckingham would have to wait for his answer.[12]

Meanwhile at 8.00 a.m. Buckingham ordered Richard Graham, his gentleman of horse, with a small servant boy and six musketeers to go ashore to see if the chosen landing place, Pointe de Sablanceaux, was clear of any French forces. Probably to avoid suspicion only the boy was sent ashore to scout the surrounding countryside. He came across three horsemen who chased his back to the shore, where he scrambled on board the boat again and returned to the fleet.

On hearing the news that the shore was deserted; Buckingham held a council of war where it was decided that the army should land that afternoon. At about 2.00 p.m. with the ships 'within a musket shot' of the shore the soldiers began to disembark, while the musketeers on four ships would cover the smaller boats while they ferried the soldiers ashore. However, many of these boats had been lost during the storms while being towed by the larger ships, so it was a slow process. As one eyewitness put it, in 'the space of three hours landed…[were] landed near upon twelve hundred men'.[13]

It was planned that Sir John Burgh's and Sir Charles Rich's Regiments would be landed first, followed by Conway's and Courtney's Regiments, then Beatie's weak regiment and Morton's so that Buckingham's infantry would form five battalions of foot, although only three battalions were landed that day. Each soldier was issued with enough provisions and ammunition for two days.[14]

Some soldiers were reluctant to leave the safety of the ships while others were eager to get ashore either from excitement or to relieve their sea sickness, but 'all things were done in confusion and tumult [and] there was no time to put the soldiers in order'. This is confirmed by another eyewitness who recorded that Burgh's and Rich's Regiment, 'being unranked and scarce stepped ashore' when the French appeared. Presumably the three riders who had chased the boy that morning had reported the incident to their superiors.[15] According to the *Journal of the Proceedings of the campaign*, Burgh's and Rich's Regiment were forced back in disorder by French cavalry, onto the soldiers of Conway's Regiment who were disembarking from their boats.

However, according to Conway's account, Burgh's, Rich's and his own regiments were landed and deployed in good order 100 yards apart, with Brett's regiment still disembarking when:

---

12 Lockyer, *Buckingham*, pp.380–381.
13 BL Add ms 9,298.
14 Quoted in Lediard, *Naval History of England* vol. 2, p.503.
15 Cherbury, *Expedition to the Isle of Rhè*, p.29; BL Add ms 72391, Trumbull papers Journal of the campaign to the Isle of Rhè, 12 July 1627–28 September 1927.

THE ÎLE DE RHÉ EXPEDITION

The enemy's horse charged us, I am sure without any fear and I think without any discretion, for they were to pass either through us, or between the sea and us, the other ground unto the landward being so uneven that it was not well to be passed.

According to our battalions I stood with my flanks unto them so that I was fair to wheel my regiment. They were so near us before they came behind the hills that they took me in some disorder, but we stood them reasonably well and the other regiments facing about paid them with their musketeer. The companies then coming on shore we were put into great disorder and many were drowned and so were some of their horse being not able to break through the divisions nor to go forwards they took the water for their best refuge.

At this encounter I was hurt with a pistol in my thigh, a little below my belly, which although it be not dangerous to my life.[16]

Whether Conway arrived during the attack of the French cavalry or before, there was enough English foot still in formation to hold the bridgehead and forced the cavalry to retire. But now it was time for the French infantry to come to grips with the English; one account recalls:

The [French] foot seeing the ill entertainment of their horse came on very unwillingly (their captains waving their hats to them) and were glad to quit us for after two or three volleys of shot and stones finding our pikes too long for them they betook themselves to flight.[17]

A Mr Graham records that the English musketeers:

Discharged their muskets and after they fell to it with swords and push of pike until they were breathless on both sides. The French finding our pikes to be longer than theirs threw away their pikes and went to it with stones, and so did our men but ours beat them out with stones and make them fly away very disorderly.[18]

Meanwhile Sir William Courtney's Regiment refused to leave the safety of the ships, until they were personally ordered to disembark by Buckingham. Whether this regiment took part in the action is not recorded, but by now the French were finally being forced back. Among the French infantry was the Regiment of Champagne, which according to an English account had 'never before been beaten'.[19]

The strength of the French force varies greatly from account to account, being from 150 to 400 horse, and 1,000 to 1,800 foot, but most place the strength at about 300 horse and 1,200–1,600 foot, who together lost about 120 horse (23 of whom were noblemen) and 100 foot soldiers.

---

16 HRO 44M69/G5/48/122/1 Edward Conway to Lord Lieutenant of Hampshire concerning the landing at La Rochelle, 16 Jul 1627.
17 BL Add ms 72,391, Trumbull papers Journal of the campaign to the Isle of Rhè, 12 July 1627–28 September 1927.
18 TNA SP 16/71/60 Journal of the Expedition, 20 July 1627.
19 S R Gardiner, *A History of England* vol. ii, p.132; BL Add ms 72, 391, Trumbull papers; Journal of the campaign to the Isle of Rhè, 12 July 1627–1628 September 1927.

One of the criticisms aimed at Buckingham was that he stayed on board ship all this time, refusing to disembark until the last Frenchman had withdrawn, but Sir John Burgh as senior colonel was perfectly capable of commanding the English on the shore.[20]

Another criticism was that no pursuit was ordered once they had routed the French, but Sir John Burgh did not have any cavalry and it would be some time before they could disembark. Moreover, since the cavalry were the eyes and ears of the army there was no way of knowing if there were any additional French forces in the vicinity.

Meanwhile, the English continued to unload their men and provisions and fortify their bridgehead. On 13 July, the French governor of the Island, the Marquis Jean du Caylar de Saint-Bonnet Toiras, sent a trumpeter several times to Buckingham to arrange a truce so that both sides could bury their dead, but this was not granted until the following day. So that the French would not gain intelligence of the bridgehead, the trumpeter was blind folded and English soldiers carried the French dead 'a good distance without [from] our trenches', to the carts which had been supplied to carry them away for burial.[21]

On 14 July, due to Sir William Beecher's diplomacy, Monsieur Soubise arrived at the English camp from La Rochelle with about 500 soldiers, but this was far short of what Buckingham had hoped for, and with his diplomatic mission having failed Beecher returned to England.

The following day, Buckingham marched to St Mary's leaving Sir Peregrine Beatie's Regiment to guard the fleet and prevent the enemy landing any additional forces. The English are known to have worn green boughs in their hats to distinguish them from the French. Buckingham has also been criticised for staying so long at the landing place, but not only did he have to consolidate his bridgehead, but his army had suffered heavy losses. According to the *Journal of the Campaign* recorded 'some 35 common soldiers drowned, but none killed'; whereas Edward Lord Herbert of Cherbury, who Buckingham asked to write an account of the campaign, states that Buckingham lost about 200 men although a French source puts the English casualties at between 500 and 600. Whatever the casualty rate among the soldiers the sources agree that among the officer corps the losses were heavy, with 19 being killed, including Sir William Haydon, the General of the Ordnance, and 12 wounded, including Colonels Rich and Conway and Lieutenant-Colonel Sir Thomas York. Sir George Blundell would die of his wounds in La Rochelle two weeks later.[22]

The army marched to La Flotta, while Sir John Burgh's Regiment was given the task of capturing the fort of Pallas, a job he failed to do. That night the English had to shelter in open fields without shelter despite the heavy

---

20 BL RP 8006 Anon account of Île de Rhé Expedition.
21 Quoted in Lediard's *Naval History of England* vol. 2, p.505.
22 Anon. *Journal of all proceedings of the Duke of Buckingham in the Isle of Rhè*. Cherbury *The Expedition to the Île de Rhé* p.36.

rain and a thunderstorm. It would not be until noon that the sun came out so the soldiers were able to dry their clothes.[23]

However, by not securing the Fort de la Prée, which was situated near to where the English had landed, Buckingham had committed a serious error, because not only did it guard a small port and he could have secured his bridgehead, but it would also have protected his lines of communications with La Rochelle. It would also be used as a rendezvous for the French relieving force.

On 17 July, the English arrived at Saint-Martins, the main town on the island, 'When we were not far off', wrote Cherbury,

> about 100 or 120 horse did discover themselves, three appearing a little before the rest. Sir William Cunningham presently taking notice hereof steps forth with as many and in a loud voice challenged any of them to single fight, but they instantly turned back withdrew themselves to their fellows.[24]

After several shots from Buckingham's artillery the French horse retired to the fort, while the inhabitants of the town had no choice but to surrender, but not before the church roof had been destroyed so that the English could not place a piece of ordinance on it to bombard the fort.

For three days Buckingham's army camped before the walls of the citadel before they began digging any fortifications, in which time the French were able to complete their preparations. Critics of Buckingham said that if he had attacked during this time, he could have easily captured the fort, but Sir John Burroughs who surveyed the fort on their arrival, said it could not be taken by assault and therefore it was decided to starve the garrison into surrender. It was only later that Buckingham discovered from several French deserters that if he had attacked the fort then it would probably have surrendered.[25]

On 19 July, seven pieces of artillery were landed from the ships to form three batteries to bombard the fort and Buckingham issued a proclamation to the inhabitants of Saint-Martins that all Catholics within the town were to be transported to mainland France in boats. How many left is unknown, but on 9 August about 50 French women and children were ordered into the fort to join their husbands. This was no act of charity on the English's part, but a way to consume the garrison's provisions faster.[26]

However, the garrison did not fall for this ploy because according to an anonymous sea captain,

> All women (papists) that had husbands in the fort were put out of the town and set between the enemy's trenches and ours. A trumpet of ours being sent with them to give notice that they were to be received by them otherwise to starve.

---

23   Anon, *Journal of all proceedings*; Grenville, *A true and exact journal*, p.7.
24   Cherbury, *Expedition*, pp.51–52.
25   BL Add ms 26,051 Journal of the Voyage of Rease.
26   Quoted in Lediard, *Naval History of England* vol. 2, p.507.

The enemy though very unwillingly received them. This was done to have their victuals devoured by these unnecessary hands.[27]

What became of these women is not recorded.

Meanwhile, Sir Peregrine Beatie's and Sir Henry Spry's Regiments, with some English cavalry, were ordered to occupy La Flotta, which would hinder two other forts on the island. Meanwhile Mountjoy's troop of horse patrolled the island searching for any straggling French soldiers.

On 23 July, the anonymous sea captain records,

Our battery played furiously upon the citadel, and they on us with as much eagerness… This 24th day having occasion to speak with our general I went to view our battery where I found some of our cannon dismounted by those of the enemy and our gabions much shaken and ruined.[28]

Another account also mentioned this incident, the guns commenced firing at the fort, 'had not shot above two or three shots but one of them was dismounted and two more almost spoiled by them of the fort'. The English appear to have done little damaged to the French gunners who sheltered behind the fort's ramparts.[29]

Neither were the soldiers having any better luck with their fortifications, as Buckingham records:

The trenches go on as fast as the hardness of the stoney ground will permit and I hope ere long to bring that work to an end… That being done I hope they will be so well pinned in both by sea and land, that they will receive no assistance from either… [because] the shortest way to take it is by famine. The ground [the fort] stands upon is rocky and of such a continued and hard kind of rock as the pickaxe will hardly fasten in it, which rakes of all possibility of making of mines.[30]

A mine was begun but had to be abandoned because of the unsuitable ground.

Unfortunately for the English, Buckingham appears to have brought with him three incompetent engineers; a Mr Rudd who was 'but a simple overseer of pioneers', a Mr Audley who was, 'so unknown a one as he cast up the trenches the wrong way', and a Mr Tap who had been, 'saved from divers hangings'. In fact, the most competent engineer turned out to be Buckingham's gardener who he had brought with him.[31]

Throughout the summer the siege dragged on, with the French making sallies from the fort. One English source records that, 'our men severally do their duties, both officers and common soldiers, three or four days in a week by watching in the trenches, approaches and batteries'. Buckingham is said to have toured the trenches, 'and if he finds any man negligent in his duty,

---

27 BL Add ms 26051.
28 *Ibid.*
29 BL Add ms 72391, Trumbull papers.
30 TNA SP16/288, Buckingham to Lord Conway, 28 July 1627.
31 BL Add ms 26051

he causeth him to suffer according to the quality of his offence for the good example of the others'.[32]

Meanwhile, of the garrison of Saint-Martins' fort, 'They are strong in number both of horse and foot', Buckingham wrote:

> Their horse consisting mostly of gentlemen and their foot of the Regiment of Champagne, which in this kingdom is called the *Invincible*. Ordnance they have great store… [and] very good and excellent gunners to us them. Corn, salt fish and wine they have in abundance and for a long siege and of all manner of ammunition and arms more than they can use or spend. And to conclude a governor that had made this the seal of honour and fortune… before he will yield up the place he will make it his death bed.[33]

On 30 July it was reported that a French 'deserter' tried to assassinate Buckingham with a poisoned knife, but the plot was discovered just in time. However, one source states that the besiegers laughed when they heard of this assassination attempt, since it was ludicrous. Certainly, Sir Richard Grenville also mentions this assassination incident, 'as some say'.[34]

Whether true or not, this was not the only time that the French tried to kill Buckingham, because on 4 August Henry De Vec wrote that,

> The enemy in the citadel makes every day diverse shot upon his [Buckingham's] lodging (and myself have been many times with his Excellency when the shot hath come through the chamber over which he was) especially about those times they imagines him to be within.

De Vec continues:

> Our trenches go on leisurely, but surely and by Sunday night we hope to bring them to the water side unless we be hindered by the enemy, who hitherto hath been reasonably quiet. All the hurt and discommodity that we have received (which God be thanked hath been very small) from them hath been from their ordnance with which that make wondrous good shots, and for ought I can see better then with their muskets… Our trenches being now run something near them at most not above half a musket shot and our men and theirs play continually upon one another… with small shot. Among the redoubts we have one of reasonably good strength wherein we intend to plant ordnance and the platforms are now ready for it. It will be of great use to us to defend and flank our trenches and will greatly incommodate [inconvenience?] the enemy in his outworks, but will chiefly annoy him in that little space of passage that remains open, which he sometimes visits.[35]

When his army had embarked at Portsmouth it was 5,934 strong, but on 20 July it mustered 5,525, a loss of 409 men in three weeks. Therefore, to

---

32   TNA SP16/288, Buckingham to Lord Conway, 28 July 1627.
33   *Ibid*.
34   BL Add ms 26,051, Grenville *A true and exact Journal* p.11.
35   TNA SP 16/73/38, Henry De Vic to Secretary Conway, 4 August 1627.

make up this loss Buckingham had 500 seamen landed under the command of a Captain Meddall, who was promoted to colonel. However, such was the ridicule they received from the soldiers that they soon returned to their ships.[36]

On 5 August three French ships tried to run the blockade, but two were sunk and the third was boarded and the crew were either killed or thrown overboard, apart from the commanders, one of whom was the king's standard bearer, who managed to escape about a week later.[37]

Despite being there almost a month Buckingham does not seem to have fortified his quarters properly and it was not until 13 August that he ordered the inhabitants of Saint-Martin to fortify the town and a further nine days passed before he ordered trenches to be dug nearer the citadel. On the other hand, there were rumours in the English camp that the French were undermining their positions despite the stony ground in order to blow them up in their trenches.[38]

On 14 August Buckingham wrote to Beecher appealing for supplies of food, men and money because, 'our provisions grow low and our men decrease'. However, the day before Charles himself had written to Buckingham:

> I have made ready a supply of victuals, munitions, four hundred men for recruits and £14,000 ready money, to be brought to you by Beecher… [who] shall set sail within these eight days. Two regiments of a thousand men a piece, victualled for three months, shall be embarked by the 10th September. I have sent for as many officers of the Low Countries as may be had… I hope likewise you shall have two thousand men out of Scotland under the command of my Lord Morton and Sir William Balfour.[39]

However, Beecher was still in England on 25 August along with the two regiments. Furthermore, the Earl of Morton had only just received a commission to raise a Scottish regiment on 7 August and it would not be until a month later that Colonels Ramsey and Ratcliffe were also commissioned to raise their regiments.

Meanwhile in France a new battery was erected by the English, which was completed on the 20 August and a boom of masts linked together by chains and a cable was begun to prevent any ships bringing provisions to the fort. The following day Buckingham called upon the citadel to surrender, which was refused.[40]

On the same day (24 August) De Vec recalls:

> Our trenches are now at length after many trials[?] and more faults come to the water side, so that the enemy hath no option left him to get out of the citadel by

---

36 BL Add ms 26,051, Grenville, *A true and exact Journal*, p.12.
37 Lediard, *Naval History of England*, p.507.
38 BL Add ms 26,051.
39 Sir Charles Petre *The Letters, Speeches, and Proclamations of King Charles I* (Cassell, 1935) p.51.
40 Cherbury, *Expedition*, pp.93–94.

land, but at low water, where he must needs pass by a redoubt of ours made at the end of our works to keep that passage.[41]

The navy blockaded the fort on the seaward side, so that Saint-Martin was now surrounded on all sides.

Meanwhile, the French army had not been idle, they had been gathering their forces together; and early in the morning of 1 August Charles de Valois, Duke of Angouleme, arrived at gates of La Rochelle with a party of horse and foot demanding to know which side the city was on: Louis XIII's or Britain's? The city assured Angouleme that they were 'the king's good servants and had no hand in the coming of the English to the Île de Rhé'. Not wishing to offend him the council of the city invited Angouleme to dine with them and showed him every courtesy. After dinner Angouleme left La Rochelle and toured the city walls with an engineer.[42]

However, despite his reception, Angouleme was not convinced and on 3 August wrote to the mayor and corporation of La Rochelle demanding that they cease supporting Buckingham's army or he would lay siege to the city. According to Peter Meruault, one of La Rochelle's inhabitants, the city council discussed the matter, but they seem to have taken too long because Angouleme appears to have taken their silence as a sign of their intent to support the English. Therefore, he sent his force, which comprised four infantry regiments, 10 troops of horse, a detachment of the Guards and 'a good quantity of cannon', to quarter in the surrounding towns of Estree, La Moulimette, Bongrenne and Courcille and began erecting fortifications. Anyone found in the neighbourhood of La Rochelle carrying provisions into the city were to be punished and have their goods confiscated. Meanwhile, King Louis himself was marching towards La Rochelle with a large force intending to rid France of the Huguenots' stronghold once and for all. This prompted the city to recall its force from the Buckingham's army; however they were sent back to the Duke a few days later.[43]

With the French digging fortifications around the city, on 10 August the city's mayor ordered 700 soldiers and some inhabitants during the night to begin improving their own defences. This lasted several days, but still no official hostilities broke out. It was not until 1 September that the population could no longer ignore the besieging force, whose lines were creeping ever nearer towards the city. Therefore, its guns opened fire on these entrenchments. The besiegers' guns replied and so at last Buckingham had an ally, albeit a reluctant one.

Buckingham received news of La Rochelle's declaration of war the same day and on 2 September Bingley's and Crosby's regiments arrived from Ireland along with three additional companies of Bertie's Regiment, bringing much-needed reinforcements of 1,899 men to Buckingham's weakened army.

---

41   TNA SP 16/75/12, Henry De Vic to Secretary Conway, 25 August 1627.
42   Peter Meruault, *The Last famous siege of the city of Rochell* (London, 1680), p.17.
43   Ibid., pp.18–19 The regiments were Navarre, Chappes, Bergerie and a detachment of Guards, with 10 'cornets' of cavalry and 'a good quantity of cannon'. The regiments of foot comprised Navarre, Champagne, Bergent and Moity.

These reinforcements disembarked two days later and appear to have had orders to capture the Fort de la Prée.

On 4 September Lord Conway's son wrote to him that:

> The weakness of our troops hath been such that we could not make approaches to force the enemy; now that the Irish troops are come we shall go resolutely to work with them and seek to beat them [the French] out of the fort de la Prèe. My Lord Duke doth let us know that there is a great courage in him which his understanding doth equal. For without him we should certainly have been discouraged.[44]

However, the order to besiege the fort de la Prèe appears to have been countermanded, because according Cherbury:

> Because many of our soldiers were dead, and others sick, our numbers were thought insufficient for both purposes. Therefore my Lord Duke of Buckingham continued his first resolution....Therefore he had compassed the citadel with a trench and redoubts, he [Buckingham] raised a kind of moveable mount in the sea, made of the keels and bottoms of ships; and seven great pieces of ordnance [were] placed on the top of it, with gabions to defend them, was conducted towards the fort, as well for the shelter of our longbows.

These measures appear to have resulted in a French pinnace being captured trying to run the blockade. The crew were thrown overboard and drowned, except for the master of the ship who alone was taken prisoner.[45]

Buckingham again summoned the citadel to surrender, which was again rejected and so the siege dragged on. According to one pamphlet, on one occasion Sir William Courtney while inspecting the trenches, 'was shot by a musket [ball] from the fort upon the belly but the bullet lighting upon some money in a little packet lost its force and did no farther hurt'. Unfortunately, Sir John Burgh was not so lucky, on 12 September he was 'shot in the belly through the guts about 5 o'clock in the evening with a musket [ball] which wound proved mortal for within four or five hours after he gave up the ghost'.[46]

The pamphlet continues:

> Our soldiers, but especially those of his regiment, were so much provoked with the unexpected accident that in revenge thereof they poured diverse volleys of small and great shot upon the French with the storm, whereof the governor's brother was slain and some other men of quality.[47]

---

44 BL Egerton ms 2533 Letter to Secretary Conway from his son, 4/14 September 1627 f.46.
45 Quoted in Lediard's *Naval History* p.508; Cherbury *the Expedition* p.508.
46 BL C.114 c.59, *A continued journal of all the proceedings of the Duke of Buckingham in the Isle of Ree*, pp.6–7.
47 Ibid.

THE ÎLE DE RHÉ EXPEDITION

However, according to Colonel William Fleetwood, Burgh met a more sinister end, and that Buckingham had a hand in his death, because while, 'Surveying his own trenches and being clear of all danger of the fort was instantly struck dead in the place with a musket shot by an unknown hand and so gave up his spotless soul unto the joys of Heaven that had never done but good on earth'. The motive was jealousy on Buckingham's part because the army looked to Burgh rather than him as their leader and there had been tension between the two men for some time.[48]

Whatever the truth, some in the army and back in England certainly believed he was murdered by the Duke or one of his faction. In England Simon d'Ewes wrote about Burgh's death, 'whether by the Duke's secret procurement or not, I am not able to affirm'.[49]

Unfortunately, we will never know who fired the shot that killed Sir John Burgh, but as Sir Richard Grenville's records, 'with him died all the hopes of our good success'. His body was taken back to London, where on 23 October the trained bands with colours, pikes and muskets trailed, dressed in black cassocks escorted his body to Westminster Abbey, where he was buried along with his own ensign, the staff being broken in two. The musketeers then gave three volleys as a salute.[50]

By now both sides were in a desperate situation, as one eyewitness records:

> The autumn now growing on, the earth was so moistened with frequent rains that the soldiers on either side had no ground but mire to do their duties on. This indeed was incommodious for the garrison unless they were covered with huts or planks, but altogether grievous for us who being a more tender and delicate constitution did hardly endure to watch in dirty places and in the open air. Hence were engendered ill habits of their bodies which had their conclusion in catarrhs, diseases of the lung, burning fevers and dysenteries our numbers hereupon were so diminished that they could scarcely be made up again by supplies both from England and Ireland.

The British were probably cheered on 20 September when between five and seven boats were captured trying to run the blockade with badly needed supplies, and once again all the French crews were killed. Four days later 13 sail of ships arrived with victuals and munitions from England. However, the British were not the only ones to receive provisions because according to the Venetian ambassador in Savoy, a soldier in the Regiment of Champagne, named Pierfort:

> With two barrels under his arms and his letters in a horn tied to his heart, he ventured to swim the sea to take them to the king [of France]. He remained 10

---

48 BL RP 8006, Anon account of Île de Rhé Expedition.
49 *Ibid.*, D'Ewes' quote printed in James Doelman 'John Earle's Funeral Elegy on Sir John Burroughs' *English Literary Renaissance* 2011, p.492.
50 BL Add ms 72,391; Birch *The Court and times* p.281. He was buried on 23 October 1627 in the St Michael's Chapel in Westminster Abbey. Close to Sir Francis Vere who he had learned the trade of soldiering.

hours in the water, pursued for four of them by two English shallops, which kept losing sight of him owing to the waves, and could never catch him. He finally reached land, but with his legs ruined by the bites of the fishes. In the letters the governor represents that he will have to surrender if he is not relieved soon, declaring he has not received a loaf since the English landed.[51]

However, for the garrison of Saint-Martin help was at hand. On the 28 September, some French boats managed to slip through the blockade during the night, the English ships not noticing them in the darkness until it was too late. These boats brought with them badly needed provisions of food and ammunition, which was a great blow to the besiegers, as one eyewitness recalls:

> Our hopes were all defeated by the coming of 15 or 16 boats from the main [land] which brought them [the French] at least two months provisions of victuals and munitions which they stood in great need of, for a fortnights time before we had not a great shot come from the fort and our soldiers marched to their guards in open view of the fort (the trenches being full of water) without receiving any hurt by their small shot for want of powder. And now to add to our miseries our soldiers began to sicken apace, having neither good lodging, but the trenches nor meat, but which stunk, nor drink, but water, all the provisions of wine in the island being spent. Yet the Rochellers would not be sensible to our wants but for that small provisions they sent they sold it at a treble of the value of its worth.[52]

The garrison of Saint-Martin fort also stuck chickens, capons and turkeys on their pikes which they waved at the besiegers to show that they had enough provisions.

By now the army had had enough and on 2 October the job of dismantling the guns from their batteries began, however, the following day Soubise returned from La Rochelle with a delegation from the town promising support, which gave Buckingham courage and the order to withdraw was cancelled, although the guns were not returned to their batteries and the sick and wounded continued to be evacuated to the ships in preparation for a retreat.

By coincidence at the same time the news of La Rochelle's support was received, intelligence arrived that a French force was on its way. On 12 October Buckingham held a council of war, where the officers argued with the Duke to abandon the siege and return to England, due to the army's weakened state. Buckingham on the other hand wanted to take the fort by storm before the French arrived, the officers argued against this stating that it was 'absolutely unassaultable'. According to Sir Richard Grenville, Sir Henry Spry is said to have:

> Used a soldier like freedom of speech in showing to my Lord the necessary of our departure; and that it would be honourable, having sufficiently so long time in an

---

51 Hinds, *CSP Venetian 1626–1628*, pp.380–389.
52 BL Add ms 72,391.

THE ÎLE DE RHÉ EXPEDITION

enemy's country, with such a handful of men… but he was answered with a court like scorn that 'he talked as if he had come from an alehouse'. And for the rest of the colonels, who were all of the same opinion… instead of harkening unto them, by whom he ought to be governed, he abused them in words.[53]

One anonymous account records that some of these 'words' the Duke used was:

> That they were all cowards in general and Sir William Courtney in particular (who was then colonel general and price of that opinion). And that he would send him prisoner into England, who answered that for his part he cares not in this account for his sake nor his love of him, but for the service of his country.[54]

Grenville continues, that 'my Lord Duke was swayed by some insinuating sycophants and others, who told them it would be dishonourable to go away, having two months victuals'.[55]

This was not the first time that Buckingham and his officers had had a heated debate at a council of war and news of these disagreements within the council soon got back to England. Denzil Holles records that there was, 'nothing but discontent betwixt the general [Buckingham] and his most understanding of his soldiers, as Burgh, Courtney, Spry'.[56]

Buckingham ordered preparations to be made for an attack the following day. An anonymous officer records that:

> Our whole army was drawn up in the fields where my Lord took a view of them. His grace made a speech to the soldiers telling them he had now certain news from England the fleet was coming. He assured them they should be eased of their duty and be supplied all with provisions both necessaries and clothes.

Fortunately for the army because of foul weather the assault was cancelled, but the storm sank at least one collier which had about 80 sick men on board, and all were drowned, and a further 20 longboats were lost.[57]

On 21 October hearing that the French had landed on the Île de Rhé, the British were ordered to quit their siegeworks to meet this new threat. However, the garrison of Saint-Martin sallied out of the fort and captured their trenches, 'pulling down our courts of guards in our redoubts'. Not wishing to be caught between two forces, Buckingham gave the order to recapture these trenches the following day.[58]

---

53 Grenville, *A true and exact journal*, p.17.
54 BL Add ms 26,051.
55 Grenville, *A true and exact journal*, p.17.
56 Letter from Denzil Hollis to Sir Thomas Wentworth, 19 November 1627 printed in William Knowler's *The Earl of Strafforde's Letters and Dispatches* (London, 1739) vol. 1 p.41.
57 Quoted in Lediard's *Naval History* p.508; *Cherbury Expedition* p.510.
58 Grenville, *A true and exact journal*, pp.18–19. On 15–16 September 1627 the French Guards (1,800 men) Suisse (1,000), Piedmont (800), Chappes (800) Bergeries (500), Chasteliers Bartiaut (500) Rambure, Plessis de Joigny, and Vaubecourt were selected to relieve St Martins Le Papiers de Richelieu vol. 2, p.493.

After retaking the trenches, the British remained inactive for the next five days, all the while the French were landing more and more forces near the little fort of la Prée. Buckingham's colonels appealed to him to abandon the Île de Rhé before it was too late. However, Soubise argued that the French force was just a detachment of recruits rather than the main French army and that the Earl of Holland was due from England with reinforcements any day. Furthermore, without their support the garrison of La Rochelle would have to surrender, which would be a 'great dishonour to the king', and not to mention Buckingham himself.

Therefore, it was not until the 27 October that Buckingham gave the order to storm Saint-Martin. After the soldiers had sung the psalm 'Let God arise and let his enemies be scattered' which according to Cherbury was 'their wonted custom' they attacked the fort:

> Our men being desperate for fear of starving and went bravely, but we quickly found (contrary to the relation of some villains who ran away from the fort that they had not 400 fighting men) a resistance fit for 4,000 and we were scarce 4,000. They killed above one hundred men with stones and they might have done the like by our army, if they had come on without shooting small shot. They blew up three mines, but did little hurt with them. In all we lost and slain 400 foot and horse, they took some 30 prisoners in traps which they made in their trenches.[59]

Another account of the assault by Sir Piers Crosby, recalls:

> The assault was given in five places without any battery or mine sprung. The four places nearest to the town were assigned to nine English regiments and certain Rochellers drawn thither by Monsieur Soubise some few days before about 500 strong…my charge, being the point on the east side. We gave on altogether, and I may say without arrogance that it was performed with as much courage as any French men we saw in any action.[60]

An English sea captain recalls:

> That many of the enemy's out works and some of their half-moons were scaled and the enemy beaten out of them but being to advance upon their bulwarks and other of their main works they were found so exceeding high and every way so well flanked and fortified as that to be further attempted with hope or reason for that after one hour and half of this being in action wherein the volleys of musket shot were never ceasing intermingled with many cannon shot and in many parts our men being gotten up with the enemy to the push of pike who always kept themselves in a covert and from the height of the works continually casting down great stones and other the like materials they were at last enforced to an orderly and soldierlike retreat and so again put themselves into their own trenches.[61]

---

59 Quoted in Lockyer Buckingham p.400; Cherbury *the Expedition* p.204; BL Add ms 72,391; Trumbull papers Journal of the campaign to the Isle of Rhè, 12 July 1627–28 September 1927.
60 TNA SP16/84/78 Sir Piers Crosby to? Lord Conway? 14 November 1627.
61 BL Add ms 26,051.

One source stated that 400 men and several officers were killed in this assault, whereas Crosby records, 'for although we lost not above 200 upon the place, yet the greater number of our humble men being the flower of the army, did both weaken and retard retreat'. One of the reasons the assault failed was because the scaling ladders were too short and there was no artillery to provide cover for the attackers, the guns having already been sent away to the ships.[62]

The assaulting force withdrew to their trenches, and the following day some troopers of Sir William Cunningham's horse reported that shooting could be heard in the distance and that the French were occupying the town of Saint-Martin. After receiving advice from Soubise, Buckingham finally decided to retreat. The only way they could be evacuated by the navy was from the Île d'Oye, but it was only connected to the Île de Rhé by a bridge and a causeway. Therefore, the British army were divided into three divisions, the regiments of Sir Edward Conway, Sir Peregrine Beatie and Sir Henry Spry formed one division and Sir Charles Rich, Sir Alexander Brett and Sir Thomas Morton's regiments formed another, while the regiments of Sir William Courtney, Sir Edward Hawley, and Sir Ralph Bingley along with the horse formed the rearguard. Crosby's Regiment appears to have formed the advance guard and occupied an earthwork near the bridge.

As Buckingham's army retreated, they had not gone a 'musket shot out of the town, but we saw the enemy at the heels of us'. The French were about 6,000 foot and 200 horse strong, compared with the Buckingham's 3,000 foot and between 60 and 100 horse. At first the French did not attack but just followed the British, who now and again stopped to face them before continuing their retreat. The French refused to give battle until the first two English regiments were already over the bridge.

The English cavalry had been divided into two bodies to cover the retreat, one commanded by Montjoy, and the other under Cunningham. However, being attacked by French cavalry:

> Montjoy's troops… presently retreated, giving fire over their shoulders, and rode in among our ranks, and routed us so, that the most began to shift for themselves. The other division of the [French] horse fell upon Sir William Cunningham's troops, but they most bravely fought it out unto the last man. Had my Lord Montjoy done the like we… [would] have not lost one-quarter so many of our men.[63]

Colonel William Fleetwood records that the French cavalry charged and clashed with the English cavalry, which:

> Were valiantly received by our men… [but] overcome with [the] multitude they underwent a diverse fortune for some part was killed among whom was that brave (and not unrevenged) Sir William Cunningham, some part yielded, among whom

---

62 TNA SP16/84/78, Sir Piers Crosby to Lord Conway, 14 November 1627; BL Sloane ms 826, Papers of concerning the proceedings in both Houses, 1621–1629, ff.30–31.
63 Birch, *The Court and times*, p.300.

my Lord Montjoy, being one [who] was curtiously [i.e. roughly] used. Others by the shock and pressing of the enemy in a sloping ground were forced back and driven to flight; these running away in full flight.[64]

However, one English source states that it was not the French cavalry that had routed the troop, but 'My Lord Newport had the ill luck to disorder our cavalry, with an unruly horse he had'. Another eyewitness records that the French charged, 'My Lord Montjoy's troops which presently retreated giving fire over their shoulders and rode in amongst us… We had no time to give fire if we had…[we would have shot] our own men.[65]

The English cavalry galloped over the causeway and bridge, knocking down the infantry as they went and being disordered by their own cavalry the 'enemy had execution of five whole regiments which they put all to the sword except 20 officers and 100 common soldiers [who were made] prisoners and those which were drowned, which were many'. The English regiments in the rear threw away their arms and 'for the most part began to shift for themselves', that is, run away.[66]

However, Cunningham's troop does seem to have rallied and counter-attacked, charging 'into the thickest of his enemy and disordered them'. Cunningham's horse was killed under him, but 'refusing quarter and yet sore wounded', attacked the French on foot, finally being captured along with most of the English cavalry, although another eyewitness records that Cunningham's troopers fought to the last man.[67]

As the English retreated over the bridge pursued by French cavalry, they found a cart blocking their way, which 'we must climb over or leap into the river or salt pits', records Fleetwood. He continues:

> Which most of our company being unable to do were instantly hewn in pieces, as my Lord Cromwell, Sir Charles Rich, and others of greatest esteem who on the very deadly extremity were offered quarter but would not, choosing rather to die honourably then longer to live with infamy and torment.
>
> I myself perceiving the folly of resisting any longer in the company was enforced to take an infirm salt pit where both myself and horse stuck fast in the ground and when I suddenly [received] a ghastly wound in the leg with a bullet and so I lay struggling for life… diverse of our company and commanders were in the like distress.

Fortunately, during a lull in the pursuit, they were dragged from the salt pits by some English soldiers who, 'conveyed them out of danger'. Another officer to fall into the salt pits was Sir William Courtney, who had been given command of the retreat, rather than Buckingham, but, according to Sir Richard Grenville, he was saved by his servant who lost his own life rescuing

---

64 BL RP 8006 Anon account of Île de Rhé Expedition.
65 James Howell, , *Familiar Letters in Importance wrote from the year 1618–1650* (Aberdeen: F Douglas and W Murray, 1753), p.185; BL Harl 390 f.330 Anonymous eyewitness account.
66 BL Add ms 72,391; BL Harl 390 f.330.
67 BL Harl 390 f.330.

Courtney. This was not the only lucky escape Sir William Courtney had, because a bullet hit a gold coin in his pocket, which bent the coin, but 'he was preserved'. Unfortunately, many were not so lucky and being pushed from bridge, which did not have any railings fell into the water and drowned.[68]

Cherbury records a different fate for some of them, after the English horse had:

> Cast our men on either side into the ditches and salt pits adjoining. The way being thus made open the French followed with their wonted alacrity and here our men received a great blow; for the French with their pikes did at their pleasure now kill those men (lying in the water and mud) whose eyes before they durst scarce behold…[The enemy] resolving not to spare any that came to their hands. And here the greatest slaughter by much [by far] was made.[69]

Luckily for the English runaways Sir Piers Crosby's Irish Regiment was at hand. It was still in good order and was manning a half-moon-shaped earthwork which had been erected to defend the bridge, although as one eyewitness put it, it was not complete and that it was 'made so ill that it was not terrible [to the enemy].' With the assistance of some musketeers commanded by Sir Thomas Fryer, they managed to put up a fight.[70] Although, according to Crosby, his musketeers were unable to fire for fear of hitting their own men.

The English soldiers, who had thrown away their arms in a desperate attempt to get away, tried to scramble over the bridge, but there were no rails and so many were pushed into the river below and swept away some pikemen which Crosby had placed at the end of the bridge. With the fugitives having passed it was only now that Crosby's musketeers were able to fire two volleys at the pursuing French, who had begun to cross the bridge. The French were caught in a crossfire which killed many, and some of Crosby's pikemen had managed to rally, which slowed their advance.

Crosby with a part of his regiment manage to keep the earthwork, which had been badly damaged by the fleeing English, that night supported by some soldiers, Sir Thomas Fryer, lieutenant-colonel to Sir Alexander Brett and Major Hackliut, of Sir Henry Spry Regiment, about 100 paces from Crosby. These two bodies managed to beat off the French attacks and later when two other regiments were sent to reinforce this rearguard action they were able to go onto the attack and according to Crosby, 'we fell on and beat the enemy out of the works, and beyond the bridge, many of them being put to the sword and the rest in great disorder'.[71] Crosby's men were also able to recover some of the wounded. Finally, under cover of darkness the rearguard withdrew, and about 3.00am the Crosby with 100 men was the last to leave the earthwork and retired to l'Oyes after having burnt the bridge.

---

68 Grenville, *A true and exact journal,* p.25.
69 Cherbury, *The Expedition,* p.244.
70 BL Add ms 72,391.
71 TNA SP 16/84/78, Sir Piers Crosby to? Lord Conway, ?14 November 1627.

Colonels Rich, Brett, Hawley, and Bingley were killed fighting in the front line of the army. Sir Richard Grenville three times refused quarter and was shot in the body by a musketeer, but despite an eyewitness claiming that he died as he desired 'with his face towards the enemy', he survived and would later fight in the Civil War. Not all officers were as brave, 'Colonel Gray fell into a salt pit and being ready to be drowned, he cried… 'a thousand crowns for my ransom', the Frenchmen hearing that, preserved him, though he was not worth a hundred thousand pence'.[72]

Where was Buckingham during the battle? It was rumoured within the army that Buckingham not only withdrew two of the best regiments to protect himself, even though he had a strong lifeguard with him, but also changed into soldier's clothing so as not to be recognised and 'got himself a ship and not only left us ignorant of the bloody intent towards us, but also made us not capable to prevent it'. Furthermore he is alleged to have dressed one of his servants in his own clothes, and with his personal standard acted as a decoy and fought bravely and 'got a gash or two in the shoulder'.[73] Unfortunately this accusation cannot be verified, although not an eyewitness himself, Denzil Holles, records Buckingham, 'carefully got himself on shipboard that night to prevent the worst, and to take order for [the] boats for the shipping of the army'.[74]

With the British army cooped up in the Ile l'Oye and while they waited for the French onslaught, where according to Crosby, a council of war was called, 'whether to fight or not'. It was decided that they should be evacuated by the navy and that Crosby's Regiment would form the rearguard. The colonels of the other regiments would, 'cast the dice who should ship first, which being done everyone took his turn and our men were shipped in good order'.[75]

What Crosby does not say was that between 30 October and 2 November a truce was called, during which time Buckingham tried to negotiate the release of the prisoners, but this was refused. By now there was little to be done and the British boarded their ships, but even now nature appeared to be against them, because a severe storm caused several boats to sink, so that it was not until 8 November that the fleet finally sailed for England.

This time there could be no excuses for Buckingham. On 27 November Joseph Meads wrote to a friend that Buckingham:

> Affirmed upon his honour there were but 400 slain in the last defeat and that he had brought 5,000 of our men home alive. But some Irish that are returned thence do report that of the English (besides Irish) were slain 2,644 and that of those 2,100 Irish that went thither there came off but 600. And certainly the loss must

---

72 BL Sloane ms 826 ff.30–31; *Familiar Letters* James Howell to Lord Clifford, p.185. Howell claimed he heard this from an officer at the battle, curiously the date given for this letter is 24 September 1627, which may be a misprint because it cannot have referred to another event during the campaign.
73 BL RP 8006, The author of this account heard this rumour on the return voyage to England.
74 Denzil Hollis to Sir Thomas Wentworth 19 November 1627 printed in Knowler's *The Earl of Strafford's letters and Despatches*. Vol. 1 p.42.
75 TNA SP 16/84/78, Sir Piers Crosby to? Lord Conway? 14 November 1627.

be great and lamentable when I saw a list this morning of colonels, lieutenant colonels, sergeant majors, corporals of the field and captains of 60 in number. And the French have hanged up in Our lady's Church at Paris 42 ensigns, the greatest dishonour our nation underwent.[76]

A Mr Beaulieu echoed Meads comments believing it was 'the greatest and shamefullest overthrow the English have received since we lost Normandy'.[77]

On 26 October Buckingham's army had mustered 6,884 men, but the following month it was just 2,671 strong. In another account, dated the 29 November1627, it was reported that 2,898 soldiers were billeted in Devon and Cornwall and 'above 1,200 sick and hurt with 300 officers'.[78]

However, it would not be until 25 May 1635 that Captain John Mason finally reported that of the 7,833 men who had served in the campaign, 409 had been either killed or drowned on 12 July when the force landed, 320 were killed during the siege and 100 'recovering our trenches from the enemy'. A further 3,895 were killed during the assault on the citadel and during the retreat and a further 120 died 'upon the bloody flux by eating grapes', leaving just 2,989 men who returned to England. Unfortunately, Mason's report does not record how many of these were wounded.[79]

Despite this humiliation, Buckingham 'was most joyfully and graciously received by the king', as if Buckingham had won a great victory. The Italian Ambassador, Alessandro Antelminalli records:

> Notwithstanding that Buckingham is upheld by the king, he never has been more unpopular with the people, and an order has been issued to preachers commanding them to refrain from speaking of him and of the retreat from Rhè in their sermons.[80]

An inquiry was set up to find out what had gone wrong during the recent campaigns. The report found that there were four points on why they had failed:

> That the army and navy had been badly victualled.
> There had not been enough sailors to crew the ships.
> There had been no overall strategy and so the campaign was conducted 'upon ill grounded and uncertain designs'.
> The officers had been too inexperienced 'especially its commander'.[81]

As well as the fear that a Spanish army might invade Britain, in January 1628 news arrived that a French army had moved into Normandy from where it would attempt to capture the Channel Islands, before launching an invasion

---

76   BL Harl ms 390, Collection of letters of Joseph Mead, Jan 1626–Apr 1631 f.322.
77   BL Add ms 26,051; Forster *Sir John Eliot* p.405.
78   TNA SP 16/84/94; *CSPD 1627–1628* p.450, Sir James Bag to Buckingham, 29 November 1627
79   TNA SP 16/289 no.39.
80   *HMC Skrine* Letter dated 22 December 1627 p.133; Birch, *The Court and times*, pp.285, 289.
81   BL Add ms 41,616.

of Britain itself. It was not until September that this force withdrew from Normandy and the threat of an invasion subsided.[82]

Strengths of Buckingham's Army, December 1627

Regiment	Mustered	Recruits	Total
Colonel Grenville	158	142	300
Sir Thomas Morton*	189	111	300
Colonel Fryer	140	160	300
Sir Edward Conway	139	161	300
Sir William Courtney	102	198	300
Sir Peregrine Bertie	111	189	300
Sir Henry Sprye	56	244	300
Sir Piers Crosby	360		360
Sir Ralph Bingley	145	155	300
Colonel Ramsey	911		911
Colonel Radcliffe	758		758
Earl of Morton	?	106	106+
	3069	1466	4535+

---

82 TNA SP 101/10.

THE ÎLE DE RHÉ EXPEDITION

Return your match

13

The Siege of La Rochelle

While Buckingham's Army was licking its wounds in England, the French Army continued to besiege La Rochelle, Louis XIII having decided to rid himself of the Huguenot threat once and for all. Among its garrison were some English soldiers who had been sent into the city to recover from the wounds they had received during the Île de Rhé Expedition, plus about 80 English soldiers who had arrived on 18 November 1627 under Captain Bourgis. A further 400 to 500 soldiers in Devon and Cornwall, who had not taken part in the Île de Rhé Expedition, were ordered to prepare to be sent to La Rochelle. However, they were still in England at the end of March 1628, and it is doubtful whether they were ever sent.[1]

Meanwhile the inhabitants of La Rochelle watched as their city was slowly encircled, and on 7 December the cannon positioned in the 'Royal battery' open fire on the city, killing a man, a woman and a child. While on the 16 December three beggars were killed and two others were wounded while playing cards from the fire from this battery.

However, not all of the inhabitants were loyal to the Huguenots; on 29 December a plot was discovered that a Captain Salle was to betray Fort Tadon to the besiegers, so he 'was immediately taken, racked [tortured] and hanged', on the 2 January. His head was placed on a pole at the fort as an example to other Catholic sympathisers.[2]

The garrison did their best to disrupt the besiegers, and according to Peter Mernault, on 3 January an attack was made on:

> The redoubt of Beatriel betwixt Bongrenne and the house of Courceille, where there was a squadron of thirty of John Sac's Regiment, who after some resistance, were forced and cut in pieces, especially by the English, in revenge for their companions which had been killed in the Île de Rhé, in such sort that not above

---

1 J. V. Lyle, APC 1627–1628, A letter to the Commissioners for soldiers at Plymouth, 17 December 1627, minute of the Privy Council, 29 March 1628 pp.181, 364.
2 Mernault The last Famous Siege pp48, 49, 50. The dates from this diary have been converted into the Old Style of calendar, which was still used by Britain and 10 days behind the new calendar.

two or three of them were saved and with the loss of no more than one killed and three wounded.³

The siege dragged on through the winter and after four months in front of La Rochelle the weather was taking its toll on the besieging force. With sickness running through its ranks the French Army was reduced to about 18,000 men by the beginning of February 1628. Many were in no condition to fight and Louis XIII himself had fallen ill and returned to Paris to convalesce, leaving Cardinal Richelieu in charge, having been promoted to lieutenant-general.

Meanwhile booms had been constructed at sea, and a fleet of 58 ships prevented any enemy ships from entering the city. This fleet included 36 Spanish ships; despite being a traditional enemy of France, King Philip IV of Spain was just as eager to crush any Protestant opposition in Europe inasmuch he had sent supplies of food to the beleaguered Fort of Sainte-Martins the year before.

With Britain still in fear of a Spanish invasion in April 1628, John Sampson, a sailor, brought news that while in Spain he had seen 'much pressing and mustering of soldiers', and that 'many of their colonels, commanders and captain are Irishmen…all these soldiers and forces were taken up for [the invasion] of England or Ireland'. Two months later Richelieu was brought the news that Spain was preparing to send 60 ships to attack England.⁴

Philip IV could have easily stayed out of the matter and let France be torn apart by a religious war which had kept her out of European politics. Why this change of heart? In August 1627 Philip IV had fallen dangerously ill and it appeared that he might die. The historian J. H. Elliott states that it 'was of a major political significance' because his death might not only have meant the end of Hapsburg dynasty in Spain, but also the end of his 'tyranny'. At the beginning of September, he began to recover and by 10 September he was out of danger. However, this recovery marks the same period as Spain's involvement in the Île de Rhé campaign. So, could it be that Philip feared that he had not done enough in God's eyes to impose the Counter-Reformation, inasmuch that he would ally Spain with her enemies to destroy Protestantism once and for all?⁵

Early in September news arrived in East Anglia that the people had been dreading, a force of 6,000 'Dunkirkers' had landed at Wakering in Essex and set fire to the town. Two beacons had been lit which caused panic among the local population, one woman with her child even fled the 16 miles to Graces to warn of the invasion. The local trained bands were mustered to repel the invasion and a Captain Humphries was sent to reconnoitre the situation. However, on his arrival at Wakering he discovered that the invasion fleet consisted of a single fishing ship which had anchored in the harbour, and one of its crew had a foreign accent.⁶ Such was the tension in England at this

---

3   *Ibid.*, p.51.
4   TNA SP 16/101/47i, Examination of John Sampson of Fowey, 19 April 1628.
5   J H Elliott writes about the importance of Philip's illness in Richelieu and Olivares pp.90–94.
6   TNA SP 16/116/67i Em Downing to Sir Robert Naunton, 6 September 1628.

time, and the wait for the expected invasion went on. Little did the people of Britain know that the Spanish ships were helping the French fleet blockade La Rochelle, while the Spanish army was fighting a covert war in Italy against France. This war would become active in 1629 after France's Huguenot problem had been dealt with and it would not be until 1630 that the Pope negotiated a ceasefire between these two countries.[7]

On 1 March, La Rochelle was informed that Richelieu had ordered scaling ladders to be built for an assault on the Cabal bastions that night, and petards would be set at the gates of Salines, Neuf and St Nicholas. It was also rumoured that about 800 to 900 inhabitants who were loyal to Louis would also rise up. The defenders waited for the attack all night, but it did not materialise. However, the assault came the following night, but it was directed against the Fort Tadon, which was held by seven companies, including one composed of English troops. A forlorn hope was composed of between 150 and 200 men and a second body, about 800 strong, and was commanded by Marshal Schomberg which, according to Peter Mernault, were the 'most sprightful, bold and vigorous soldiers in the army, with the flower of the gentry'. A reserve would follow this body, flanked by cavalry. A diversion of 30 soldiers was sent to the seashore and a soldier went to the Port of St Nicholas to ask the defenders not to fire, which caused some confusion.

The musketeers in the forlorn hope had 'a cover, that they might not be discovered', but a musketeer within the fort saw a shadow and fired his musket. The flash from the barrel highlighted the attackers who were already in the approaches of the fort, but after fierce fighting they were forced to retreat. On 4 March, to mark this success the inhabitants held a thanksgiving service, but it would be the last celebration for some time.

Louis returned to the siege in April 1628 with more French troops bring the besieging force up to about 25,000 men. While the besieged were informed that the English fleet would arrive in March or April to break the siege, 'which rejoiced much the city throughout'. but as the weeks passed there was still no sign of the fleet, which prompted them again to write to Charles for assistance. This despatch is probably the one the Privy Council received on 21 April 1628 from the mayor and council of La Rochelle, which said, 'In the name of God come with all speed'. They warned their food was exhausted and that the enemy's fortifications were being strengthened each day.[8]

In December 1627 on their arrival back in England the 'broken companies' of Buckingham's Army mustered just 3,069 men, not including the Earl of Morton's Scottish Regiment, which was about 1,500–2,000 strong. That month a further 1,466 recruits arrived to reinforce the army, so that when another muster was held in December 1627 it was reported that the army now mustered 7,557 men. However, of this figure only 3,255 were said

---

7 TNA SP 12/237/63, List of ships before La Rochelle, February 1628; TNA SP 101/10, pamphlet News from Paris dated 4 September 1627 which reported that Spain had offered to help victual Sainte-Martins.

8 Mernault The last Famous Siege p.57; TNA SP 16/101/47iii Mayor and Council of La Rochelle to their Deputies in England, 4 April 1628.

THE SIEGE OF LA ROCHELLE

Plan of the fortifications at La Rochelle, 1628 (Author's Collection)

to be healthy and quartered in the West Country, plus 1,302 sick, 1,500 Scots and about 1,500 quartered in Hampshire and other counties. When Captain John Mason held another muster early in the following year, he found that the army was just 4,535 men strong and was not strong enough to attempt another landing in France.[9]

Therefore, it was decided to send the English fleet back to France with provisions on board to re-supply La Rochelle. This time the fleet was to be commanded by the Earl of Denbigh, but when he arrived at Plymouth at the end of February, he found that there were no supplies and many of the ships were not seaworthy. Those that were, about 24 to 25 ships, were not strong enough to break the siege. However, at the end of April intelligence was received that the besieging force at La Rochelle had been reduced to 8,000–10,000 men and there were now just 12 ships blockading the city. Moreover, the boom placed in the bay which had been erected to prevent any ships replenishing La Rochelle from the sea had been destroyed in a storm.

With this intelligence, the English fleet finally sailed on the morning of 27 April and during the afternoon of the 1 May the inhabitants of La Rochelle saw the English fleet's approach, which according to Peter Mernault consisted of 11 king's ships, 30 or 40 lesser men-of-war and as many other vessels, laden with corn and provisions. The morale of the inhabitants of La Rochelle increased at the sight of these ships.

Unfortunately, the inhabitants of the city's hopes were dashed, because the intelligence that had been received in England was wrong and the barrier had been rebuilt. According to Captain Jacob Williams, 'all men who saw the palisade thought it an impossibility to break through it'. Williams was not the only one to believe this. Moreover, there were many more French and Spanish ships within the bay than they had expected. Richelieu ordered that these ships were on no account to leave the bay, but withdrawn further into the bay so that they could be protected by the besieger's artillery.[10]

Faced with this opposition Denbigh was forced to order his fleet to anchor at the head of the bay at between 4.00–5.00 p.m. but the besiegers had set up a battery near to where the English fleet had anchored, which forced Denbigh to order his ships to anchor further off. The inhabitants sent word pleading with Denbigh to fight his way through to the city, but he refused, saying that he did not have any orders to fight. According to the Venetian ambassador the 'La Rochellese' had deliberately misled Charles' government over the strength of the besiegers' defences, but neither fleets were prepared to give battle, and they just exchanged a few shots now and again. However, the Venetian ambassador believed that if Denbigh had pushed on, then nothing could have been done to prevent him relieving La Rochelle.[11] Nevertheless on 5 May Denbigh sent several fireships, 'full of fireworks, in the fashion of petards into the enemy fleet to set fire on them, but taking fire before its time,

---

9   TNA SP 16/87; SP 16/18; SP 16/98.
10 TNA SP 16/105/85; SP 16/103/57, William Earl of Denbigh to Buckingham, 9 May 1628; SP 16/103/61 Captain Thomas Kettleby to Sec. Nicholas, 10 May 1628.
11 Hinds, CSP Venetian Zorsi Zorsi's letter to the Doge and Senate, 20 May 1628 p.94; TNA SP 16/101/47ii Report of Mons de Sainte-Martin who came out of La Rochelle 4 April 1628.

the fireships and those that were in it perished miserably without any being saved'.¹²

This incident is similar to one recorded in a ship's logbook, but for the 7 May, that at about 9.00–10.00 p.m. Captain Allen with Captain Williams placed 'their floating engines so near to the enemy upon the tide of flood as conveniently they could'. Unfortunately, for Captain Allen his 'engine' accidentally touched the boat's side while he was putting it overboard and blew himself up along with six men. Three other men were blown into the air and were 'sorely bruised' and rescued by Captain Williams' boat:

> The corpse of Captain Allen was taken up but much diffused and torn. But the coxswain of the scallop with his mate and four men were never heard of. At the firing of it, it gave a double report like unto two whole cannon making our ships shake.

Three other mines were laid, but either failed to find a target ship or were faulty.¹³

Meanwhile the besiegers feared that the garrison of La Rochelle would take the opportunity to make a sortie and they would be caught in a two-pronged attack, but no attack materialised and according to Andrew Le Brunn, the captain of the *Mary Rose*, Denbigh boasted that he would 'sink the French ships when the water increased and the wind came west; but when those circumstances happened he would not fight, but came away'.¹⁴

However, Captain Thomas Kettleby, referred to the 'impossibility' of the enterprise, which would have been 'madness' to attempt to relieve the town and that the English fleet did not receive any word from La Rochelle of their intentions, so Denbigh was forced to withdraw and on 8 May the fleet weighed anchor and sailed away. The fleet had not been able to deliver its provisions to La Rochelle, which according to Peter Mernault, 'greatly dejected the La Rochellers and put them in great pain and perplexity'. According to the Venetian ambassador, this left Louis and Richelieu 'singularly relieved', and they sent word to Paris. Without these provisions the ambassador believed the city would not be able to hold out for another month, while Richelieu boasted that he would hold Mass in La Rochelle at Whitsuntide.¹⁵

The day after the fleet left the bay Captain Allen was buried at sea, his body being saluted by the rest of the fleet.¹⁶ A ship was sent to report the failure of the expedition to Charles, who on hearing the news wrote to the inhabitants of La Rochelle on 15 May:

> Gentlemen, Be not discontented though my fleet be returned, hold out unto the last, for I am resolved that all my fleet shall perish rather than you should not be relieved, and to this end I have countermanded it and have sent ships to make

---

12   Mernault, *The last Famous Siege*, p.112.
13   BL Egerton ms 2533, Account of the English Exhibition to La Rochelle, 1628 ff.51–52.
14   TNA SP 16/106/3i Examination of Andrew le Brunn, 16 May 1628.
15   TNA SP 16/103/61, Captain Thomas Kettleby to Secretary Nicholas, 10 May 1628.
16   BL Egerton ms 2533, 'Account of the English Exhibition to La Rochelle, 1628 ff.51–52.

them change their design that they had taken to come back. I shall shortly send you some number of ships to reinforce it and with the help of God, the success will be happy for your deliverance.[17]

This letter would not arrive at La Rochelle until the 1 June, which made the inhabitants expect 'succour within a few days'. The following day another letter arrived from Charles, which was dated 17 May, again reassuring them that he would not abandon them. However, the fleet would remain off the French coast until 20 May when it sailed to the Scilly Islands arriving on 24 May, and from there it sailed on to England.

According to William Whiteway on his arrival back in England, Denbigh excused his failure for relieving the city by saying that 'he had no commission to hazard the king's ships in a fight and so returned shamefully to Portsmouth'. Nevertheless, this voyage must be one of the earliest incidents of the use of mines in naval warfare and Samuel R. Gardiner is probably correct when he says that Sir Francis Drake or even Nelson could not have broken through the blockade.[18]

On 30 May 1628 Denbigh was ordered to assemble his fleet once more and prepare to return to La Rochelle, but this order was cancelled the following day because the 'ships were found to be so defective as that the whole fleet could not be in readiness to go together'.[19]

The inhabitants of La Rochelle waited for the promised relief and at the end of June or early July, according to Peter Mernault, they began to:

> Kill horses, asses, mules, cats and other creatures, the flesh of which was sold for ten or eleven Sols the pound, that of horse flesh was above all savoury, there being little difference betwixt it and beef… the famine so increased, every one reserving themselves their provisions, that the greatest part were in great want and bread failing they had recourse to Brazil sugar, Colworts, Frigase with a little tallow and such other nourishment.

To make matters worse some provisions which had gone bad had been destroyed in May when it was believed that the English fleet had brought fresh supplies. Therefore, to help supplement their rations, Peter Mernault, continues:

> [They] scattered themselves upon the Fens where the salt pans were, to make war with eels and other little fishes and on the coast to fish for cockles and after that eat all sorts of herbs…good and bad, boiling them in two or three waters to take away the bitterness and ill taste.

---

17   Mernault, *The last Famous Siege*, p.120.
18   Whiteway Diary of William Whiteway pp.96–97; S R Gardiner History of England vol. 6, 1625–1629 pp.291–292.
19   Lyle, *APC 1627–1628* Privy Council minutes, 30 May 1628, p.481.

For those that could afford it, sugar was used to sweeten the taste, but soon scurvy broke out in the city.[20]

Those inhabitants who tried to flee were to be shot, many preferred this quick end rather than:

> Return home to die there miserably from famine and many women and maids of the common people… were violated and beaten with forks [i.e. musket rests] and shafts of halberds, then stripped naked as when they came from the womb of their mothers and so sent back into the city.[21]

At the end of June Richelieu summoned the city to surrender, to which the inhabitants replied that they expected to be relieved in eight to 15 days, but the weeks and then months passed and still the English fleet did not come.

Meanwhile the preparations in England were going on apace. Having seen the French fortifications in July, 10 ships had been commissioned to carry 'several warlike engines', but they were still not finished. These engines may be the ones referred to in the Office of the Ordnance's account book which records a payment of £40 to Lewis Tayte for 'eight engines of iron for the sterns of ships with stays, stirrups and spikes', while another payment was made for 'six engines for fireworks'.[22]

Around the middle of August Charles' Privy Council sent a letter to the Archbishop of Canterbury, George Abbot, to 'cause a form of prayer to be conceived to be used in all churches throughout the kingdom for the good success' of the fleet. On 15 August Buckingham left London, but on his arrival at Portsmouth two days later about 300 sailors surrounded his coach demanding their arrears of pay. One mariner is even said to have pulled the Duke from his coach. He was arrested and on 22 August hung on the gibbet between Portsmouth and Southsea Castle.[23]

Buckingham stayed at Captain John Mason's house, known as the *Greyhound* in Portsmouth's High Street, where on the morning of the 23 August Buckingham entered a room which was crowded with people. No one noticed Lieutenant John Felton as he walked up to Buckingham and stabbed him in the chest. 'Gods wounds the rogue has killed me', the Duke is allegedly to have said, while another source states that Buckingham just said 'Villain!' before falling to the ground dead. While another says, 'villain, thou hast slain me', and that Buckingham had kicked Felton, because he had not taken off his hat while talking to him.[24]

Whatever Buckingham said, in the confusion that followed Felton managed to escape. He had been wounded in 1627 and ever since had been 'troubled by dreams of fighting'. and held a grudge against Buckingham, who

---

20 Mernault, *The Last Famous Siege*, p.125.
21 *Ibid.*, p.133.
22 Hind, *CSP*, Venetian Alessandro Antelminelli to the Grand Duke of Florence, 12 July 1628 NS, p.157; TNA WO 49/59 Debenture Book, 1627 f.157.
23 Lyle, *APC 1628–1629*, Privy Council to the Lord Archbishop of Canterbury, 15 August 1628, Privy Council minutes, 30 May 1628, pp.103, 481 Oglander His Observations pp.33–34.
24 Dyfnallt Owen Calendar of the Cecil Papers in Hatfield House, 1612–1668 (1971) vol. 22, pp.239–249; Cooper, J P (ed) Wentworth Papers, 1597–1628 (London, 1973), p.303.

he blamed for being overlooked for promotion. A mob scoured the streets looking for Felton, who they believed to be a French assassin. As the story goes when he heard the cries 'a Frenchmen!' he gave himself up, believing that they were saying 'Felton'. Believing that he would be killed while attempting to assassinate Buckingham, Felton had sewn a note into his hat which read:

> That man [Buckingham] in my opinion is cowardly and base, and deserves neither the name of gentleman nor a soldier, that is unwilling to sacrifice his life for the honour of God and the good of his king and country. Let no man commend me for it, but rather discommend themselves, for if God had not taken away their hearts for their sins, he had not gone so long unpunished.[25]

Felton was charged with treason and taken to London for trial, but by this act he had become a hero to the people. At Kingston-on-Thames in Surrey an old women said to him, 'God bless thee little David!'. In Dover they drank to Felton's health and one Irishman, James Farrell, is alleged to have said that 10,000 in England wished the King was dead also. On the other hand, Charles allegedly wanted to have Felton 'racked' (tortured) and his hand which had held the dagger cut off. However, Charles was told that this would be illegal and on 29 November 1628 Felton was hung at Tyburn in London. His body was returned to Portsmouth and hung in a gibbet to rot like a common criminal.[26]

On the other hand, Buckingham's body received a very different welcome. Instead of a sombre silence it was accompanied by cheers as it passed through the streets of London. At his funeral only about 1,000 members of the London Trained Bands turned out to escort his body to Henry VII's chapel at Westminster. Even then the pikemen did not wear their armour and carried their pikes at the 'advance' rather than the 'trail', the traditional posture for a funeral march. The musketeers also refused to fire a volley as a salute. Of course, the King was saddened by the death of his friend and ordered that all his servants wear black for 14 days, although as one courtier put it, that although he obeyed this order, 'his sorrow did not last longer'.[27]

In August it was rumoured that the English fleet had been wrecked in a storm and that according to the Venetian ambassador to Rome, Anzolo Contarini, the Pope was overjoyed at the news claiming, 'that God was working marvels for the Catholic faith', but this joy did not last long because another ambassador pointed out that the fleet was not due to sail until September.[28]

---

25 *Ibid.*, pp.239–249.
26 Quoted in C Hill, *The World Turned upside Down* (London, Penguin Books, 1991) p.111; TNA SP 16/114/46, 51 Examinations of John Foard, Walter Roades and John Waller 29–30 August 1628; SP 16/116/28 Dr Issac Bargrave, Dean of Canterbury to William Weld, Secretary to Lord Conway with confession of James Farrell, 2–3 September 1628.
27 National Archives of Scotland GD237/25/1 letter from Sir David Cunningham to the Laird of Robertland, undated.
28 Hind, *CSP,* Venetian Anzolo Contarini letter to the Doge and Senate, 23 September 1628 pp.303–304.

However, in the besieger's camp things were also going badly. As in all sieges it was a matter of who could consume the provisions available the slowest. The inhabitants of La Rochelle were suffering greatly from starvation and disease, but the besiegers were also stripping the country bare of victuals. On 1 September Zorsi Zorsi, the Venetian ambassador to France, wrote:

> Everything seems to be going to ruin. The king and cardinal remain away from the camp. The troops diminish daily and food is scarce. I feel sure that by the end of this month there will be no army whatever. There is neither powder nor shot in the camp and the guns mostly lie dismounted in the fields and the works are falling into decay. God grant that the English do not come now, as they could succour the place without the slightest difficulty.[29]

Some soldiers had even fled to the city thinking that conditions would be better there. However, Louis could not afford to abandon the siege because he knew that this would encourage the Huguenots in other parts of France to rise up.

On 7 September, the English fleet put to sea again, this time under the command of the Earl of Lindsey in a desperate attempt to break the blockade. Lindsey had only been made an admiral in 1626, having only commanded regiments in the army before that. Nevertheless, he was much more aggressive than Denbigh had been and despite reports in May that the barrier was impossible to break, it appears that many believed that the defences were much weaker than was rumoured.[30]

On 18 September Lindsey's fleet, which was composed of about 60 ships anchored off Saint-Martin's Road, but it was not until 23 September that he attacked the French fleet, which was anchored off Charlebois Point. However, the merchant ships fired their guns at too great a range and although Lindsey ordered them to engage the enemy within 'caliver shot' upon pain of death they refused. The crews of the merchant ships again disobeyed this order the following day when Lindsey again engaged the French fleet. It was essential that Lindsey defeat the French fleet before he could attempt to destroy the barrier which blocked the entrance to La Rochelle.

Both fleets continued to bombard each other for the next few weeks and during the night of 11 October the French sent some fireships to disrupt the English ships. Over the coming week Lindsey planned to attack the French, but there was no wind to propel the ships and so neither side could come to grips with each other.

By now the conditions in La Rochelle were desperate. One report stated that 16,000 people had died during the siege, and an English eyewitness recorded:

> The rest endured a wonderful misery most of their food being hides of leather and old gloves. Other provisions which were scarce were at an expensive rate… a bushel of wheat was at £120, a quarter of mutton at £5… a pound of bread at 20s, a

---

29 *Ibid.*, Zorsi Zorsi's letter to the Doge and Senate, 1 September 1628 p.259.
30 *Ibid.*, Alessandro Antelminelli to the Grand Duke of Florence, 20 October 1628 NS, p.164.

pound of butter at 30s and an egg at 8s… Yea the famine was such that their poor people would cut off the buttocks of the dead that lay in the churchyard unburied to feed upon. All the English that came out thence looked like anatomies they lived two months with nothing but cowhides, and goat skins boiled dogs, cats, mice and frogs being all spent before. And this a world of misery did they suffer in hope of our relieving them.[31]

This is not the only account of their misery, Peter Mernault an inhabitant of the town records on 15 October:

The little corn that was found was reserved for the nourishment of the soldiers, who with the inhabitants were like anatomies and by little by little died away. And it is much observable what befell two English soldiers, who finding that they could [do] no more went together to the house of a joyner to bespeak each their coffins for the next day by eight in the morning; he at first refused, believing they mocked him, as thinking himself more wasted with famine than the soldiers; 'have you not the strength' said they 'to work?.' And pressing him and paying him in advance for gain made him undertake it, and before them and in their presence began to work and finish these two coffins and came at the time appointed… [That] same evening one died and the other the next day betwixt ten and eleven in the morning.[32]

Four days later Peter Mernault wrote in his diary:

Now famine increased dreadfully, nothing being left, the greatest number having in three months' time not known what bread was, nor anything of ordinary provisions, flesh and horses, asses, mules, dogs, cats, rats, mice, were all eaten up. There were no more herbs or snails left in the fields so that the recourse was to leather, hides of oxen, skin of sheep, cinnamon, cassia. Liquorish out of Apothecaries shops,… bread of straw made with a little sugar, flower of roots, Irish powder, jelly of skins of beasts and sheep, horns of deer beaten to powder, old buff coats, soles of shoes, boots, aprons of leather, belts for swords, leather points, parchment, wood beaten in a mortar, plaster, earth, dung (which I have seen with my own eyes), carrion and bones that dogs had gnawed and indeed all that came into sight, through such food rather gave death than substance.[33]

On 16 October, an English messenger reached La Rochelle bringing news that there was nothing more to be done and that Lindsey had begun negotiating with Richelieu for a peace treaty. On hearing this news, the Mayor and Council of the city had no choice but to begin negotiations themselves.[34]

The inhabitants of La Rochelle had never wanted a war with Louis, but had been forced into it by Buckingham when he had landed the previous year at the Île de Rhé, and had been encouraged to hold out by Charles on

---

31 BL Harl ms 390, Collection of letters of Joseph Mead, Jan 1626–Apr 1631 f.455.
32 Mernault, *The Last Famous Siege*, pp.161–162.
33 *Ibid.*, p.163.
34 *Ibid.*, p.167.

the understanding that they would soon be relieved by the English fleet. La Rochelle, once described as a 'state within a state', surrendered and on 24 October Louis entered La Rochelle and ordered its defences slighted.

The following day with nothing else to be done Lindsey gave the order for the fleet to return to England, 'as soon as the wind would allow', but on the 28 October the calm of the sea was interrupted by a storm which damaged many ships. Therefore, it was not until 31 October that Lindsey himself set sail for England.[35]

According to one source of the 600 English soldiers who had formed part of the garrison 62 are said to have survived, although the Venetian ambassador in England, Alvise Contarini reckoned 200 English soldiers were evacuated from La Rochelle, a figure also suggested by Captain William Button who was captain of one of the whelps in Lindsey's fleet. Like the other soldiers within its garrison, they were allowed to leave with just their swords and 'with white staves in their hands, after the names and surnames of everyone [were recorded] and their oaths taken never to bear arms more against the service of His Majesty [Louis XIII]'.[36]

---

35 TNA SP 16/120/17i, Relation of Captain William Button. 28 October 1628.
36 Louis le Juste XIII de nom p.500; Hind, *CSP*, Venetian Alvise Contarini to Zoriz Zorzi 21 November 1628, p.404; Mernault The Last Famous Siege p.178.

THE FIRST BRITISH ARMY 1624-1628 (REVISED EDITION)

Clear your pan

14

Homecoming

For three weeks the soldiers on board Lindsey's fleet had been short of provisions, and most had not had any beer for 10 to 12 days by the time the fleet arrived back in England on 9 November. However, they were more fortunate than those soldiers who were at La Rochelle. On 13 November, the Mayor of Barnstable informed the Privy Council that a ship had arrived off the town with 15 soldiers who 'by reason of the famine in Rochelle are so weakened that they are not able to travel'. A week later the Council ordered £600 to be paid for provisions for the expected 1,200 soldiers from La Rochelle, who were to be transported to Scotland.[1]

With the surrender of La Rochelle there was no longer any need for an army, so Captain John Mason was ordered to muster the men. At this muster, the corporals and soldiers were ordered to hand in their weapons to the officers of the ordnance and be discharged. They received their pay, plus eight pence per day that it was expected to cost for them to return home, which was calculated at 15 miles per day. They also received a pass stating that they had been discharged from the army, which was probably similar to the one issued to the following sailor:

> 'Whereas the bearer hereof Thomas Pyning being lately impressed for His Majesty's Service from Pomfret in the county of Yorkshire and having served under the command of Sir James Scott in the fleet sent from Portsmouth and is now arrived with thereof his company at Portsmouth. And being so sick and weak that he is not like to be fit for his Majesty's Service upon any further employment we whose names are hereunder written the members of the park[?] part of His Majesty's army as in the county of Southampton do these present give license unto the said Thomas Pyning to depart from the said service and to return unto Pomfret aforesaid from which he was impressed and have allowed him for his time and travel home 28 days. And we do pray and require all constables, tithingmen and others [of] His Majesty's offices to be aiding and assisting unto him in the said journey and so see him provided of necessary lodging. He behaved

---

1 TNA SP 16/120/10, Kenrick Edisbury to the Lords Commissioners, 9 November 1628 SP 16/121 f.389.

himself orderly and according to the Laws of the realm. Given at Southampton the 28 January 1626'.[2]

They were told to, 'employ themselves in their honest vocations, until his Majesty shall have some other occasion to use their service'. However, since there was no amnesty for soldiers who had deserted, this caused the county authorities a large headache in distinguishing a discharged soldier from a deserter, both being rounded up until their position could be clarified. [3]

The Earl of Morton's regiment was to march back to Scotland travelling about 12 miles per day for which each soldiers also received eight pence per day. It arrived in Scotland in December, where it was dispersed into winter quarters. On 21 December 1628 Morton wrote to the Earl of Carlisle requesting that the regiment should not be disbanded and it was suggested that the regiment should be taken into Dutch service, because 'such a regiment they never had from our country'.

By 9 March 1629 Sir George Hay appears to have taken over the regiment, which was now 2,000 men strong. It is believed to have been present at the siege of Bois-le-Duc, however in May 1630 the regiment returned to Scotland where it appears to have been disbanded.[4]

Although he was writing in 1578 Barnaby Rich summed up the attitude of civilians to the soldier returning home after risking their lives for England:

> The wars being once finished and there is no need of them, how be they rewarded, how be they cherished, what account is there made of them, what other thing gain they than slander, misreport, false imposition, hatred… or anything preferred for the service they have done.[5]

There was probably no hero's welcome for the discharged soldiers who returned to their parishes. The local Justices of the Peace had to find them work, although many parish officials were probably not too pleased to welcome them back into the community again since many had been a burden on the parish before they had left. This was particularly true of Crosby's Irish Regiment, many of whom appear to have been Catholics, and so many Protestant Irish were glad to be rid of them. They were even more horrified when in July 1628 they were told that the regiment would be incorporated into the Charles' Irish army, 'under the same captains and officers that now command them', since it had 'well approved its sufficiency and fidelity' at the Île de Rhé.[6] The Lord Deputy of Ireland had had nothing but praise for

---

2   HRO 44M69G5/37/29.
3   Lyle, *APC 1627–1628*, Letter to Captain John Mason, 26 October 1628, pp.208–209; Proclamation declaring His Majesty's Royal Pleasure touching the English Soldiers late employed at sea (EEBO)
4   TNA SP 84/138, Earl of Morton to the earl of Carlisle, 21 December 1628, Duplin to the earl of Carlisle, 21 December 1628, ff.200, 202; E403/2747, Order Book, 1628, ff.111, 127, 134, 137, 140, 144.
5   Rich, *Allarme to England*, p. E ivb.
6   Lyle, *APC 1627–1628*, the Privy Council to the Lord Deputy of Ireland, 27 July 1628, p.58.

Crosby's Regiment when it had left, but on 11 August 1628 he appealed to the Privy Council to change their minds because:

> We must say unto your Lordships that there is a mistaking in esteeming the regiment of Sir Piers Crosby to be wholly volunteers, for if he had not been assisted by the power of me the deputy intaking many men out of the gaols and receiving of divers malefactors and rebels that stood upon their keeping and pressing of other persons of notorious ill fame in their countries, he had passed over with a very weak regiment, and now these people [are] to be returned hither again with enablement to do mischief and with weapons in their hand in these times. We humbly leave it to your Lordships grave consideration whether it be expedient or no. The wisdom of this state having ever been to purge the realm of such ill members and known disturbers, by venting them into foreign parts, for other services of the crown where they may be of stead, which hath proved a great safety and preservation of the peace of this kingdom.[7]

On 27 August, the regiment was still in Bristol, waiting for a 'convenient wind', for them to sail for Ireland, but by the 18 September the regiment had finally landed in Munster. By early January 1629 Crosby's company, at least, had been sent into King's County, where the inhabitants of Phillipstown, protested to Sir Charles Coote, 'The soldiers have repeatedly drawn their swords and threatened to kill gentlemen of worth in the county. They have not paid the company the £13 which they were ordered to pay as the soldiers have already stolen so much'.[8]

Nothing had improved by 2 May 1629 when the Lord Deputy of Ireland wrote to Lord Dorchester, 'I think there will be no rebellion now, unless it be by reason of the army and especially of the Irish Regiment which is here universally murmured against. I fear they will goad the people into rebellion unless some money is sent'.[9]

Whether these accounts are exaggerated or not is not known, but on 21 May 1630 the Committee of the Privy Council for Irish Affairs finally thought Crosby's Regiment should be disbanded because, '[it] costs more money than any other part of the army, yet is totally unarmed and being natives and bred in loose and uncivil manners'. The date set for the regiment's disbandment was to be the 30 June 1630 when half the soldiers' pay was to be paid by the Irish government and the other half by the Crown. [10]

Despite this, the regiment was still in existence on 17 July 1630 when Lord Esmond wrote to Lord Dorchester:

---

7 TNA SP 63/247/1119, Lord Deputy and Council to English Privy Council, 11 August 1628, ff.97–98.
8 TNA SP 63/248/1281, Petition of King's County, to Sir Charles Coote, 10 January 1629.
9 CSP Ireland 1625–1632 Lord Deputy to Lord Dorchester, 2 May 1629 p.450.
10 Lyle, *APC, Jan-June 1627* Minutes of the Privy Council, 24 May 1627, p.294; Clarke, Aidan 'Sir Piers Crosby, 1590–1646, Wentworth's Tawney Ribbon' *Irish Historical Studies* (Nov. 1988) vol. 26 no. 102 (Dublin, Hodges, Figgis & Con, 1988) pp.152–153; TNA SP 63/250/1683, Report of Committee of the Privy Council on Irish Affairs, 21 May 1630; TNA SP 63/250/1716, Draft of the king to the Lord Justices concerning the housing of the Irish Regiment, 28 June 1630.

> There is a strange rumour here of cashiering the Irish regiment. I think it would be much better to send them to serve one of the king's foreign allies, for there are a great many gentlemen of the mere Irish amongst them as well of Moores and Connors as of Carties, Kavanaghs and Byrnes who have nothing to live on and must be in difficulties next winter.[11]

Finally, with too much opposition to Crosby's Regiment remaining in Ireland, it was proposed to send it to join Gustavus Adolphus whose army had just landed in Germany. However, despite the king raising regiments to reinforce his army, he declined the regiment, because, it was claimed, he 'did not trust the Irish'. Unfortunately for Crosby's Regiment it appears not to have found another employer and so was finally disbanded.

If the Protestant officials in Ireland disliked the soldiers of Crosby's Regiment, they hated the idea of paying pensions to the maimed soldiers from it. Sir William St Leger was given the task of overseeing the petitions and payments of wounded Irish soldiers. On 13 April 1630 he wrote to the Council, complaining that the cities, towns and villages where the soldiers had been raised refused to pay their pensions, claiming that there was no legal basis to support the maimed soldiers in Ireland and that they had 'put themselves voluntarily upon the action', and therefore did not deserve a pension.[12]

On the 29 April 1628, the 1601 *Act For The Necessary Relief Of Soldiers And Mariners* had been extended to Ireland. This Act was 'to provide relief and maintenance to soldiers and mariners that have lost their limbs and disabled their bodies in the defence and service of Her Majesty'. In England, this Act was to be administered by the county's Quarter Sessions under the Treasurer of Maimed Soldiers, who was chosen annually, or the Lord Presidents of provinces Connaught, Ulster, Munster and Leinster in the case of Ireland.[13]

For those soldiers who had been pressed into service they had to return, 'if he be able to travel', to the county where they were enlisted, while those who had volunteered had to return to their county of birth or where they had settled for three years. However, before they could claim a pension the soldier had to obtain a certificate from the army or garrison commander, but if they were caught begging then they would lose their pension.

Some of these petitions give a great deal of information about a soldier, such as that of Lieutenant Robert Bellott, who served in Viscount Valentia's Regiment in the Cadiz Expedition, and afterwards landed in Ireland, where his company was reduced into the Colonel Bertie's Regiment, before being sent to the Île de Rhé. It was here, while assaulting the fort, that he was shot in both hands and lost the use of one of them. Despite this handicap he was able to make clay pipes in Ireland, until he lost the use of the other hand. He visited five or six surgeons who were unable to cure him, because of

---

11 TNA SP 63/251/1745, Lord Esmond to Lord Dorchester, 17 July 1630.
12 TNA SP63/250, Sir William St Leger to Lord Conway and Lord Kullultagh, 13 April 1630, ff.182–183.
13 Lyle, *APC 1627–1628*, Privy Council Minute concerning petitions of Irish maimed soldiers from expedition to Rhè, 29 April 1628, pp.389–390.

a fracture 'in the bone which causeth it to break out often whereby your petitioner is utterly disabled'. Whether Robert Bellott was granted a petition is not recorded.[14]

Unfortunately, it was not just the Irish counties who resented paying pensions to maimed soldiers. The JPs of Wiltshire and Hampshire complained that they had more soldiers' pensions to pay than the poor law could afford. In July 1631, the Somerset Quarter Sessions ordered that:

> Whereas we find by due examination that by reason of the many late presses of soldiers out of this county… the number of maimed soldiers do daily increase, whereby the collection or sum of money which hath been heretofore usually raised… is not sufficient to satisfy them for their pensions and other necessary disbursements for relief of them. Ordered that £50 should be yearly raised over and above the usual rate.

The first £50 was to be paid at the next Session. However, other parishioners refused to pay and parishes such as Holywell in Oxfordshire were said to be 20 years in arrears by 1627 and in 1630 the Hundred of Aylesford in Kent was eight years in arrears.[15]

Among those soldiers who applied for a pension was William Spry, a gentleman who had been a sergeant in the Cadiz and Île de Rhé Expeditions, where he received several 'dangerous hurts'. However, it was not until 30 October 1635 that he actually applied to the Privy Council for assistance, who recommended that his case should be looked upon favourably. Many who applied for assistance were described as 'gentlemen', but others came from the lower classes including Thomas Chace of Captain Goring's company who was discharged in 1626 because he was 'distract[ed] of his wits'.[16]

One officer to petition for a pension was Ensign William Ware who 20 or so years before had been an impressed soldier, and had served with Colonel Morgan in the Danish army until a shot had blinded him and also made him deaf. With a wife and four children to support, the Privy Council petitioned the Justices of Hereford to grant him a pension.[17]

Sometimes a one-off payment was given to an individual until he received his pension, as in the case of Ralph Crewe of Sherborne in Dorset, who 'was impressed to serve the king in the war in the Low Countries under Count Mansfeldt. This has left him so disabled that he cannot work'. Crewe was awarded 10 shillings for his 'immediate relief' and then 40 shillings per year. In July 1638 he would petition to get his pension increased but to no avail, although he did receive a one-off payment of £3. On the other hand it was

---

14 TNA SP 63/269/322, Petition of Robert Bellot, nd c.1630.
15 1601 Eliz c.3. This Act superseded the 1592–1593 35 Eliz c.4; *Somerset Quarter Session records* p.156 Crossley, Alan, (ed); Oxford Quarter Sessions Order Book (Woodbridge, Boydell, 2009) pp. Xxvi-xxvii, 143; Kent Archives Q/FM/5 Aylesford was two years in arrears at the rate of £3 10s 4d and six at 35s 2d.
16 TNA PC2/45 f.192; HRO 44M69/G5/48/4 Account of Acts of Commission of Martial Law at Winchester, 10 January 1627.
17 TNA SP 16/179/30, Petition of Ensign William Ware, nd c.1630.

not until 1677 that Richard Thomas, a husbandman of Ashcombe applied to the Dorset Quarter Sessions for a pension:

> Your petitioner served his late Majesty, of famous memory in the several expeditions at St Martins and Cadiz actions. And that at the fight at St Martins aforesaid was grievously wounded and maimed by a shot of a bullet in the head which went into his mouth and came out at the hinter part of his head, tearing and shattering all his teeth and jaw bone; so as he hath not been able to eat his bread or any solid mean ever since, but has been feign to live upon broth and liquid meats which gave him not that nourishment which solid meat might have done by reason whereof he is grown imbecile and weak in body in this his old age of four score and two years. That the petitioner refrained to make his addresses to this worshipful Bench in all the time past since his maiming because he could not meet with any that could prove the premisses [?presumption] by oath. But not by providence hath men with one William Crooke, who was a soldier with him in the same action, who is now in the city and is ready to be deposed before your Worships of the truth of the premisses [?presumption] and the petitioner being now grown very weak for wanting of [the] ability to eat proper meats for his better nourishment and not able to work for his living formerly.
>
> He therefore humbly prayeth that in tender consideration of the premisses your worships would be graciously pleased to grant him the King's Majesty's pension to be yearly paid him out of the Treasury for Maimed Soldiers for his subsistence during his life.[18]

Richard Thomas was lucky that he had survived for 50 years with this injury because surgeons could do little for fractured jaws and the chance of infection was great. A surgeon could try bandaging the jaw, but these were likely to slip and displace the jaw once more. If he was lucky then the jaw would not be broken all the way through the bone so that there would be some healing over time.

Unfortunately, the petition does not record whether he was granted a pension or not, but he was probably the Richard Thomas whose burial is recorded in the Ashcombe parish register as taking place on 29 January 1683.[19]

Pensions ranged from £1 6s 8d to £2 a year. Until they received their pension a maimed soldier had to live as best he could, like William Rodner of Buckland Newton who was a maimed soldier under Mansfeldt and found employment as a cowherd by the tenants of the manor. However, if being dismissed from his employment was not bad enough his house also fell down. At the July 1633 Dorset Quarter Sessions it was ordered that his house was either to be rebuilt by the overseers of the poor of his parish or find him a new one. It was also at this Session that he was finally given a pension.[20]

---

18   Devonshire Record Office, QS128/4 f.4.
19   South Western Heritage Trust, 462A/PR/1/2.
20   Terry Hearing and Sarah Bridges (eds) *Dorset Quarter Sessions, Order Book, 1625–1638* (Dorset Record Society, 2006) pp.11, 19, 20, 55, 77, 81, 116, 117, 225, 249, 426; Crossley, *Oxford Quarter Session Order Book, 1614–1637* p.145; Willis Bund, J W (ed) *Worcestershire Quarter Sessions* (Worcestershire County Council, 1900) p.457.

Pensions would be paid in quarters and distributed when the Quarter Sessions met, although Henry Cottrell received his payments twice a year from the Worcestershire Quarter Sessions since he had to travel from London at 'a great charge' four times a year. Cottrell was lucky because not all pensions were paid. John Mathews who had been awarded a pension of £2 in 1624, had only received 50 shillings up to July 1638. The Dorset Session ordered that he should be paid £10 immediately and £4 a year until the arrears had been paid, then he was to receive £2 again. The entry does not record how he had been living up to then, but like many he probably had to throw himself on the poor relief.

Despite the 1601 Act for maimed soldiers ordering that all maimed soldiers should have a pension, most counties appear to have adopted a waiting list, so in 1635 the Hertfordshire Quarter Sessions awarded Nicholas Lawrence, 'a maimed soldier impressed from Royston to have the next pensioner's place which shall become void and which is not already granted in revision'. Richard Cressey of Harding was to be paid 10s and have the next pension after Lawrence. When Lawrence or Cressey finally received their pensions is not recorded, but another pensioner would have to die before they did so. The Western Division of Kent had just 20 pensions to offer its maimed soldiers, while the Justices of the Peace of Buckinghamshire went further and cruelly informed the county's pensioners that Charles had repealed the Maimed Soldiers Act in the last Parliament and so the Justices had ordered the constables of the parishes to return the money to the tax payers, which brought in a flood of petitions to the Privy Council.[21]

On 15 March 1629 Sibel Alley, whose husband and two sons had drowned on the return of the Earl of Holland's Fleet and was described as 'destitute of maintenance, 'was paid just £24, which was her husband's arrears of pay, for herself and 'four fatherless children'.[22] Widows could also apply for pensions. In an undated petition, (probably 1628), Jane Watkins a Dutchwoman petitioned the Privy Council on behalf of herself and two children because her husband, Major Watkins of Sir Thomas Morton's regiment had been killed on the Île de Rhé. She had accompanied him to England and was now destitute in London and appealed for her husband's arrears to be released to her. It was agreed that she should receive £100 immediately and the remainder in February 1629 when the pay of those officers killed during the campaigns would be paid. Lady Sara Brett was luckier, in June 1629 she was granted a pension of £200 a year for her and her two children. However as late as September 1637, the family of Captain Thomas Gates, who had been killed at La Rochelle was still petitioning the Privy Council over the settlement of his estate.[23]

---

21 William Hardy (ed), *Hertfordshire County Record Session Book,* (Hertford County Council, 1961), p.200; Kent Archives Q/FM/1-8, maimed soldiers accounts, West Kent, 1626–1635, 1640; TNA SP 16/88/44, Declaration of William Coleman, Thomas Kettlewell, Edward Eaton and Robert Hill, maimed soldiers of the county of Buckingham.

22 TNA SP 16/28, Petition of Sibel Alley, wife of Lieutenant-Colonel Alley, 15 March 1629, ff.69–70.

23 TNA SP 16/154 f57, Petition of Jane Watkins, TNA E407/13, Ordnance Quarter Book, January-December 1625; Noel Sainsbury, W *Calendar of State papers Colonial* vol.9, Order of the Privy

What of their children? On 12 September 1628 the Signet Office petitioned the Trustees of Thomas Sutton's Hospital to admit William Acton, whose father was an ensign and had died in captivity after being captured on the Île de Rhé. Sutton Hospital had been established at Charterhouse in Middlesex in 1611 with an aim to care for elderly people and as a school for poor children over the aged of 11. What the fate of William's six brother or sisters was is unknown, even though they are described as having 'little or nothing to maintain them'.[24]

However not all soldiers went home. In January 1630, the Dorset Quarter Sessions ordered that Richard Bindall, a soldier from Cambridge, who had been billeted at Cranborne was 'living incontinently' with Margaret Mathew of that parish. He was ordered that if he did not return to Cambridge he would be put in the house of correction, whipped as a vagrant and then forcibly sent back to Cambridge.[25]

With the disbandment of Buckingham's army the majority of soldiers disappeared into obscurity. Some would take up arms again in the Civil War while others would continue soldiering, like Sir Richard Grenville. Grenville had been reported as being killed in 1627, but later served in Ireland returning to England in late 1643 to become Sir William Waller's Lieutenant-General of Horse. However, early in 1644 he changed sides, which earned him the enmity of Parliament who dubbed him 'Skellum Grenville'. Another veteran of 1627 was George Monck, who first served the king and later Parliament and would bring about the Restoration in 1660.

Council re Margaret and Elizabeth Gates, 30 September 1637, p.82.
24   TNA SO 1/1, Petition to William Acton, 14 May 1629 Ensign Acton had been reportedly captured on 29 October 1627.
25   Hearing (ed.), *Dorset Quarter Sessions*, p.116.

Appendix I

Regiments 1624–1628

Regiments of Mansfeldt's Army

In 1624 the following were delivered to Count Mansfeldt:[1]

Company	Raised	Mustered
Colonel Earl of Lincoln's Regiment of Foot:		
Earl of Lincoln	Lincolnshire	250
Lieutenant Colonel Allen	Lincolnshire	250
Major Boughton	Devon	200
	Dorset	50
Captain Sir Edward Fleetwood	Oxfordshire	200
Captain Wirly	Buckinghamshire	100
	Hertfordshire	100
Captain Reynolds	Devon	100
	Somerset	100
Captain Babington	Yorkshire	150
	City of York	50
Captain Sir Matthew Carey	Norfolk	200
Captain Barlee	Dorset	200
Captain Cromwell	Bedfordshire	100
	Huntingdonshire	100
Colonel Viscount Doncaster's Regiment of Foot:		
Colonel Viscount Doncaster	London	250
Lt Colonel Sir James Ramsey	London	250

---

1 HRO 5M50/1954 f.81.

THE FIRST BRITISH ARMY 1624-1628 (REVISED EDITION)

Major Alexander Hamilton	London	250
Captain Archibald Douglas	London	200
Captain Zouch	Northants	200
Captain John Douglas	London	200
Captain Bell	Warwickshire	200
Captain William Douglas	London	200
Captain George Kelwood	London	200
Captain Andrew Heathly	London	200

Colonel Lord Cromwell's Regiment of Foot:

Colonel Lord Cromwell	Staffordshire	150
	Derbyshire	100
Lt Colonel Dutton	Wiltshire	250
Major Gibson	Gloucestershire	250
Captain Bassett	Berkshire	200
Captain Lane	Northumberland	100
	Bedfordshire	100
Captain Vincent Wright	Hampshire	200
Captain Jenner	Suffolk	200
Captain Vaughan	Buckinghamshire	200
Captain Ousley	Worcestershire	150
	Rutlandshire	50
Captain Crave	Hertfordshire	200

Sir Charles Rich's Regiment of Foot:

Colonel Sir Charles Rich	Essex	250
Lt Colonel [Ralph] Hopton	Somerset	250
Major Killigrewe	Somerset	150
	Hampshire	100
Sir W[alter] Waller		
Captain Sir Marchant St Leger	Kent	200
Captain Burton	Essex	200
Captain Francis Hammon	Kent	200
Captain Winter	Essex	200
Captain Goring	Sussex	200
Captain Fowler	Gloucestershire	150
	Monmouthshire	50

Sir Andrew Grey's Regiment of Foot:

Colonel Sir Andrew Grey	Yorkshire	100
	Nottinghamshire	150

APPENDIX I

Lt Colonel Boswell	Yorkshire	100
	Suffolk	150
Major Doburne	London	250
Captain Daniel Murray	Norfolk	200
Captain Forbois	Norfolk	200
Captain Carey	Suffolk	200
Captain Ramsey	Surrey	200
Captain Williams	Herefordshire	150
	Monmouthshire	50
Captain Seaton	Salop	200

Sir John Burrow's Regiment of Foot:

Colonel Sir John Burrows	Kent	250
Lt Colonel Brett	Middlesex	250
Major Willoughby	Cambridgeshire	200
	Surrey	50
Captain William Lake	Hampshire	150
	Surrey	50
Captain Roberts	Yorkshire	200
Captain Webber	Essex	150
Captain Shipwith	Leicestershire	200
Captain Thomas Woodhouse	Oxfordshire	50
	Warwickshire	50
	Wiltshire	10
Captain Gorge	Hampshire	200
Captain Mustain	Derbyshire	50
	Northamptonshire	50
	Wiltshire	100

Four Regiments Raised 1624[2]

Earl of Essex's Regiment:

Sir Charles Rich
Captain Swinton
Sir Walter Devereux

---

2   BL Add ms 46,188, Jessop Papers vol. 1 Miscellaneous official and private correspondence and papers of Robert Devereux, 3rd Earl of Essex, seventeenth century, ff.32–33.

Captain Sir Robert Knollies
Captain Thornhurst
Captain Sir Sigismund Zinzay
Captain Terringham
Captain Wainesman
Captain Davis
Captain Higham

Earl of Southampton's Regiment:

Sir John Burlace
Captain Astley
Lord Wrothersly
Lord Montjoy
Captain Conners
Captain Sir Thomas Conners
Captain Sir Thomas Littleton
Captain Barclay
Captain Cromwell
Captain Hubbert
Captain Jucks
Captain Goring

Lord Willoughby's Regiment:

Sir Edward Conway
Sir Thomas Conway
Sir Peregrine Bertue
Captain Sir John Ratcliffe
Captain Fane
Captain Clapham
Captain Hunck
Captain Terrett
Captain Aspersnone
Captain Ramsey

Earl of Oxford's Regiment:

Sir James Pevesteine
Captain Seaton
Captain Lord Delaware
Captain Sir Dudley North
Captain Wainworth
Captain Halley
Captain Harry Croft
Captain Sir William Brunkard
Captain Goodwick
Captain Sir William Heydon
Captain Sir John Wentworth

APPENDIX I

Regiments Raised for the Cadiz Expedition:[3]

Unfortunately, there are several lists of the ten regiments, but not all of them agree on the officers who served in them, all have October 1625 pencilled in. What is known is that each regiment had ten companies. Prowde's, Gifford's company landed in Ireland after the Cadiz Expedition.

Duke of Buckingham's Regiment:

Lt Col Sir John Prowde; Lt Bromingham (killed), Ensign Owen, later Lieutenant Ensign Quarles.
Major Sir Thomas Thornhurst, Lt Proude, (killed); Ensign Russell
Lieutenant Clarke, Ensign Dighton
Captain Philip Gifford, Lieutenant Potts, Ensign James Bradsey or Brasey
Captain Knolles, Lieutenant Henry Neville, Ensign Green
Captain Elphenston, Lieutenant Richard Tremaine, Ensign Moore
Captain Paddon, Lieutenant Colwell, Ensign Pennant or Pettit
Captain Reynolds, Lieutenant Whitehead, Ensign Fearne or Lieutenant Cheverton, Ensign Muskeley,
Captain Kirton, Lieutenant Joseph[?] Donne, Ensign Nathaniel Otby later Lieutenant Ensign Paddon
Captain John Courtney, Lieutenant Brett, Ensign Ward or Were
Captain Robert Preston, Lieutenant Lee, Ensign Bagg

Colonel Sir Henry Bruce's Regiment:

The regiment was raised from recruits from Worcestershire, Yorkshire, Middlesex, Shropshire, Cornwall, Surrey, Cambridgeshire, Carmarthenshire and Breckon.[4]
Colonel's company Lieutenant St Paul, Ensign Robert Gibbs later Ensign Bruce
Lt Col Sir Henry Killegrew, Lieutenant Stephen Broadrepp, Ensign Andrew Bruce
Major Sir James Scott, Lieutenant Cowley, Ensign Boswell or Ensign James Stevens or Stephenson
Captain Patrick Wood, Lieutenant John Sandilance, Ensign Willoughby
Captain Thomas Cornwall, Lieutenant James Coffin, Ensign Lowe or Clarke
Captain Gilpin, Lieutenant Fox, Ensign Vaughan
Captain Ashley, Lieutenant Honeywood, Ensign Robinson
Captain John Glenne, Lieutenant Powell, Ensign Hobby
Captain Meauties, Lieutenant Bathurst, Ensign Williams
Captain Walter Norton, Lieutenant Thomas Jarvis, Ensign Nicholas Webber
Captain Edward Yates, Lieutenant W Houghton, Ensign Greene

---

3   Dalton, *The Life and Times of General Sir Edward Cecil*, pp.391–394, TNA 16/126, A list of the officers of the army employed in Cadiz, October 1625, f.657.
4   Hants RO 44M69/G5/37/29.

Colonel Sir John Burgh's (or Burrough's) Regiment:

The regiment was raised in Cornwall, Bedfordshire, Hertfordshire, Gloucestershire and Radnorshire.
Colonel Company, Captain Lt Jefferies or Gifford, Ensign Fanshaw
Lt Col Sr Alexander Brett, Lieutenant Turney, Ensign Bluddell
Major Sir Edward Hawley, Lieutenant Watts, Ensign Watnam
Captain John Brett or Betts, Lieutenant Yates or Gates, Ensign Waterman
Captain Terwhit or Territt, Lieutenant Atchison or Lee, Ensign William Gibbs
Captain Anthony Hill, Lieutenant Benjamin Outridd or Utricke, Ensign Foliatt
Captain Richard Bond, Lieutenant Searles, Ensign Knowles, later Lieutenant then Ensign Bond
Captain Thomas Lindsey, Lieutenant Jones Cludd, Ensign Foy or Scott
Captain Grove, Lieutenant Dodsworth, Ensign Foy or Thorpe
Captain Richard Grenville, Lieutenant Pollard, Ensign Thorpe
Captain John Parkinson, Lieutenant Long, Ensign Ayleworth

Colonel Sir Edward Conway's Regiment:

Colonel's company, Capt-Lt James Dawson, Ensign Pinchbeck
Lt-Col Sir Francis Willoughby, Lieutenant Chaworth, Ensign Ottley
Major Sir Sheffield Clapham, Lieutenant Browne, Ensign Wells or Thomas Witham Cole
Captain Goring, Lieutenant Shelley, Ensign Maxey
Captain Thomas Pelham, Lieutenant Powell later Ensign Thomas John Bamfield Kettleby
Captain Sir Francis Rainsford, Lieutenant Morgan or Huson, Ensign Bartlett or Birkley
Captain Michael Williams, Lieutenant Huson, later Ensign Cross Wake Starkey
Captain Alford, Lieutenant Heigham or Shelley, Ensign Hudson
Captain Dixon, Lieutenant Melcomb, Ensign Ayres
Captain John Hammond, Lieutenant Mark Robinson, Ensign Netherton or Champerhoone
Captain William Ogle, Lieutenant Plessington, Ensign George Hudson
William Hammond, preacher to the regiment from 3 July 1625 until 12 February 1625 or 1626.

Colonel General Robert Devereux, 3rd Earl of Essex's Regiment:

Colonel's Company, Lieutenant Frogmorton, Ensign Pelham
Lt-Colonel Sir Thomas York, Lieutenant Hinton, Ensign Fry
Major Hackliut, Lieutenant Hackluit, Ensign Gwynne
Captain Carleton, Lieutenant Ottley or Kelke, Ensign Kelke or York
Captain George Took, Lieutenant Edward Spring, Ensign Watts later Captain-Lieutenant, Ensign Atkins

APPENDIX I

Captain Peter Hone, Lieutenant Barrington, Ensign Henry Smith
Captain Thomas Shugburg, Lieutenant Culvert or Ensign Balleigh or Shugborough or Holcombe or Barleigh
Captain Peter Alley, Lieutenant Paul Quarles, Ensign Heigham or John Gore
Captain Nicholas Crisp, Lieutenant Edward Jarman, Ensign Potts or Le Grice, or German Terrick
Captain Samuel Leake, Lieutenant Goodridge, Ensign Matthews or Godfrey Lucas Dodson
Captain Bowles Jr, Lieutenant Vernon, Ensign Tennison
Captain Babbington, Lieutenant Lewkerne, Ensign Grimes

Colonel Sir Edward Harwood's Regiment:

Colonel's company, Captain-Lt Alcock, Ensign Arkeld or Stanton
Lt-Col Sir Thomas Morton, Lieutenant Charles Dawson, Ensign Betnam
Major Watkins, Lieutenant Humphries, Ensign Stewart or Wells
Captain Jackson, Lieutenant Tillier, Ensign Stanton or Peake
Captain Thomas Abraham, Lieutenant Lewkin, Ensign Champernowne
Captain Gitthorpe, Lieutenant Bridges, Ensign Lucas
Captain Hartley or Heatley, Lieutenant Anderson, Ensign Hunt
Captain Sir Archibald Douglas, Lieutenant Woodward, Ensign Saltingstone
Captain Richard Seymour, Lieutenant Westcott, Ensign Fortescue
Captain Thomas Masteron, Lieutenant Arthur Love, Ensign Stevens later John Potts
Captain William Morgan, Lieutenant Thomas James, Ensign Eden
William Cross, preacher to the regiment 3 July 1625 to 12 February 1625 or 1626

Sir Charles Rich's Regiment of Foot:

Colonel's company, Lieutenant Richard Leigh, Ensign Thomas Frith
Lt Col Sir John Ratcliffe, Lieutenant Drury, Ensign Colte
Major Francis Standish, Lieutenant Waller, Ensign Huncks or De La Hay
Captain Steward, Lieutenant Crisp Ensign Bowyer
Captain Robert Gray, Lieutenant Francis Grover, Ensign Ramscroft, later George Reading
Captain Sir Ralph Skelton, Lieutenant Gray Ensign Story
Captain Leighton, Lieutenant Williams, Ensign Price
Captain Walter Waller, Lieutenant Brand, Ensign Dudley
Captain Cook, Lieutenant Parry, Ensign Jarvis or Winter
Captain Thomas Staverton, Lieutenant Edmund Chadwell, Ensign Wormwood
Captain Sir Warham St Leger, Lieutenant Holdham or Hanson Ensign Robert Wright later Johnson
Captain Brand, Lieutenant Dudley, Ensign Steed

Major General Sir William St Leger's Regiment:

Colonel's Company, Lieutenant Judd; Ensign Whitney; later Lt, then Parker
Lt Colonel Sir John Gibson, Lieutenant Abraham, Ensign Hull
Major Sir Thomas Fryer, Lieutenant Stephens Ensign Spilling or Weston
Captain William Courtney, Lieutenant Prideaux or Farrar, Ensign Trefuse or Jordon
Captain Richards, Lieutenant Grace or Grove, Ensign Bockard
Captain Anthony Matthews, Lieutenant Roger Powell, Ensign Thomas Parker
Captain Richard Mostyn, Lieutenant Caspar Ward, Ensign Hugh Hookes
Captain John Read, Lieutenant Cole, Ensign Francis Maddison;
Captain Bowles, Lieutenant Sherrock, Ensign William Bowles
Captain Brutus Buck, Lieutenant Cooper Ensign Matthews later Robert Brereton
Captain William Molesworth Lieutenant Sydenham, Ensign Molesworth
Captain Sir Thomas Piggot, Lieutenant Powell, Ensign William Browne
Captain Judd, Lieutenant Cole, Ensign Broome
Provost Marshal, John Brookes

Lord Valentias Regiment, Master of the Ordnance:

Two entire companies of this regiment were drowned on 13 October 1625 when their ships were lost.[5]
Colonel's Company, Lieutenant Henry Frodsham, Ensign Francis Bowyer
Lt Col Sir Henry Spry, Lieutenant Searle or Greenfield, Ensign Greenfield or Spry
Major George Kennithorpe, Lieutenant Judge, Ensign Bennett
Captain John Hammond, Lieutenant Bowyer, Ensign Markham
Captain Thomas Brett, Lieutenant John Appleyard, Ensign Thomas Appleyard
Captain Nathaniel Taylor, Lieutenant Wilton or Ensign Leigh or Philip Hougedge Gates
Captain Henry Fisher, Lieutenant Brooke, Ensign Ogle
Captain Hackett, Lieutenant Bemersyde, Ensign Bullock
Sir Thomas Bruce, Lieutenant John Reynolds, Ensign William Fullerton
Captain Anthony Porter, Lieutenant Matthews, Ensign Veale
Captain John Tolkerne, Lieutenant Barnard or Barnett, Ensign Ogle
Captain John Hammond, Lieutenant Markham, Ensign [Blank]
Captain Michael Mathews, Lieutenant Porter, Ensign Neale
Captain Richard Vaughan, Lieutenant John Reynolds, Ensign William Fullerton
Captain Woodhouse, Lieutenant [blank], Ensign [blank]

Lord Marshal Edward Cecil, The Lord Wimbledon's Regiment:

Colonel's Company, Captain Lt Powell, Ensign Humphrey Hawkins

---

5   TNA SP 16/12 f.81.

Lt Col Sir George Blundell, Lieutenant Booth or Browne, Ensign Marbery or Gwynne
Major Robert Farrar, Lieutenant Bassett, Ensign Carlile or Eaton
Captain Alexander Crofts, Lieutenant Grimshaw, Ensign William Halls or Hungate
Captain Christmas, Lieutenant Sheverton, Ensign Dobson
Captain Nicholas Crisp, Lieutenant Wormwood, Ensign Lindsey
Captain John Paprill, Lieutenant Richard Burthogg, Ensign Dixon or Paparell
Captain William Bridges, Lieutenant Tristiam Horner, Ensign Carew; later Thomas Rowse
Captain Robert Gore, Lieutenant Browne, Ensign Pagitt
Captain Edward Leigh, Lieutenant John Felton[6], Ensign Dedham
Captain Anthony Leigh, Lieutenant Talbot, Ensign Bagnall
Captain George Blundell, Lieutenant Whitehead, Ensign Blundell
Captain John Talbot, Lieutenant Bagnall, Ensign Malbery
Captain Henry Huncks, Lieutenant Musgrave, Ensign Wood
Captain James Powell, Lieutenant Francis, Ensign Hall

Regiments during the Isle of Rhè Expedition

Colonel Peregrine Beatie's Regiment of Foot:

Raised in May 1627 from the companies from the Cadiz expedition who had landed in Ireland and Irish recruits.

Ensign Richard Morrison; an ensign on 2 November 1627
Lieutenant Colonel Vaughan; killed on 29 October 1627
Lieutenant Colonel William Molesworth; regiment's lieutenant colonel by May 1628
Lieutenant Robert Molesworth; lieutenant on 2 November 1627
Major Sir Francis Willoughby; on 12 July 1627 'at night by chance treading upon a pike of a pallisadoe was run through the foot'
Captain Vaughan; John Reynolds, Ensign Thomas Fullerton
Captain Pelham
Captain Cooke
Captain Gilpin
Captain Henry Ashley, Lieutenant? killed on 29 October 1627, then Lieutenant Alexander Ramsey
Capt Richard Bolles or Bowles, Lieutenant ? killed on 29 October 1627, then Lieutenant John Powell.
Ensign Charles Hyerin
Captain Browne
Captain York; he was succeeded by Captain Seymour

---

6   Felton would assassinate Buckingham in 1628.

Captain Arthur Brett; a captain on 2 November 1627
Lieutenant Richard Cook
Captain Parr; still a captain on 2 November 1627, Ensign Benedict Winter
Captain John Powell; a captain by 2 November 1627, Ensign Anthony Thelwall
Robert Beal, Quartermaster
John Marshall, Provost Marshal

Colonel Sir Ralph Bingley's Regiment of Foot:

Raised in May 1627 from the companies from the Cadiz expedition who had landed in Ireland and Irish recruits. On 30 August 1627, the Lord Deputy of Ireland wrote to the English Privy Council:

> We the council have done our utmost to help Sir Ralph Bingley and Sir Piers Crosby to raise men. They are good and serviceable troops as you will see by the enclosed and sailed on the 25 [August] so that they should by now have joined His Majesty's Army.

It sailed from Waterford on 27 August 1627.[7] In January 1627/28 the regiment was ordered to quarter in Kent.

Colonel Sir Ralph Bingley; killed on 29 October 1627.
Captain John Roberts; killed on 29 October 1627, Lieutenant Robert Broughton
Captain Lewis Williams; killed on 29 October 1627.
Captain Barrett; killed on 29 October 1627.
Captain Starkey; killed on 29 October 1627, Lieutenant Willoughby; killed on 29 October 1627.
Captain Wood, Lieutenant Thomas Rawlins
Captain Dawson, Ensign Edward Thynne
Captain Marvyn, Ensign Andrew Mynn
Captain Owen, Ensign William Bingley
Captain Archibald Campbell

Colonel Sir Alexander Brett's Regiment:

Raised March 1627 and from 10 May until 21 June 1627 the regiment, which was 1000 strong was quartered in the Isle of Wight. When the regiment returned from the Île de Rhé expedition from January 1627/28 the regiment was ordered to quarter in Dorset.

Colonel Sir Alexander Brett; He was killed on 29 October 1627.
Lieutenant Colonel Sir Thomas Thornhurst; killed on 12 July 1627.
Sergeant Major Thomas Fryer, Ensign John St Leger
Captain Richards; killed in the Isle of Rhè as a major.

---

7   Lyle, *APC 1627*, p.294.

APPENDIX I

Captain Glen; killed on 12 July 1627
Captain Molesworth
Captain Robert Preston; killed on 29 October 1627.
Captain Babington; killed on 12 July 1627.
Captain Sir Alexander Brett junior; killed on 29 October 1627.
Captain Galpin; gone by 2 August 1627.
Captain John Hamon or Hammond
Captain Robert Clarke, Ensign Daniel Sheldon
Captain Jeremy Brett, Lt John Appleyard, Ensign Lionel Batty
Captain Turney, Ensign Thomas Hobby
Captain Thomas Brett, Lieutenant John Ferrar, Ensign Henry Honeywood
John Brookes, Provost Marshal

Colonel Sir John Burgh's (or Burrough's) Regiment:

Raised March 1627. In January 1628, the regiment was ordered to quarter in Somerset.

Colonel Sir John Burgh; killed during the siege.
Lieutenant Walter Owen
Lieutenant Colonel Sir Edward Hawley; He took over the regiment but was killed on 29 October 1627.
Sergeant Major Sir Richard Grenville; later lieutenant colonel and took over the regiment.
Lieutenant Arundell Thorp Ensign George Monck
Captain Grove
Captain Thirwhit; killed on 29 October 1627, Ensign William Gibbs
Captain Robert Hammond, Lt Robert Markham, Ensign Henry Markham
Captain Haytley or Heetley; killed on 12 July 1627.
Captain Shugbury; killed during the siege.
Captain Betts; killed on 29 October 1627.
Captain Blundell; killed 12 July 1627.
Captain Bolles
Captain Watts; killed on 29 October 1627.
Captain Whitehead or Whitmore, Lieutenant Christopher Burgh
Captain Fanshaw, Ensign Wainwright
Captain Gates, Lieutenant Thomas Gates

Colonel Sir Edward Lord Conway's Regiment:

In January 1628, the regiment was ordered to quarter in Hampshire. On 26 August 1628, his officers were:

Colonel's company, Lieutenant Henry Shelley, Ensign Thomas Sheldon
Lt Col Sir James Scott, Lieutenant Thomas Cowley, Ensign Robert Dixon
Major George Kenethorpe, Lieutenant James Stephenson, Ensign John Langridge

Capt Henry Hanks or Huncks, Lieutenant Thomas Feameson, Ensign William Luttrill
Captain Sir Francis Rainsford, Lieutenant George Hewitson, Ensign Maximilian Bartlie
Captain Goring; He was killed on 12 July 1627.
Captain Thomas Pelham, Lieutenant John Colt, Ensign Knight
Captain Ashley
Captain Sir William Ogle, Lieutenant Daniel Scott, Ensign Harvey Charnlollony
Captain Thomas Dixon, Lieutenant William Welcombe, Ensign John Aires
Captain Lancelot Alford, Lieutenant James Stephenson, Ensign John Griffin then James Bennett
Captain Thomas Alcock, Lieutenant Thomas Pinchbeck, Ensign Henry Keyes then John Massey then John Middleton
Captain Tolkerne
Thomas Arnold, Provost Marshal

Colonel Sir William Courtney's Regiment:

On 22 November 1627, the regiment was in Sussex, when was ordered to be sent to Kent. However, in January 1628 the regiment was ordered to quarter in Sussex.

Colonel Sir William Courtney; After Sir George Blundell was killed he succeeded him as Major General and then Colonel General when Burgh was also killed. Ensign Thomas Hamlyn
Lieutenant Colonel Sir Thomas York.
Major Ferrar; captured on 29 October 1627.
Captain Sir George Blundell; mortally wounded on 12 July 1627, after being shot through both hips.
Captain Paddon; killed on 29 October 1627.
Captain Cornwall; killed on 29 October 1627.
Captain Country; killed 12 July 1627.
Captain Powell
Captain Reynolds was killed on 29 October 1627
Captain Peter Mewfus, Ensign Lewis Williams
Captain Donne; captured on 29 October 1627.
Captain Thomas Brett
Captain Watkins
Captain Sir Francis Willoughby, Lieutenant John Manley Ensign Fane Beecher
Captain Gilpin his ensign was killed on 29 October 1627.
Captain Browne; by 2 August 1627 had taken over Captain St Paul's company, but was killed on 29 October 1627. Ensign John Tarbuck
Captain Pennant; killed on 29 October 1627.
Lieutenant Willoughby; killed on 29 October 1627.
Ensign Sparr; killed on 29 October 1627.
Ensign Willoughby; killed on 29 October 1627.

APPENDIX I

John Manley, Quartermaster
John Austin, Provost Marshall
Mr Brewer, Surgeon

Colonel Sir Piers Crosby's Regiment of Foot:

Raised from Irish recruits in May 1627 and formed into 10 companies. Despite being paid £3,191 13 shillings and 4 pence for victualling, clothing, arming and transporting the regiment, it the Irish council still needed £200 'for the ten Irish captains to furnish colours, drums and halberds'.[8]

In January 1627/28, the regiment was ordered to quarter in Essex. Captain Carey's company was at first billeted at Malden, but constant complaining of the town's inhabitants resulted in his company moving to Witham, where on 17 March 1628, St Patrick's Day, the company rioted where 'many townsmen and soldiers were dangerously wounded'. This time it does not appear to have been the soldiers' fault. Wearing red crosses in their hats to celebrate St Patrick's Day, 'an untoward boy' tied a red cross to a dog's tail and to a whipping post. It was reported that 30 or 40 were killed on both sides, unfortunately the parish registers for this period have not survived to confirm the number. However, the soldiers were disarmed by their major.[9]

The regiment was quickly ordered to Kent, where six companies were ordered to quarter at Sandwich and Hythe in Kent and 300 at Canterbury.

Colonel Sir Piers Crosby, Lieutenant Purcell, Ensign Butler
Lieutenant Colonel John Butler, Lt James Keating, Ensign Michael Butler
Major Thomas Esmonds or Osmond
Lt Patrick Welch, Ensign Alexander Herbert
Captain Sir Rose Carew or Carey
Captain Cave
Captain John Crosby, Lt Thomas Taylor, Ensign William Crosby
Captain Shortall
Captain John Stafford, Lt Matthew Goldingham, Ensign James Batty
Captain Roger Donoghue, Lt Edward Furlough, Ensign Brian Crosby
Captain Richard Crosby.
Captain Sir Morgan Cavenaugh, Lt Turlough Mohun, Ensign Morgan Canenaugh

Colonel Sir Thomas Morton's Regiment:

On 18 June 1627 William Handson of Colonel Morton's Regiment signed for six pairs of breeches, 30 shirts, 30 pairs of breeches and 30 pairs of shoes for St Leger's company which he had received from the quartermaster of Hampshire. Another Hampshire warrant also provided 191 soldiers' coats,

---

8   *CSP Ireland 1627*, p.240.
9   Aylmer, 'Communication, St Patrick's Day 1628 in Witham, Essex' *Past and Present* vol.61, (Kettering: Oxford University Press, 1973), pp.139–148.

30 pairs of stockings, 30 pairs of shoes and 30 pairs of shirts, at a cost of £119 16s 7d, which may have been the clothes which Handson had signed for.[10]

Colonel Sir Thomas Morton
Lieutenant Colonel Sir Warwick St Leger.
Major Watkins; killed on 29 October 1627.
Captain Abraham; killed on 29 October 1627.
Captain Masterson; killed on 29 October 1627.
Captain Anthony Hill
Captain Taylor
Captain Bond
Captain Edward Spring; killed on 29 October 1627.
Captain Judd
Captain Thomas Sherley
Lieutenant John Knolles; His right leg was broken 'all to shivers' at Cadiz; was shipwrecked at Baltimore, to the loss of all he had, even his shirt; received two musket wounds in the Isle of Rhè.[11]
Captain Dawson
Captain Handson; By 2 August 1627 he had taken over Captain Morton's company.
Lieutenant Outradd; killed on 29 October 1627.

Colonel William, Earl of Morton's Regiment of Foot:

On 7 August 1627 Morton was commissioned to raise 2,000 men in Scotland. The recruits were ordered to muster at Edinburgh on 15 September. The regiment was to consist of 22 companies each with 10 officers and 80 men, although Morton had permission to reduce the number of companies and increase their strength. This he seems to have done because a commission dated 10 September 1627 requires an unnamed captain to raise a company of 100 men.[12]

The regiment appears to have received the following arms at Portsmouth in October 1627:

Muskets with bandoliers and rests	1,000
Armour	1,000
Swords, girdles and hangers	2,000
Long pikes	1,000[13]

---

10 Hants RO 44/M69/G5/48/113, Receipt by William Hanson of Colonel Morton's Regiment to Sir Thomas Jervoise for parcels of clothes, nd [c1627] 44/M69/G5/48/36/1–3, Accounts of pay, conduct, clothes etc., to soldiers, 30 March 1627.
11 *CSPD 1629*, p.489.
12 *Acts of the Scottish Privy Council, 1627*, pp.38, 50–51. National Archive of Scotland, GD248/458/12.
13 WO 55/1684, Surveys of stores and arms ships returning from Rhè, 1627.

APPENDIX I

The regiment could not join Buckingham in time for the Isle of Rhè and from 20 November 1627. Oglander describes the regiment at this time:

> There were 1,500 Scotch soldiers… billeted in our island… There were 12 commanders and every captain was a knight, but commanders and common soldiers were most inexperienced in martial discipline. They lay here a long time to the great sorrow, loss and undoing of the whole island – a people base and proud…
>
> They were such a burden to us, for with rest and high feeding these poor knaves grew so basely peremptory, relying more on hope of their fellows taking part with them than on their own valour, that they not only committed divers murders but also became a terror to the inhabitants.[14]

On 14 August 1628 with the 'Island being miserably oppressed with the Scotch Regiment' the Islanders complained to the king about the 'murders, rapes, robberies, burglaries, getting of bastards and almost the undoing of the whole island' committed by the regiment. During this time the regiment mustered 1494 men and cost the island £8136 3s 6d. Petitions were sent to the King and his Privy Council complaining of their conduct, but despite appeals to the officers, the soldiers continued their abuses.[15]

On 1 September 1628 with the army once more mustering to relieve La Rochelle the regiment was ordered to Portsmouth and two days later left the Isle of Wight, much to the relied of the islanders.

On 3 September Oglander wrote:

> We were freed from our Egyptian thraldom, or like Spain from the Moors. For, since the Danish slavery, never were these Islanders so oppressed as in January 1627, and so till the Michaelmas following, when the regiments of Scots ate and devoured the land.[16]

In 1628 the officers were:

Colonel William Douglas, Earl of Morton
Lieutenant Colonel Sir John Balfour
Major Sir John Meldrum; succeeded Balfour as lieutenant colonel
Major Stuart; Succeeded Meldrum as major and was still in the regiment in 1629.
Captain Lord Gray, Lieutenant John Gray, Ensign Gilbert Gray
Sergeant Alexander Caddell
Captain Sir George Hayes; took over the regiment.
Captain Sir John Hamilton
Captain Sir William Balladine or Ballantine
Captain Sir William Carr
Captain Sir James Jefferies

---

14   Oglander, *His observations*, pp.26–27.
15   IoW RO OG/14/1, Oglander, *His observations*, pp.36–37.
16   Oglander, *His observations*, pp.45–46.

Captain Sir David Home or Hume of Wedderburne
Captain James Hayes
Captain Daliell
Captain Sir Archibald Campbell
Captain Magnaities
Captain Lesley
Captain Ramsey
Captain William Hay
Captain Kammell
Captain M Kammell
Captain Uttison
James Melville, Quartermaster under Morton and Hay
Robert Carmichael, Provost Master under Hay by 1629

Colonel Sir Charles Rich's Regiment:

Raised March 1627. In January 1627/28 the regiment was ordered to quarter in Wiltshire.

Colonel Sir Charles Rich 'dangerously hurt in the shoulder' on 12 July and killed on 29 October 1627.
Lieutenant Colonel Sir John Ratcliffe; killed on 29 October 1627.
Major Francis Standish; killed on 29 October 1627.
Captain Sir Warren St Leger
Captain Sir Ralph Shelton; killed on 29 October 1627.
Captain Hone
Captain Cartleton; killed on 29 October 1627.
Captain Paparell
Captain Morgan
Captain Gifford
Captain Cooke
Captain Gellard
Captain Brands; killed on 29 October 1627.
Captain Leigh; killed on 29 October 1627.
Captain Drury?; killed on 29 October 1627.

Colonel Sir James Ramsey Regiment:

Raised 8 September 1627 and mustered 1,000 strong, which was part of the 2,000 recruits, which were to rendezvous at Plymouth. Each company was to be 83 men strong. In January 1627/28 the regiment was ordered to quarter in Berkshire.
    The following were appointed to be his officers:[17]

Colonel Sir James Ramsey, Lieutenant Roger Powell, Ensign Andrew Bruce
Captain Robert Le Gris or Grice, Lt Edward Jarman, Ensign John Terrike

---

17   APC 1627, pp.15–16.

APPENDIX I

Captain Reginald Mohun, Lt Banfield Leigh, Ensign Henry Morgan
Captain Christopher Jackson: By 2 November 1627 Captain John Fowler had taken over the company.
Lt Henry Frodsham, Ensign Edward Frodsham
Captain Humphrey Hawkins, Lieutenant Hugh Hookes, Ensign Jacob Lovell
Captain John Longworth, Lieutenant John Potts, Ensign Gwinn
Captain George York, Lieutenant John Fox, Ensign Francis Bowyer
Captain Sir William Tresham or Tresson
Lt Thomas Witham, Ensign Willoughby
Captain John Pell, Lt Thomas Parker, Ensign Henry Baskerville
Captain Bartholomew Jukes or Jewkes
Lt Theodore Stephens, Ensign John Wood
Captain Francis Williams, Lt John Felton, Ensign Edward Salter
Captain John Read, Lieutenant Saunderline, Ensign Graham
Captain John Parkinson, Lt Thomas Long, Ensign Robert Wright
Quartermaster John Sutton

Colonel Sir John Ratcliffe's Regiment:

Raised 8 September 1627 and mustered 1000 strong, which was part of the 2000 recruits, from Derbyshire, Warwickshire, Cambridgeshire, Northamptonshire, Radnor, Buckinghamshire, Bedfordshire, Herefordshire, Leicestershire, Staffordshire and Oxfordshire.

The following were appointed to be his officers:[18]

Colonel Sir John Ratcliffe, Lieutenant Dawson, Ensign Pigot
Captain Peter Alley; formed by 39 men
Lieutenant Quarles
Ensign Gove
Captain Peter Hone, Lieutenant Coffin, Ensign Thomas Kettleby
Captain Edward Yates, Lt Stephen Broadripp, Ensign Jelfe
Captain Alexander Crofts, Lieutenant Halsey, Ensign Burley
Captain William Bridges, Lieutenant Horner, Ensign Thomas Rowse
Captain Sir Thomas Pigot, Lt William Browne, Ensign Scott
Captain Robert Gray, Lieutenant Grover, Ensign Peyton Cooke
Captain Thomas Linsey or Lyndsey
Lieutenant Linsey, Ensign William Mathews
Captain John Parkinson, Lt Francis Madison, Ensign Wright
Captain Michael Mathews, Lt Edward Porter, Ensign Andrews
Captain Thomas Staverton, Lt Edward Chadwell, Ensign Richard Andrews
Quartermaster Anthony Witherings

---

18 APC 1627, pp.15–16.

THE FIRST BRITISH ARMY 1624-1628 (REVISED EDITION)

Sir Charles Rich's Regiment:

Raised March 1627 when the officers were:

Sir Charles Rich
Lieutenant Colonel Sir John Ratcliffe; later a colonel of his own regiment.
Major Standis
Captain Warren St Leger
Sir Ralph Sheton
Captain Hone
Captain Carleton
Captain Paperell
Captain Morgan
Captain Philip Gifford.

Colonel Sir Henry Spry's Regiment:

The regiment, 1000 strong, was quartered in the Isle of Wight from 27 May until 21 June 1627. When it returned from the Isle of Rhè Expedition it was quartered in Sussex. On 22 November 1627, the regiment was when was ordered to be sent to Kent. However, in January 1628, the regiment was ordered to quarter in Gloucestershire, but in August 1628 it returned to Hampshire to taken part in the La Rochelle Expedition.

Colonel's Company:

Lieutenant Colonel Sir John Tolkern; killed on 29 October 1627.
Major Philip Hackluit; succeeded Tolkern
Lieutenant John Walsh, Ensign Robert Grant
Captain Richard Seymour, Lt Sebastian Westcott, Ensign John Copley
Captain Courtney
Captain Leake; killed on 29 October 1627.
Captain Mostian by 2 August 1627 had been succeeded by a Captain Ward.
Captain Brutus Buck; succeeded Hackluit as major Lt John Simonds
Ensign Nicholas Rea
Captain Owen, Lieutenant Hull, Ensign Harries
Captain York; killed on 29 October 1627.
Captain Constance Ferrar, Lt Francis Norwood, Ensign William Mann
Captain Arthur Spry, Lt Edmund Hackluit, Ensign William Mintridge
Captain Thomas Ogle, Lt William Pomroy, Ensign Nicholas Barrett
Captain Peter Aylworth, Lt Henry Stanford, Ensign Thomas Langford
Captain Thomas Jernegard, Lt William Fortescue, Ensign John Bowers

Lord Montjoy's (later Earl of Newport's) Troop of Horse:

Served in the Isle of Rhè campaign. His troops consisted of three corporals and 19 gentlemen in 1628.

APPENDIX I

Sir William Cunningham's Troop of Horse:

Raised May 1627 and mustered only 63 troopers. Served in the Isle of Rhè campaign where Cunningham was killed. Only 10 troopers survived the campaign.

The following are known to have served at the Île de Rhé, but their regiments are not known:

Captain Michael Woodhouse killed 12 July 1627.
Lt Powell; killed 12 July 1627
Lieutenant Sydenham; killed 12 July 1627.
Captain St Paul; slain in the siege.
Captain Houghton; slain in the siege.
Lt Lower; slain in the siege.
Lieut Waller; slain in the siege.
Ens Coitz; slain in the siege.
Captain Murton; died on Isle of Rhè.
Capt Roger Greenfield died on Isle of Rhè.
Captain James Bradsey formerly ensign to Captain Philip Gifford in the Cadiz Expedition, On his return in 1325 he was forced by foul weather to remain in Ireland 13 months. During the Isle of Rhè he received 40 wounds and was left for dead.[19]
Captain Fanshaw; captured on 29 October 1627
Captain Hinton; captured on 29 October 1627
Captain Whitehead; captured on 29 October 1627
Captain Owen; captured on 29 October 1627
Captain Norton; captured on 29 October 1627
Lieutenant Jessop; captured on 29 October 1627
Lieutenant Bassett; captured on 29 October 1627
Lieutenant Lee; captured on 29 October 1627
Lieutenant Kelke; captured on 29 October 1627
Lieutenant Jaffar; captured on 29 October 1627
Lieutenant Huckley; captured on 29 October 1627
Ensign Rend; captured on 29 October 1627
Ensign Stomp; captured on 29 October 1627
Ensign James; captured on 29 October 1627
Ensign Dimmock; captured on 29 October 1627
Ensign Newcomte; captured on 29 October 1627
Ensign Ackton; captured on 29 October 1627
Ensign Moyle; captured on 29 October 1627
Ensign Sledde; captured on 29 October 1627
Ensign Brett; captured on 29 October 1627
Ensign Quinet; captured on 29 October 1627
Ensign Yates; captured on 29 October 1627

---

19 CSPD 1629, p.487.

THE FIRST BRITISH ARMY 1624-1628 (REVISED EDITION)

Artillery:

Between 13 August and 30 June 1627, the following made up the artillery train:[20]

Sir Thomas Love, Master of the Ordnance
10 halbertier to guard the Master of the Ordnance
Ralph Sampford, ensign to the train
Henry Johnson, clerk of the Ordnance
Robert Morace and Edward Smith, assistant clerks
Henry Ielt Master Gunner
John Allen and George Parsons, Master Gunners mates
John Andrews, Jeremy Spratt, Robert Squire and Joseph Dudley, four quarter gunners.[21]
Thomas Williams, George Allen, Robert Parsons, William Yawling, Libias Craft, Henry Hook, Thomas Day, Richard Briggs (quarter gunner by end of 1627), Thomas Piborn, Alexander Stringer, Edward Hopcroft, Robert Drayton, John Allen, Thomas Raymond, Cuthbert Ethrington, John Mausby, Thomas Young, Gunners.
William Green, drum of the train.
Richard Green, fife of the train.
Richard Howse, Master Carpenter
William Petley 'one of two of his servants'
William Abbot, Master Smith
William Mylland 'one of four of his servants'
Richard Usted, Master Wheeler.
Henry Penfold 'one of two of his servants'.
John Mason, Tent keeper, Nicholas Harbart, his servant
Thomas Gore, Master Armourer
Thomas Gust, his servant
James Crewes, Master Ladlemaker
John Brett, servant to James Crewes
Roger Phillips, Master Gunmaker
William Rafe and Thomas Lee, his two servants
Shadrack Miller, Master Cooper
Thomas Griffin and James Starr 'his two of his servants'.
William Devereux, Barber Surgeon
Alexander Fair and Isaac Thorneton his servants
Robert Norton, Chief Petarder

---

20   TNA AO 1/299/1136.
21   Only three are mentioned in a list compiled at the end of 1627, Jeremy Spratt, Richard Briggs and Robert Seaborne.

APPENDIX I

1628 Regiments[22]

Lord General's The Earl of Lyndsey's Regiment of Foot:

Colonel's Company, Capt Lt Monk, Ensign Maynard
Lt Col. Robert Hammond, Lt Markham, Ensign Skinner
Major Thomas Groves, Lt Mayes, Ensign Essington
Captain Long, Lt Treswell, Ensign Trebicon
Captain Waitmore, Lt Burgh, Ensign Matthews
Captain Gaiter (later Collins), Lt Groves, Ensign Collins
Captain Skudamore, Lt Hayman, Ensign Senior
Captain Gibbs, Lt Leach, Ensign Hibbins
Captain Markham, Lt Walsh, Ensign Subbury
Captain Whitehead, Lt Tremayne, Ensign Lower
Chaplain, Mr Carey
Quartermaster, Mr Withering
Provost Marshal, Mr Mitchell

Major General Sir Francis Willoughby's Regiment of Foot:

Sir Francis Willoughby, Capt Lt Owen, Ensign Beecher
Lt Col. Walter Norton, Lt Lea, Ensign Williams
Major James Dawson, Lt Moon, Ensign Thymm
Captain Gorge, Lt Cason, Ensign Wilson
Captain Bagg, Lt Rogers, Ensign Ward
Captain Mervyn, Lt Chantcell, Ensign Merez
Captain Bardsey, Lt Castle , Ensign Paddox
Captain Broughton, Lt Newcombe, Ensign Day
Captain Rawlings, Lt Games, Ensign Edwards
Captain Bagnell, Lt Dowett, Ensign Kelley
Chaplain, Mr Sturton
Quartermaster, Mr Jefferies
Provost Marshal, Mr Long

Colonel Sir James Ramsey's Regiment of Foot:

Col. Sir James Ramsey, Capt Lt Bruce, Ensign Hammond
Lt Col. Sir William Tresham, Lt Gwilliams, Ensign Willoughby
Major Fowler, Lt Frodsham, Ensign Fanning
Captain Sir Robert Le Grice, Lt Jermin, Ensign Stephen
Captain Pell, Lt Baskerville, Ensign Gagoyne
Captain Langworth, Lt Gwynn, Ensign Sandford
Captain Hawkins, Lt Hook, Ensign Underhill
Captain Mohun, Lt Lovell, Ensign Potts

---

22 TNA SP 16/116 f.125.

Captain Jukes, Lt Stepcher, Ensign Wood
Captain Rowell, Lt Salter, Ensign Winde
Captain Fox, Lt Bowyer, Ensign Fitzjames
Captain Sandelyn, Leviston, Ensign Greaham
Chaplain, Mr Mighan
Quartermaster, Mr Batchelor
Provost Marshal, Mr Ashfield
Surgeon, Mr Griffin

Colonel Earl of Newport's Regiment of Foot:

Col Earl of Newport, Capt Lt Crane , Ensign Legg
Lt Col Paprill, Lt Willis, Ensign Peacock
Major Sir Sheffield Clapham, Lt Eaton, Ensign Willoughby
Captain Story, Lt Bell, Ensign Thring
Captain Smyth, Lt Gates, Ensign Saunders
Captain Foulks, Lt Drake, Ensign Coleman
Captain James, Lt Everard, Ensign Povey
Captain Neal, Lt Bold, Ensign Osborne
Captain Bell, Lt Dew, Ensign Whatman
Captain Keark, Lt Willett, Ensign Littlebury
Chaplain, Mr Theobald
Quartermaster, Mr Ellis
Provost Marshal, Mr Pollard

Quartermaster General Sir James Scott's Regiment of Foot:

Col Sir James Scott, Capt Lt Stevenson, Ensign Waller
Lt Col Kennythorp, Lt Mauihill, Ensign Langrish
[then] Lt Col Raysnford, Lt Masey, Ensign Midleton
Major Pelham (later Huncks], Lt Jennison, Ensign Lutterell
Captain David Scott, Lt Carleton, Ensign Jefferies
Captain Shelley, Lt Scott, Ensign Shelley
Capt Welcombe (later Frodsham, Lt Rellinson, Ensign Butler
Captain Bradrip, Lt Lee, Ensign Watchowse
Captain Colt (later Baker), Lt Woodman, Ensign Bacon
Captain Plessington, Lt Cooper, Ensign Padley
Captain Cowley, Lt Dixon, Ensign Heath
Chaplain, Mr Clungson
Quartermaster, Mr Ructing
Provost Marshal, Mr Arnold
Surgeon, Mr Jones

Colonel Sir Thomas Fryer's Regiment of Foot:

Col Sir Thomas Fryer, Capt Lt Stephens, Ensign St leger
Lt Col Jeremy Brett, Lt Appleyard, Ensign Batty
Major John Hammond, Lt Evans, Ensign Tyler

APPENDIX I

Captain Turney, Lt Hobby, Ensign Hen
Captain R Clark, Lt Brett, Ensign Skelton
Captain Watts, Lt Freeman, Ensign Woolnor
Captain Thomas Brett, Lt Farrer, Ensign Honeywood
Captain Thomas Powell, Lt Hortley, Ensign Williams
Captain Leigh, Lt Townsend, Ensign Gray
Captain Carne, Lt Price, Ensign Stephens
Chaplain, Mr Eliis
Quartermaster, Mr Cossard
Provost Marshal, Mr Brooks
Surgeon, Ebenzer Smyth

Colonel Sir Peregrine Beatie's Regiment of Foot:

Col Sir Peregrine Beatie, Capt Lt Bamfield, Ensign Morrison
Lt Col Moulesworth, Lt Molesworth, Ensign More
Major Arthur Brett, Lt Cook, Ensign Godfrey
Captain Sir Roger Bertie, Lt Francis, Ensign Rentz
Captain John Powell, Lt Salt, Ensign Orton
Captain Parrey, Lt Mason, Ensign Pistor
Captain Ashley, Lt Ramsey, Ensign Danbridgecourt
Captain Sawyer, Lt Jennison, Ensign Cooper
Captain Reynell, Lt BellotEnsign Noakes
Captain Gilpin, Lt Quadrin, Ensign Blankard
Chaplain, Mr Bradley
Quartermaster, Mr Beale
Provost Marshal, Mr Cuppledike
Surgeon, Mr Clarietto

Colonel Sir Thomas Morton's Regiment of Foot:

Col Sir Thomas Morton, Capt Lt Bettenham, Ensign Moyle
Lt Col Sir Warrick St Leger, Lt Johnson, Ensign Ketteridge
Major Taylor, Lt Yates, Ensign Pretty
Captain Elphingstone, Lt Atkins, Ensign Smynoe
Captain Judd, Lt Smyth, Ensign Thomas
Captain Lukin, Lt Pardam, Ensign Barnes
Captain Woodward, Lt Sherborne, Ensign Appleton
Captain Lowe, Lt Mannering, Ensign Hanmer
Captain Sherley, Lt Knowles, Ensign Mitchell
Chaplain, Mr Bennett
Quartermaster, Mr Tenche
Provost Marshal, Mr Price
Surgeon, Mr Griffith

Colonel Ferrar's Regiment of Foot:

Col Ferrar, Capt Lt Bell, Ensign Wiseman
Lt Col Connisby, Lt Donne, Ensign Hudson
Major Donne, Lt Hambledon, Ensign Davis
Captain Ottby, Lt Legat, Ensign White
Captain Bassett, Lt Apsley, Ensign Poyntz
Captain Caswell, Lt Jocelin, Ensign Parrey
Captain Francis Clark, Lt Forbes, Ensign Norton
Captain Goodrick, Lt Harwood, Ensign Holbrook
Captain Browne, Lt Muckloe, Ensign Tarbuck
Captain Thelwell, Lt Dish, Ensign Crisp
Chaplain, Mr Ansell
Quartermaster, Mr Wase
Provost Marshal, Phillips
Surgeon, Bolyer

Colonel Philip Huckliut's Regiment of Foot:

Col Philip Huckliut, Capt Lt Weshe, Ensign Grant
Lt Col Buck, Lt Symmonds, Ensign Herring
Major Bolle, Lt Powell, Ensign Rea
Captain Seymour, Lt Banks, Ensign Copeley
Captain Owen, Lt Hall, Ensign Harris
Captain Hackliut, Lt Wright, Ensign Leighton
Captain Spry, Lt Hackuit, Ensign Mintridge
Captain Ogle, Lt Pomeroy, Ensign Barrett
Captain Ferrar, Lt Marbury, Ensign Champoon
Captain Ayslworth, Lt Standford, Ensign Langford
Captain Jersegan, Lt Fortescue, Ensign Bennet
Chaplain, Mr Forbes
Quartermaster, Mr Redar
Provost Marshal, Phillips

Colonel Dodo Von Knyphausen's Regiment of Foot:

Col Knyphausen, Capt Lt Horner, Ensign White
Lt Col Peter Hone, Lt Kettleby, Ensign Barnard
Major Edward Yates, Lt Jelfe, Ensign Jackson
Captain Sir Thomas Piggot, Lt Browne, Ensign Scott
Captain Gray, Lt Grovett, Ensign Cook
Captain Staverton, Lt Chadwell, Ensign Andrews
Captain Lyndsey, Lt Sucklyn, Ensign Matthews
Captain Halse, Lt Clapp, Ensign Killinghall
Captain Christopher Dawson, Lt Piggot, Ensign Wilkinson
Captain Keyes, Lt Burleigh, Ensign Willoughby
Captain Witham, Lt Gregson, Ensign Hyppisly
Chaplain, Mr Pemberton
Provost Marshal, Mr Pattison
Surgeon, Parr

Appendix II

Pay

Rank	1620[1]	1627[2]
Army		
Lord general of the army	£10	
Lord marshal	£4	
Major general	£2	
Quartermaster general	£1	
Provost marshal	6 shillings 8 pence	
Carriage master general	6 shillings 8 pence	
Train		
Treasurer at war	40 shillings	
Muster master general	20 shillings	
Commissary general	10 shillings	
Judge marshal	10 shillings	
Chaplain	6 shillings 8 pence	
Physician	6 shillings 8 pence	
Apothecary	3 shillings 4 pence	
Secretary to the council	5 shillings	
Surgeon	6 shillings 8 pence	5 shillings
Halberdier (body guard)	12 pence	

1 BL Kings 264 ff.260–261
2 *APC 1627*, p.291, SP 16/60 ff.89–90

THE FIRST BRITISH ARMY 1624-1628 (REVISED EDITION)

Infantry		
Colonel		
Lieutenant colonel	£1	£1
Major	6 shillings 8 pence	10 shillings
Quartermaster	5 shillings	6 shillings 8 pence
Provost	5 shillings	5 Shillings
Carriage master	5 shillings	3 Shillings
Preacher	3 shillings 4 pence	
Chief surgeon	4 shillings	5 Shillings
Assistant surgeon	4 shillings	2 Shillings 8 pence
Captain (of colonel's company)		2 Shillings
	8 shillings	8 Shillings
Captain	6 shillings	8 Shillings
Captain Lieutenant	4 shillings	
Lieutenant	3 shillings	3 Shillings
Ensign	2 shillings 6 pence	2 Shillings 6 pence
Sergeant	12 pence	8 Shillings (per week)
Drummer	12 pence	5 Shillings (per week)
Gentleman	12 pence	
Corporal	12 pence [sic]	5 shillings (per week)
Soldier	8 Pence	8 Pence
Cavalry		
Lord General of horse	£4	
Lieutenant general of horse	40 shillings	
Major General of horse	30 shillings	
Quartermaster general of horse	6 shillings 8 pence	
Provost	5 shillings	
Preacher	4 shillings	
Chief surgeon	4 shillings	
Captain of horse	8 shillings	
Lieutenant of horse	5 shillings	
Cornet	4 shillings	
Trumpeter	2 shillings 6 pence	
Corporal	2 shillings 6 pence	

Quartermaster	2 shillings 6 pence	3 shillings 4 pence
Surgeon	2 shillings 6 pence	
Cuirassier	2 shillings	
Harquebusier	1 shilling 6 pence	

Train of artillery

Master of the ordnance	£3	£3
Lieutenant general	20 shillings	10 Shillings
Surveyor	6 shillings 8 pence	
Clerk to the surveyor	2 shillings	
Auditor	6 shillings 8 pence	
Clerk to the auditor	2 shillings	
Paymaster	6 shillings 8 pence	
Clerk to the paymaster	2 shillings	
Trench master general	10 shillings	'Not found necessary'
Engineer	6 shillings 8 pence	6 shillings 8 pence
Clerk to the engineer	2 shillings	2 shillings
Conductors for the works	2 shillings	'Not found necessary'
Clerk of the ordnance	6 shillings 8 pence	5 shillings
Clerk to the clerk of the ordnance	2 shillings 6 pence	2 shillings
Gentleman of the ordnance	6 shillings 8 pence	'not allowed'
Harquebusier (guard of the master)	1 shilling 6 pence	
Halberdier (guard to the master)	10 pence	1 shilling
Quartermaster	10 shillings	
Horseman to attend him	1 shillings 6 pence	
Furrier	2 shillings	
Under furrier	1 shilling	
Commissary of victuals	5 shillings	
Muster master	6 shillings 8 pence	
Clerk to muster master	2 shillings	
Purveyor general	6 shillings	'Not found necessary'
Horseman to purveyor general	1 shilling 6 pence	
Master of the carriages	6 shillings	
2 Halberdiers to attend him	1 shilling each	
Overseer to the carriages	3 shillings	
Master of mines	5 shillings	
Captain of pioneers	4 shillings	

THE FIRST BRITISH ARMY 1624-1628 (REVISED EDITION)

Lieutenant of pioneers	2 shillings	
Overseer of the pioneers' work	1 shilling 6 pence	
Chief Petardier	6 shillings 8 pence	6 shillings 8 pence
Attendant to petardier	1 shilling 6 pence	1 shilling 6 pence
Master gunner	6 shillings	4 shillings
Master gunner's mate	2 shillings 6 pence	2 shillings
Constable or quarter gunner	2 shillings	not necessary for service
Gunner	1 shilling 6 pence	1 shilling
Labourer	1 shilling	1 shilling
Provost marshal of ordnance	5 shillings	
Provost or jailor	1 shilling 6 pence	
Under jailer	1 shilling	
Founder of the brass ordnance	4 shillings	
His man	2 shillings	
Master of fireworks	4 shillings	
Chaplain	5 shillings	
Ensign	2 shillings	
Drummer	1 shilling	
Trumpeter	2 shillings	
Barber surgeon	2 shillings 6 pence	2 shillings 5 pence
Under barber surgeon	1 shilling	1 shilling 6 pence
Master carpenter	3 shillings	2 shillings
His mate	2 shillings	1 shilling
Carpenter	1 shilling 6 pence	
Assistant to the carpenter		
Master smith	3 shillings	2 shillings
His mate	2 shillings	1 shilling
Servants (six to a forge)	1 shilling 6 pence	
Master wheeler	3 shillings	2 shillings
His mate	2 shillings	1 shilling
Servant to master wheeler	1 shilling 6 pence	
Master farrier	2 shillings 6 pence	
Servant	1 shilling 6 pence	
Pioneer	8 pence	1 shilling
Conductor (one to every 160 cart horses)	2 shillings	2 shillings
Tent keeper	2 shillings	2 shillings
Servant to tent keeper	1 shilling	
Armourer	3 shillings	2 shillings

APPENDIX II

Servant to the armourer	1 shilling 6 pence	1 shilling
Furbusher for the store of arms	3 shillings	
Servant to furbusher	8 pence	
Basket maker for gabions, hurdles et cetera	2 shillings	2 shillings
Servant to basket maker	1 shilling	
Ladle maker	3 shillings	
Miner	?	1 shilling*
Surveyor of the water engines	?	3 shillings 6 pence
His assistant	?	
Servant to ladle maker	3 shillings	2 shillings 6 pence
Cooper	3 shillings	2 shillings
Servant to the cooper	1 shilling 6 pence	
Carter	12 pence	
Hired cart	6 shillings	

* Another source states 1 shilling 6 pence.

In May 1627 new daily rates of pay were introduced for the artillery, which were lower than those in 1620.

Appendix III

Clothing, 1625

In 1625 a survey was made of the troops at Plymouth for the attack on Spain the various counties had supplied their recruits with the following clothing:[1]

Bedfordshire	Captain Thomas Lynsey's company of grey cassocks lined yellow.
Berkshire	Captain Nicholas Crisp's company 'Their coats were prized to be worth 13 shillings but their coat being of the countries charge 16 shillings as the soldiers assure. All complain for shirt, shoes and stockings and most for breeches and doublets'. The muster roll for the company said the men were 'coated with coats of grey cloth lined through with white cotton and trimmed with ash coloured buttons'.
Breckon	Captain Walter Morton's company red cassocks lined blue.
Bristol	Captain Robert Gore's company in 50 in watchet coats lined yellow, 30 in blue cassocks lined green and 20 cassocks and unlined.
Buckinghamshire	Lieutenant Adrian Brooke and Ensign Banfield's companies wore blue cassocks lined white.
Cambridgeshire	Captain Yates' company, 'their coats were prized to be worth 9 shillings... All complain for want of shoes, shirts and stockings and most for breeches and doublets'.
Cardiganshire	Lieutenant Benjamin Ughted company in grey cassocks, 'very short'.
Carmarthenshire	Ensign Webber's company in dark grey cassocks unlined.
Cornwall	Captain Grenville's company, 'there is no complain of any things but want of apparel among the soldiers... It is alleged that some coats are not worth 8 shillings, some 10 shillings and some are not worth 6 shillings'.

---

1   SP 16/4/160 Report of Captain Edward Leigh, July 1625.

APPENDIX III

Devonshire	Captain John Courtney's company blue cassocks unlined.

Lieutenant Samuel Burnett's company blue cassocks.
Captain Richard Sermour blue cassocks unlined

Dorset	Lieutenant John Reynolds' company cassocks of 'several colours'.
Essex	Lieutenant John Cossin and Ensign John Pryse's company undyed cassocks with red and yellow lace.

Lieutenant Pate's company, 'Their coats are prized to be worth 8 shillings a piece and they have no lining in them… All complain for shirts, shoes and stockings and most for breeches and doublets'.

Glamorganshire	William Montgomery's company in watchett cassocks 'lined'.
Gloucestershire	Captain Parkinson's company in grey cassocks unlined.

Captain Peter Alley's company in grey cassocks unlined.
Captain Peter Aylworth's company in grey cassocks unlined.

Hampshire	Captain William Moulsworth's company of grey cassocks lined with white.

Lieutenant Ratlings' company in grey cassocks

Hertfordshire	Lieutenant John Appleyard and Ensign Thomas Appleyard's company blue cassocks lined white.

Captain Thomas Lynsey's Company mixed blue Cassocks.

Huntington	Ensign Mark Robinson's company watchett cassocks lined with yellow.
Kent	Captain Bolle's company 'Their coats are prized to be worth 10 shillings a piece… All complain for shirts, shoes and stockings and most for doublets and breeches'.

Captain Dixon's company, 'all complain for shirts, shoes and stockings and most for breeches and doublets… Their coats are prized to be worth 10 shillings a piece'.

Leicestershire	Captain Edward Leigh's company '16 [men have] run away with their coats'. No other mention of clothing. Blue coats
London	Lieutenant [unreadable] company 'their coats valued to be 8 shillings a piece.

Lieutenant Adye's company 'their coats are valued to be not worth more than 10 shillings a piece.
Lieutenant Spring's company 'there are three sorts of coats whereof some are lined with red, those are valued at to be worth but 5 shillings a piece. There be others lined with white and others with blue and those both are prized to be worth 10 shillings a piece'.
Ensign Grimshaw's company 'There are three sorts of coats in the company, the whitish grey prized to be worth 7 shillings and 6 pence; the brownish coats are worth 5 shillings and those that are lined with blue is prized to be worth 7 shillings a piece.

THE FIRST BRITISH ARMY 1624-1628 (REVISED EDITION)

	Captain Cooke's company 'Most of them are very poor in apparel their coats were prized to be worth about 8 shillings but some 7 shillings a piece'.
Middlesex	Ensign Otby's company grey cassocks lined yellow.
	Captain James Powell's Company grey cassocks lined yellow.
	Ensign Henry Smith's company grey cassocks lined yellow.
Monmouthshire	Lieutenant Richard Tremayne's company red cassocks unlined.
Northampton	Captain Anthony Leigh's company in blue cassocks with orange tawny lace.
Norfolk	Ensign Austin's company 'Their coats are prized to be worth 7 shillings and some 8 shillings a piece'.
	Lieutenant Burthogge's company 'Their coats were prized to be worth at 7 shillings some at 8 shillings a piece'.
	Lieutenant Groner's company 'Their coats is alleged that some are worth 8 shillings and some 7 shillings… All complain for shirts, shoes and stockings and the most for breeches and doublets'.
Norwich	Captain Anthony Leigh's company in blue cassocks 'with orange tawny lace'.
Oxfordshire	Lieutenant Powell's Company grey cassocks 'good and lined with yellow'.
Pembrokeshire	Lieutenant Benjamin Ughted company in dark grey cassocks.
Radnorshire	Captain Thomas Lynsey's company russet cassocks lined red.
Rutland	Captain Edward Leigh's company 'Their coats are valued to be worth not above 8 shillings a piece and they are lined through with white'.
Shropshire	Captain Leeke's and Lieutenant Carlisle's companies, 'their coats are prized to be worth 5 shillings, some 6 shillings and some 7 shillings, but it is said they cost 11 shillings… All complain for shirts, shoes and stockings and most for breeches and doublets'.
Somerset	
	Sir Archibald Douglas' company in Watchett Cassocks, their apparel partly good.
	Captain William Preston's company in grey cassocks 'buttoned and unlined'.
	Captain John Talkarne's company in red cassocks 'buttoned button unlined'.
Staffordshire	Lieutenant Frodham's company. 'Their coats are prized to be worth 10 shillings a piece… It is alleged by the soldiers that some of them were pressed and kept so long without the king's allowance as they were forced to pawn their clothes'.
Suffolk	'the soldiers' coats of the Isle of Ely were praised to be worth 9 shillings lined all through but the arms. They are very short and mard with making. Suffolk coats are praised to be worth 6 or 7 shillings a piece but they lay

APPENDIX III

	county in 11 shillings a piece… All complain for shirts, shoes and stockings and most for breeches and doublets'.
Surrey	Captain Peter Mates' company grey lined blue.
Sussex	Captain John Hermon?'s company grey cassocks lined white.

Ensign Wakeman's company blue cassocks 'very tight and unlined'.
Captain Thomas Steverley's company dark grey unlined.
Lieutenant Timothy Lobyes' company grey cassocks lined white.

Warwickshire	Lieutenant Cooper 'their coats are not worth past 10 shillings 6 pence they cost 16 shillings a piece'.
Wiltshire	Captain Anthony Porter's company in grey cassocks 'lined with several colours'.
Worcestershire	Ensign Robinson's company cassocks 'of several colours'.

Ensign Gilbert Weare's company dark grey cassocks and unlined.

On 28 June 1625, the Privy Council directed the commissioners at Plymouth to 2,803 pairs of shoes at 2 shillings a pair at £208 6 shillings, 396 pairs of breeches at 8 shillings the pair at £158 8 shillings and 1,900 shirts at 3 shillings per £286 12 shillings and 1,800 pairs of stockings at 6 pence the pair or £120 6s 8 pence. These were distributed as stockings and shoes as the followings to the companies in Devon and Cornwall.[2]

Company	Shirts	Shoes	Breeches	Stockings
Capt John Tallakarne	25	27	5	25
Capt. Robert Gore	25	25	5	25
Capt William Preston	25	25	5	25
Lt Samuel Bernard	30	30	6	30
Capt. Brutus Buck	6	8	-	6
Sir Archibald Douglas	50	50	10	50
George Kennelthorpe	25	25	5	25
Sir William St Leger	30	40	5	12
Capt George Tuck	25	25	5	25
Lt Richard Burthogg	25	25	5	25
Capt Edward Yates	22	22	4	22
Capt Edward Spring	25	40	5	18
Sir Richard Grenville	30	30	6	30
Capt William Molesworth	32	32	6	32
Capt John Courtney	27	27	6	27
Capt John Hammond	40	40	8	40
Capt Anthony Leigh	37	37	7	37

---

2   TNA AO 1/300/1138.

THE FIRST BRITISH ARMY 1624-1628 (REVISED EDITION)

Capt Edward Leigh	42	42	8	42
Lt Thomas Pote	44	44	9	44
Capt Walter Norton	85	85	17	85
Capt Francis Standish	23	31	4	18
Capt William Morgan	37	37	7	37
Capt Thomas Lindsey	45	45	9	45
Lt John Reynolds	30	30	6	30
Lt Henry Neville	25	25	5	25
Lt Thomas Rawlins	23	23	4	23
Ensign Nathaniel Itby	24	25	4	24
Capt Thomas Staverton	24	25	4	25
Lt Brent Adie	45	50	9	40
Capt Porter	60	60	12	60
Capt Cross	45	42	9	35
Ensign John Austin	35	42	5	-
Lt Henry Frodsham	25	25	5	25
Capt Benjamin Owtred	5	2	-	-
Lt Richard Tremaine	35	37	7	37
Capt Cooke	25	25	5	25
Capt Francis Rainsford	17	18	3	17
Capt Buck	27	27	6	27
Cap Buck	6	2	-	-
Lt John Appleyard	45	45	49	39
Ensign Henry Smith	22	22	4	22
Capt Nicholas Crisp	47	77	15	48
Capt Thomas Lindsey 2nd supply	16	20	6	15
Lt Francis Grover	28	28	5	28
Capt George Hacquett	25	25	5	25
Capt Samuel Leake	25	25	5	25
Capt Thomas Pelham	40	32	6	36
Capt John Glen	25	25	5	25
Capts Bolles & Dixon 2nd supply	55	123	8	30
Capt William Ogle	25	25	5	25
Capt Steven Country	25	25	5	25
Col Harwood	25	25	5	25
Capt Richard Seymer	30	30	6	30
Capt Gore, 2nd supply	20	20	1	20
Capt Peter Newtree	25	25	5	25

APPENDIX III

Capt James Scott	25	25	5	25
Capt Francis Rainsford 2nd supply	-	6	-	-
Capt Peter Alley	25	25	5	25
Capt George Slingbury	25	25	5	25
Capt Daniel Powell	30	25	5	25
Capt John Bettes	25	25	5	25
Capts Cornwell and Leighton	50	50	10	50
Capt John Parkinson	25	25	5	25
Capt Henry Fisher	25	25	5	25
Capt Gore, 2nd supply	8	8	-	-
Capt John Tallakerne, 2nd supply	4	4	-	-

Appendix IV

Instructions given to Cecil, 1625

Instructions given by Sir Edward Cecil Knt, Baron de Putney and Viscount Wimbledon, Admiral of the fleet, lieutenant general and marshal of his Majesty's land forces, now ready to go to sea, to be daily performed by all the commanders and their companies, Masters and other interior officers both by sea and land for the better government of his Majesty's fleet. Dated in the sound of Plymouth aboard his Majesty's good ship, The Ann Royal, the 3 October 1625:

> That above all things you shall provide that God be duly served twice every day, by all the land and sea companies in the ship, according to the usual prayers and liturgy of the Church of England and shall get a discharge every watch, with the singing of a Psalm and prayer usual at sea.
> You shall keep the companies from swearing, blaspheming, drunkenness, dicing, cheating, picking and stealing and the like disorders.
> You shall take care to have all the companies live orderly and peaceably and shall charge the officers faithfully to perform the office and duty of his or their places ; And if any seaman or soldier shall raise tumult, mutiny or conspiracy or commit murder, quarrel, fight or draw weapon to that end, or be a sleeper at his watch, or make a noise, or betake himself to his place of rest after his watch is out , or shall keep his cabin cleanly or be discontented with the proportion of victuals assigned unto him or shall spoil or waste them, or any other necessary provisions in the ship, or shall keep clean his arms or shall go ashore without leave, or shall be found guilty of any other crime or offence you shall use due severity in the punishment and reformation thereof, according to the known orders and custom of the sea.
> For any capital or heinous offence that shall be committed in the ship by the land or seamen, the land and sea commanders shall join together to take a due examination hereof in writing and shall acquaint me therewith to the end I may proceed in judgment according to the quality of the offence.
> No sea captain will meddle with the punishing any of the land soldiers, neither shall the land commanders meddle with the punishment of the seamen.
> You shall with the master take a particular account of the scores of Boatswains and carpenters of the ship examining their receipts, expense and

remains, not suffering any unnecessary waste to be made of their provisions or any works to be done which shall not be needful for the service.

You shall every week take the like account of the purser and steward, of the quantity and quality of victuals that are spent and provide for the preservation thereof without any superfluous expense; and if any suspected persons be in that office for wasting and consuming of victuals you shall remove him and acquaint me herewith and shall give me a particular account from time to time of the expense, goodness, quantity and quality of the victuals.

You shall likewise take a particular account of the Master Gunner for the shot, powder and munition and all manner of stores contained in his indenture.......suffer any part hereof to be sold, embezzled or wasted, nor any piece of ordnance to be shot without directions, keeping an account of every several shot in the ship to the end I may know how the powder spends.

You shall suffer no boat to go from the ship without special leave and upon necessary cause to fetch water or some other needful thing and then you shall send some of the officers or men of trust for whose good carriage and speedy return you will answer.

You shall have a special care to prevent the dreadful accident of fire and let no candles be used without lanterns, nor any at all in or about the powder room. Let no tobacco be taken between [i.e. below] decks or in cabins, or in any part of the ship but upon the forecastle or upper [deck] where shall stand tubs of water to throw the ashes into and to empty pipes.

Let no man give offense to his officer, not strike his equal or inferior aboard and let mutinous persons be punished in [a] most severe manner.

Let no man depart out of the ship wherein he is first entered without leave of his commander, nor let any captain give him entertainment after he is listed upon pain of the severity of the law in that case.

Appendix V

Martial Law

Martial laws ordained and instituted by His Majesty with the advice of the Council of War for the government of good ordering of the troops in this Kingdom either being in an army or in regiments or in single companies to punish the malicious and wicked and to defend the innocent according to the custom of all well governed kingdoms. [1625?]:[1]

1. He that blaspheth or is known to be a common swearer to have his tongue bored through with a hot iron.
2. He that is known to be a drunkard is first to be admonished and for the next to be put imprison and to be kept there at bread and water seven days. And his means that while that is ever plus to be stowed on some sick or poor soldier.
3. That if any soldier be drunk on guard or on the day he should exercise if he be an officer he is to be cashiered if a soldier he is to be seven days at bread and water as is before said. And if that do no good he is to have the strappedo, either more or less.
4. That if any shall absent himself from the church to have the same punishment of bread and water.
5. That if any soldier or officer do abuse either man or woman the party grieved shall go to the officer that commands in chief let him be captain or more than captain or lieutenant or ensign if none of these then to the next justice of the peace if the cause require, and desire him to have him forth coming and to write to the officers of the offender to let him know of the offence then that chief officer must crave justice of a council of war according to the offence either to ask the party forgiveness or to punish him with imprisonment or the strappedo or with more or less as the fault requireth.
6. If any soldier do shed blood or maim any of his fellows or any other that is no soldier he must be more or less punished till the party offended be satisfied according to the punishment before named and at their direction of their officers if there be no martial court in the place he must be brought where there is one.

---

1   NA SP 16/13/42.

APPENDIX V

7. If any soldier do force any women or maid he is to suffer death.
8. If he shall steal anything that is of note he is to suffer death, if of small value he must be more or less punished and imprisoned until he make satisfaction and to stand with a paper on his breast where there shall be written the offence and if it be done in a market place or nigh then to it on a market day.
9. That if any soldier do give any hindrance to market folks passage to be severely punished but if they took anything from them to be punished with death without mercy.
10. That no soldier be seen without his sword and if he will take no warning then to be put in prison.
11. That no soldier be found a mile from his garrison or lodging without leave of his officer in writing upon pain of imprisonment.
12. If a soldier have leave and do not return within the time written without some sufficient excuse and then to bring some sufficient testament of it he must be punished as before.
13. If any soldier do sell any part of his clothes or arms to another or to any other he shall have the aforesaid punishment and the party that doth receive it shall loose his money for they are not his but his Majesty's.
14. That if any officer of a town shall see a soldier pass without asking from whence his pass and forelass is and if the soldier have no pass then is he [is] to send him from constable to constable till he come to his officer if he do not his duty herein to the Majesty he must be presented to the lords.
15. That any soldier that doth strike his officer or any other or doth but make show of it as by laying his hands upon his sword or give him ill language shall suffer death without mercy.
16. That if any soldier do use any uncivil speeches either against his majesty or any other person in authority or any other ways is to be punished without mercy.
17. That no man shall do of his guard without leave of his chief officer that there commandeth without due punishment which have been fore mentioned.
18. That no sentinel be found sleeping upon his guard upon pain of death.
19. All other disorders whatsoever are to be punished as these formerly mentioned, though not formerly nominated.
20. That all soldiers and officers do take an oath as the custom is in all countries to be faithful to his Majesty's service, both in thought, words and deeds, and not to conceal anything that they shall hear or see, that is not for his Majesty's service and honour and to obey all his commandments.
21. That there be authority given to any three or more of the commissioners to call a martial court, and sit in commission to hear, judge and determine any fact done by soldiers, but to have no power to put to death till they have advertised the general, that shall have authority of life and death, for such troops as he shall command.
22. To have in every market place a gibbet and strappedo for the sight of such a remembrance will do good in a wicked mind.
23. That there be a provost appointed in every regiment where the regiments shall be full and a prison ordained for soldiers a part from any other.
24. The provost must have a horse allowed him and some soldiers to attend him and all the rest commanded to obey him and assist him otherwise the

service will suffer for he but one man and must correct many and therefore cannot be beloved and he must be riding from one garrison to another to see that their soldiers do no outrage nor scath about in the country. And these orders we hold in all counties where true discipline is esteemed.

Instructions for the Execution of Martial Law in his Majesty's Army, 3 December 1626:[2]

1. He that shall take God's name in vain and blaspheme God shall for his first offence be kept in prison with bread and water and for the second time of offending, so he shall have a hot iron thrust through his tongue and be stripped and so banished [from] the troops.
2. The like penalty shall be inflicted on those who shall either do or say ought in despite or derision of God's word or the ministers of the word.
3. All willful murders, rapes firing of houses, robberies, outrages, unnatural abuses and such like, shall be punished with death.
4. No man shall beat, threaten or dishonestly touch any women or children on pain of punishment according to the quality of their offence.
5. Whosoever shall raise a mutiny shall be punished to death without mercy.
6. Whosoever shall make any unlawful assembly shall be punished with death.
7. Whosoever shall be present at such an assembly and call or stir up and entice any to increase it shall be subject to the same punishment and officers more than any.
8. He that shall speak words tending to sedition, mutiny or disobedience or having heard any such and shall not presently acquaint the superior officer therewith shall be punished with death.
9. In like manner they shall be punished who rehearse any such words in the presence of private soldiers without order.
10. Whosoever shall go out of his quarter from the Colours or garrison further than shall be limited without his captain's leave shall be punished with death.
11. Whosoever shall neglect his watch or any other service commanded him by his officers [or] shall be found sleeping upon his watch or shall depart from it when he hath been placed by his officers unless he hath been called thence or relieved by his officers shall be punished with death.
12. No man shall make known the watch word by order, nor give any word other than is given by the officer upon pain of death.
13. Whosoever shall absent himself out of the corps of Guards without his officer's leave shall be punished with death.
14. No man shall make alarm or discharge a piece in the night or presume to draw his sword after the watch is set without lawful cause on pain of death.
15. Whosoever shall strike his fellow soldier shall be punished according to discretion.

---

2   Sources BL Lansdown Mss 844 f.309, printed in C Dalton, *Life and times of General Sir Edward Cecil…* pp.386–388.

APPENDIX V

16. No man shall demand money by unlawful assembly or by way of leading to mutiny upon pain of death.
17. No man shall quarrel or fight in a private quarrel or call any to his help in such a quarrel upon pain of death.
18. Whatsoever soldier shall challenge another in the field shall be punished with death.
19. The soldier wronged either by word or deed shall repair to a commissar for satisfaction wherein there shall take order or if he take his own course he shall be punished by discretion.
20. Whosoever shall refuse to perform the command of his officer according to the discipline of war shall be punished with death.
21. Whosoever shall by word or deed interrupt any officer in the execution of his office without order shall be punished with death.
22. Whosoever shall not come fully armed to his Colours or being to march or exercise shall be punished according to discretion.
23. Whosoever shall be drunk during the time of his watch shall be severely punished.
24. Whosoever shall diminish or pawn his arms shall be punished according to discretion.
25. Whosoever shall take away [a] soldier's arms [weapons] shall be punished according to discretion.
26. Whosoever shall be found with his arms fowl or unserviceable shall be punished according to discretion.
27. Whosoever shall take away any other soldier's provisions or arms shall be punished according to the quality of the offence.
28. Whosoever shall draw his sword against his officer shall be punished with death.
29. Whosoever shall refuse or oppose the punishment of any offender by way of attempting to rescue the offender or otherwise shall suffer death.
30. Whosoever shall break prison shall suffer death.
31. No victualler, alehouse keeper or the like shall entertain soldiers at an unreasonable hour on pain of being severely punished.
32. All other abuses and offences not specified in these orders shall be punished according to discretion.

Instructions for the Execution of Martial Law in his Majesty's Army, 17 September 1627:[3]

1. He that shall take God's name in vain or blaspheme God shall for the first offence in that kind, be kept three days in prison with bread and water. And for the second time of offending so shall have a hot iron thrust through his tongue and be stripped to his shirt and so banished [from] the army.

---

[3] HRO 5M50/1954, Volume of Southampton Martial Business memorandums, 1588–1628, ff.109–111, 114–116.

2. The like penalty shall be inflicted upon those who shall either say or do aught in despite or derision of God's word or the ministers of God.
3. All willful murders, rapes firing of houses, robberies, outrages, unnatural abuses and such like, shall be punished with death.
4. No man shall beat, threaten or dishonestly touch any women or children on pain of punishment according to the quality of their offence.
5. Whosoever shall conspire to do anything against the fleet or army shall be put to death without mercy.
6. Whosoever shall come to the knowledge of any such fact and conspiracy and not acquaint their chief officer thereof, shall be punished with death without mercy.
7. Whosoever shall raise a mutiny shall be punished to death without mercy.
8. Whosoever shall make any unlawful assembly shall be punished according to discretion: and if it shall be judged of that quality and of such danger to the service as it shall deserve it, to be punished with death.
9. Whosoever shall be present at such an assembly and call and stir up, or entice any to increase it, shall be subject to the same punishment. And officers more thn any other.
10. He that shall speak any words tending to sedition, mutiny or disobedience, or that having heard any such words and shall not acquaint his superior officer with it, shall be punished with death.
11. In like manner they shall be punished who shall rehearse any such words in the presence of private soldiers without order.
12. Whosoever shall entertain conference or hold correspondence with the enemy, or send any message or letter to the enemy, or receive any from him without the consent of the commander in chief shall be punished without mercy.
13. Whosoever shall converse with any trumpet or drum of the enemy or any sent in message from them without leave of the commander in chief shall be punished with death.
14. Whosoever shall go out of his quarter from his Colours or garrison, further than a cannon shot, without his captain's leave shall be punished with death.
15. Whosoever shall forsake his Colours shall without mercy be punished with death.
16. Whosoever shall neglect his watch or any other service commanded him shall be punished with death.
17. Whosoever shall be found sleeping upon his watch, either of sentinel or perdu shall without mercy be punished with death.
18. Whosoever shall depart from his watch when he hath been placed by his officer, unless he be called thence or relieved by his officer, shall receive punishment of death without mercy.
19. No man shall make known the watchword to the enemy, or any other, but by order, nor give any other word than is given him by the officer, on pain of death.
20. Whosoever shall absent himself out of the Corps de Garde without his officer's leave shall be punished with death.
21. No man shall make an alarm, or discharge his piece by night, nor make any noise without lawful cause, on pain of death.

APPENDIX V

22. Whosoever shall presume to draw his sword without order after the watch is set, shall be punished with death.
23. Whosoever shall strike his fellow-soldier shall be punished according to discretion.
24. No man shall command money by any unlawful assembly.
25. No captain, lieutenant or ensign shall depart from his garrison or quarter without sufficient leave, on pain of death.
26. No man shall quarrel or fight in any private quarrel, or call any to his help in such a quarrel, on pain of death.
27. What soldier so ever shall go on free booting, or comit any spoil without order, shall suffer death.
28. What soldier so ever shall challenge another into the field shall be punished with death.
29. If any corporal or other commanding the watch, shall suffer a soldier to go forth to [a] private fight, he shall without mercy be punished with death.
30. The soldier wronged either by word or deed shall repair to his officer for satisfaction, wherein there shall be order taken, or, if he take his own course, he shall be punished according to discretion.
31. Whosoever shall go out of his garrison or quarter, or come in any other way than at the ports and ordinary allowed passages, shall be punished with death.
32. Whosoever shall refuse to perform the commands of his officer according to the discipline of war shall be punished with death.
33. Whosoever shall not repair unto his Colours (unless it be upon evident necessary) when an alarm is given, or shall go to another place, without special order, or shall neglect his time in coming to his colours, shall be punished with death.
34. Whosoever shall go out of his order ir rank, where he is placed by his officer, without leave, shall be punished with death.
35. Whosoever shall run away at a battle, assault or encounter, may bwe killed by any man that meeteth him; And if he shall escape he shall be declared a villain.
36. No man that shall be appointed to the defence of any place, be he officer or soldier, shall quit that place without order from the Chief on pain of death without mercy.
37. Whosoever shall deliver any place to the enemy by betraying it, shall suffer death without mercy.
38. Whosoever shall persuade the leaving of the defence of a place without sufficient order shall suffer death without mercy.
39. Whosoever shall run to the enemy, being taken again, shall be punished with death.
40. No captain shall entice away the soldier of another captain upon pain of being punished according to discretion.
41. No soldier shall go away from the service of his captain to the service of another on pain of being severely punished unless he have lawful order.
42. Whosoever shall by word or deed interrupt an officer in the execution of his office without order shall without mercy be punished with death.

43. Whosoever shall not come fully armed to his Colours, being of the watch or being to watch or exercise shall be punished according to discretion.
44. Whosoever takes any prisoner of the enemy shall instantly bring the prisoner to the chief commander in that quarter, from him to be brought to the chief of the army, upon pain of death.
45. Whosoever shall take a prisoner and suffer him to depart without the order of the chief of the army shall be punished with death.
46. Whosoever shall take any prize shall presently acquaint the chief of the army with it, on pain of death.
47. All captains and officers shall observe to acquaint the chief commanding in their quarter and the chief of the army with all offenders under their charge, on pain of being punished according to discretion.
48. Whosoever shall be drunk during the term of his watch shall be severely punished.
49. Whosoever shall diminish [break] or pawn his arms [weapons] shall be punished according to discretion.
50. Whosoever shall take to pawn any soldier's arms shall be punished according to discretion.
51. Whosoever shall be found with his arms foul or unserviceable shall be punished according to discretion.
52. Whosoever shall take away other soldiers' provisions or arms shall be punished according to the quality of the offence.
53. Whosoever shall draw his weapon against his officer shall be punished with death.
54. Whosoever shall resist or offend any provost, or his officers, in the execution of his office by way of attempting to rescue any offender or otherwise shall suffer death.
55. Whosoever shall break prison shall suffer death.
56. Whosoever shall make, or help to make, a false muster shall suffer death.
57. Whosoever shall spoil, sell or convey away any munitions shall suffer death.
58. No soldier shall be a victualler without the consent of the chief of the army, upon pain of being punished according to discretion.
59. No victualler shall entertain soldiers at unlawful or unseasonable hours upon pain of being punished by discretion.
60. All other abuses and offences not specified in these orders shall be punished according to the discipline of war and opinions of such officers and others as shall be called to make a council of war.

Select Bibliography

British Library
Add Roll 77,174-75	Account of Richard Graham for £1000 spent on the horses for the expedition to the Île de Rhé, 1627
Add ms 4,712	Miscellaneous papers relating chiefly to ceremonies at the English Court, 16th -17th centuries
Add ms 4,474	Miscellaneous fragments, 17th -18th centuries
Add ms 10,609	Muster rolls of the trained bands, list of military stores etc from the Privy Council, 1608-1637
Add ms 12,528	Sir Sackville Crowe's Book of accompts containing receipts and disbursements on behalf of the Duke of Buckingham, 1622-1628
Add ms 18,764	Miscellaneous Exchequer accounts, 16th -18th centuries
Add ms 26,051	Journal of the Voyage of Rease [Rhè] 1 May to 7 November 1627
Add ms 29,609B	Muster roll for the county of Surrey, 26 May 1627
Add ms 30,766	Correspondence relating to La Rochelle and the Île de Rhé, 17th Century
Add ms 35,832	Hardwick papers, vol. 484, Miscellaenous state letters, 1615-1625
Add ms 37,817	Nicholas papers vol iv, Correspondence of the Duke of Buckingham as Lord Warden of the Cinque Ports, 1624-1627
Add ms 37,819	Nicholas papers vol iv, Correspondence of the Duke of Buckingham as Lord Warden of the Cinque Ports, 1624-1627
Add ms 39,245	Muniments of Edmond Wodehouse
Add ms 41,616	Miscellaneous papers, including tracts concerning the war with Spain, 1625-1626, 1628
Add ms 46,188	Jessop Papers vol. 1 Miscellaneous official and private correspondence and papers of Robert Devereux, 3rd Earl of Essex, 17th century.
Add ms 46,189	Jessop Papers vol. 2 Miscellaneous official and private correspondence and papers of Robert Devereux, 3rd Earl of Essex, 1585-1641.
Add ms 48,182	Survey of musters of the wards of London and suburbs, containing the names of captains, devices on their ensigns and numbers, 1600
Add ms 69,907	Coke papers, accounts and papers relating to the army, 1588-1639
Add ms 72,315	Trumbull papers of Sir Horace de Vere, 1620-1622
Add ms 72,422	Trumbull papers of the council of war and the Muster master General, 1624-1635
Add ms 72,391	Trumbull papers Journal of the campaign to the Isle of Rhè, 12 July 1627-28 September 1927
C.114 c 59.	*A continued journal of all the proceedings of the Duke of Buckingham in the Isle of Ree.*

THE FIRST BRITISH ARMY 1624-1628 (REVISED EDITION)

Egerton ms 2,087	Miscellaneous papers relating to Dover and the Cinque Ports, 1566-1784
Egerton ms 2,596	Correspondence to Lord Carlisle and Henry Rich, Earl of Holland, 1622-1625
Harl ms 135	Sir John Smythe's answer to Captain Humphrey Barwike's book on military affairs, 1595
Harl ms 389	Collection of letters of Joseph Mead, Dec 1620 – Dec 1625
Harl ms 390	Collection of letters of Joseph Mead, Jan 1626 – Apr 1631
Harl ms 429	Journal of the Office of the Ordnance in the Tower of London, Aug 1626 to Feb 1630
Harl ms 1,584	Various instructions to diplomats, 1626-1657
Harl 3,638	A volume containing a variety of historical papers, 1600-1690
Harl ms 4,771	Diary of the proceedings of Parliament, 17 Mar to 27 May 1627
Harl ms 5,109	A volume of miscellaneous military matters
Harl ms 6,344	*A Short military treatise concerning all things needful in an army*
Harl ms 6,807	Miscellaneous tracts including charges by the Earl of Essex against Sir Edward Cecil and a journal or diary of the most martial passages happening at and after our landing at the Isle de Ree 1627.
Harl ms 7364	A book of tactics in Charles I's time
Kings ms 265	Transcripts of military and naval papers, chiefly temp. Charles I
Lansdowne ms 498	Miscellaneous papers relating to parliamentary papers between 1620 and 1628
Lansdown ms 844	Miscellaneous articles, 1558-1726
Maps CC.5.a.343	A new map of the bay and town of Cadiz, 1702
Maps C.7.e.4 (10)	Map of Breda, 1625
Royal 18 A LXXVIII	Memorial by Sir Edward Cecil to Charles I, 1628
RP 8006	Photocopy of an Anonymous account of Île de Rhé Expedition, Discovered by William Fleetwood
Sloane ms 363	Volume of letters including William Fleetwood's account of Buckingham's actions at the Île de Rhé
Sloane 826	Papers of concerning the proceedings in both Houses, 1621-1629
TT	Thomason Tracts

The National Archives, Kew

AO	Audit Office
E	Exchequer Records
MPF 1/250/1-3	Maps of La Rochelle, 1628
MPF 1/256	Map of La Rochelle, 1627
PC/-	Privy Council Papers
PRO/30	Non Public records
PROB	Probate records
SO	Signet Office
SP 9/-	State Papers, Williamson Collection, pamphlets and miscellaneous
SP 14/-	State Papers of James I
SP 16/-	State Papers of Charles I
SP 19/-	Committee for the Advancement of Money
SP 28/-	Commonwealth Exchequer Papers
SP 63/-	State Papers, Ireland
WO	War Office

National Archives of Scotland

GD237/25/1	Undated letter from Sir David Cunningham to the laird of Robertland.
GD248/458/12	Miscellaneous Papers, 17 – 19 Century papers

SELECT BIBLIOGRAPHY

Berkshire Record Office
D/ELL/C140   Lenthall Family Papers
AX1/   Papers of Musters and Trained Bands

Essex Record Office
Q/SR 261A/2   Commission of George Took, 1625

Guildhall Library
12079   Records of the Cutlers Company
12085   Records of the Cutlers Company

Hampshire Record Office
44M69/-   Jervoise family of Herriard, 12th-20th centuries
5M50/1954   Southton Marshall Busines, 1588-1628
1626B/057   Will of William Hare, sen, of Longparish, Hampshire, 1626

Kent Archives
Q/F/M/1-8   List of maimed soldiers (west Kent), 17th century
TR2451/6/7   Transcript of St Mary's Church, Dover burial register

House of Lords record Office
HL/PO/JO/10/1/27   Main papers, Jun 1624-Mar 1625

Isle of Wight Record Office
OG/-   Oglander papers

Lancashire Record Office
QSB/1/30/37   Relief for Jenett Fradsam, wife of pressed soldier, 1627
DDKE/Box 32/63   Copy of a licence from the Earl of Warwick for Henry Rycroft soldier to 'go about his needful affairs for space of three months', 21 Apr 1628

Surrey History Centre
6729/-   Loseley Manuscripts
LM/Cor   Loseley Manuscripts Correspondence
LM   Loseley Manuscripts

Wiltshire Record Office
1178/325   Papers of William Calley, 17th century

Printed Primary Sources

Anon. *A Relation of the passage of our English Companies from time to time* (London, 1621)

Anon. *Journal of all proceedings of the Duke of Buckingham in the Isle of Rhe.* (T Walkley, 1627)

Anon. *Journal of the House of Commons* vol 1

Anon. *Instructions for Musters and Armes and the use thereof* (London, 1623)

Anon., *Calendar of the manuscripts of the Most Hon. The Marquis of Salisbury, K.G., etc., preserved at Hatfield House,* Royal Commission on Historical Manuscripts, 9 (1883-1940)

Anon., *The manuscripts of Rye and Hereford corporations, Capt. Loder-Symonds, Mr. E.R. Wodehouse, M.P., and others (13th Rep., app. iv),* Royal Commission on Historical Manuscripts, 31 (1892)

Anon., *The manuscripts of the Marquess of Abergavenny, Lord Braye, G.F. Luttrell, esq., etc. (10th Rep., app. vi),* Royal Commission on Historical Manuscripts, 15 (1887)

Anon., *The manuscripts of the Earl Cowper, K.G., preserved at Melbourne Hall, Derbyshire (12th Rep., app. i-iii)*, Royal Commission on Historical Manuscripts, 23 (1888-9)

Anon., *The manuscripts of Henry Duncan Skrine, esq. Salvetti correspondence (11th Rep., app. i)*, Royal Commission on Historical Manuscripts, 16 (1887)

Anderson, R C Examinations and dispositions, 1622-1644' Southampton Record Society vol .2, (Southampton: Cox and Shorland, 1629)

Aubrey, John, *Brief Lives edited Richard Barber* (Woodbridge, Boydell, 1982)

Barker, Thomas, *Military Intellectual in Battle* (University of New York Press, 1975) Barry, Gerrat, (translator) *The Siege of Breda, by Herman Hugo* (Ilkley: Scolar Press, 1975)

Barriffe, William, *Military Discipline or the Young Artillery man* (London: Ralph Mab 1635)

Bates Harbin, E H Somerset Quarter Session records (London: Harrison and Sons, 1908)

Bingham, John, *The Art of Embattaling Armies* (1631)

Bingham, John, *Tacticks of Aelian* (1616)

Birch, Thomas *The Court and Times of James the First* (London: Henry Colburn, 1848)

Birch, Thomas, *The Court and Times of Charles the First* (London: Henry Colburn, 1848)

Boas, Frederick, (ed) *The Diary of Thomas Crosfield* (London: Oxford University Press, 1935)

Bray, William (ed) *The Diary of John Evelyn* ed William Bray. (Frederick Warne & Co, 1818)

Bruce, John (ed) *Calendar of State Papers Domestic* 1625-1639 (London: HMSO, 1858-1871)

Buckingham, George, Duke of, A Manifestation or Remonstrance of the Duke of Buckingham, (1627)

Cherbury, Edward Baron Herbert of, *The Expedition to the Isle of Rhe* (London: Whittingham and Wilkins, 1860)

Cockburn, J S (ed) Assizes of Kent during the reign of Charles I (London: HMSO, 1995)

Cooke, Edward, *The Prospective Glass* (1628)

Cooper, J P (ed) *Wentworth Papers, 1597-1628* (London, 1973)

Crossley, Alan, (ed) *Oxford Quarter Session Order Book, 1614-1637* (Woodbridge: Boydell, 2009)

Cruso, John, *The Art of Warre* (Cambridge, 1639)

Cruso, John, *Military Instructions for the cavalry* (Cambridge, 1635)

Daniel, Howard, (ed) *Jaques Callot's Etchings*, (New York: Dover Publications, 1974)

Davies, Edward, *Military Directions and England's training plainly demonstrating the duty of the private soldier* (1619)

Douglas Hamilton, William (ed) *Calendar of State Papers Domestic, addenda* (London: HMSO 1897)

Estiene, A et al *Declaration du Roy, sur la reduction de la ville de la Rochelle* (Paris, 1628)

Grillon, Pierre (ed) *Les Papiers de Richelieu* vols. 2-3 (Paris: Editions A Pedone, 1977, 1979)

Hinds, Allen B (ed) *Calendar of State Papers Venetian* (London: HMSO,1914-1919) *Calendar of State Papers Ireland* (HMSO

Fraser, Sir William *The Book of Carlaverock, memoirs of the Maxwells, Earls of Nithsdale* (Edinburgh, 1873) 2 vols.

Gareston, Mr, *A Continuing journal of all the Proceedings of the Duke of Buckingham on the Isle of Ree* 1627 (EBBO, accessed 2014)

Gerrard, William, *The Art of Warre,* 1591,

Glanville, Sir John *The voyage to Cadiz in 1625* (London: Camden Society, 1883)

Green, M A E (ed) *Diary of John Rous, incumbent of Santon Downham, Suffolk* (London: Camden Society, 1856)

Grenville, Sir Richard, *Two Original Journals of Sir Richard Grenville* (London, 1724)

Grimmelshausen, Hans von, *The Adventurous Simplicissimus*, (Lincoln: University of Nebraska, 1962), Translator A T S Goodrich

Gunkel, Alexander and Handler, Jerome S (ed) 'A German indentured servant in Barbadoes in 1652' in *The Journal of the B.M.H.S.* vol.33 (1970)

SELECT BIBLIOGRAPHY

Helfferich, Tryntje, *The Thirty Years War, A Documentary History* (Indianapolis: Hackett Publishing, 2009)

Hearing, Terry and Bridges, Sarah (ed) *Dorset Quarter Sessions, Order Book, 1625-1638* (Poole: Dorset Record Society, 2006)

Hardy, William (ed), *Hertfordshire County Record Session Book,* (Hertford County Council, 1961)

Hexham, Henry, *The First Part of the Principles of the Art Military Practiced in the Warres of the United Netherlands* (1642)

Hume Browne, F (ed) *Register of the Privy Council of Scotland 2nd Series* (Edinburgh, 1901)

Howell, James, *Familiar Letters in Importance wrote from the year 1618-1650* (Aberdeen: F Douglas and W Murray, 1753)

Kellie, Thomas, *Pallas Armata or the Art of Instructions for the Learned: and all generous spirits who effect the Profession of Arms* (Edinburgh, 1627) (EBBO, accessed 2014)

Knowler, William (ed) *Letters and Dispatches of Thomas Earl of Strafforde with an essay towards his life by Sir George Radcliffe* (London, 1739)

Long, W H (ed) *Oglander Memoirs, extracts from the Mss of Sir John Oglander* (London: Reeves and Turner, 1888)

Lyle, J V (ed) *Acts of the Privy Council of England, 1623-1628* (London, 1932-1940) vols. 39-43

Mansfeldt, Count Ernest von, *The appollogie of the illustrious Prince Ernestus, Earle of Mansfield* (1622)

Mansfeldt, Count Ernest von, *Directions of Warre given to all his officers and soldiers in general* (London, 1624)

Markham, Francis, *Five Decades of Epistles of Warre* (London, 1622) (EBBO, accessed 2014)

Markham, Gervase, *The Souldiers Accidence* (London: Printed by John Dawson for John Bellamie, 1625)

Markham, Gervase, *The Souldiers Grammar* (London, 1626)

Markham, Gervase, *The Souldiers Grammar*, part 2 (London, 1627)

Markham, Gervase, *Souldiers Exercise* (London, 1639)

Marolois, Samuel *The Art of Fortification translated by Henry Hexham, (1638)*

Maclean, J, *De huwelijksintekeningen van Schotse militairen in Nederlands* (Walburg, 1976)

Meruault, Peter, *The last famous siege of the city of Rochell* (London, 1680)

Middlesex Session Rolls 1627, (Middlesex County records, 1888) vol. 3,

Monro, Robert *His Expedition with the Worthy Scots Regiment called Mackay's levied in 1626* (London, 1637)

Mungeam, Gerald, (ed), 'Contracts for the Supply of equipment to the New Model Army' in *The Journal of the Arms and Armour Society* 1968

Murphy, W P D, *The Earl of Hertford's Lieutenancy Papers, 1603-1621* (Devices, Wiltshire Record Society, 1969)

Noel Sainsbury, W *State papers Colonial Series, America and West Indies, 1675-1676, Addenda, 1574-1674* (London: HMSO, 1893)

Orrery, Roger Earl of, *A Treatise of the Art of War* (1677)

Owen, G Dyfnallt Owen (ed) Calendar of the *Cecil Papers in Hatfield House, 1612-1668* (London: Eyre and Spottiswoode, 1971)

Palmer, Thomas, *Bristol's Military Garden* (1635)

Peachey, Stuart (ed) Richard Symonds, *The complete military diary* (Leigh on Sea: Partisan Press,1989)

Peters, Jan *Ein Soldnerleben in Dreissigjahrigen Krieg.* (Akademie Verlag, c1993)

Petre, Sir Charles, *The Letters, Speeches and Proclamations of King Charles I* (Cassell, 1935)

Martin Philippson *Geschichte des Dreissigjahrigen Krieg* (Berlin: Historischer Berlag Baumgartel, nd c.1900)

Poyntz, Sydnam, *.A Relation of Sydnam Poyntz, 1624-1636* (Royal Historical Society, 1908)

Radcliffe, S and Johnson, H C (ed) *Warwick County records, Quarter Session Order Book* (Warwick: L Edgar Stephens, 1935)
Raymond, Thomas, *Autobiography of Thomas Raymond* (London: Royal Historical Society, 1917)
Rayner J L and Crook G T, (ed), *Complete Newgate Calendar* (London: Navarre Society Ltd., 1926)
Relf, Frances Helen (ed) *Notes of the Debates in the House of Lords* (Offices of the Royal Historical Society, 1929)
Rich, Barnabe, *Allarme to England* (1578)
Roberts, G (ed), *Diary of Walter Yonge* (London, 1848)
Rushworth, John *Historical Collections* of Private Passages of State (1659)
Rye, Walter *State Papers relating to Norfolk* (Norwich: Norfolk and Norwich Archaeological Society, 1907)
Schukking, W H *The Principal Works of Simon Stevin* (Amsterdam: Swets and Zeitlinger, 1955-1966)
Smythe, Sir John, *Instructions, Observations and Orders Mylitarie* (1595)
Tadra, Ferdinand, *Briefe Albrecht von Waldstein to Karl von Harrach* (Vienna-Oesterreichische Akademie der Wissenschaften, vol. 41, 1849)
Thompson, T (ed) *Sir James Turner, Memoirs of his own life and Times* (Edinburgh, 1829)
Tooke, Captain George, *The history of Cales passion, or, as some will by-name it, The miss-taking of Cales presented in vindication of the sufferers and to forewarne the future* (1652)
Turner, Sir James, *Pallas Armata* (London, 1683)
Vernon, John, *The Young horseman* (London, 1644)
Wallhausen, Johann, *Kriegkunst zu Fuss.* (Oppenheim, 1615)
Wallhausen, Johann, *Art militaire a Cheval, Instructions des principes et fondements de la Cavallerie* (Frankfort, 1616)
Ward, Robert, *Animadversions of Warre* (1639)
Whiteway, William, *William Whiteway of Dorchester, his Diary* (Dorchester, Dorset Record Society, 1991)
Wood, Marguerite, *Extracts from the Records of the Burgh of Edinburgh* (Edinburgh: Scottish Burgh Records Society, 1882)
Woodall, John *The Surgeon's Chest* (London, 1628)
Woodall, John *The Surgeon's Mate* (London, 1617)
Willis Bund, J W (ed) *Worcestershire Quarter Sessions* (Worcestershire County Council, 1900)

Secondary Sources
Ackermann, S, Gatti E and Richardson, T (ed) 'A 17th century pikeman's armour from Antwerp' in *Arms and Armour* 2000 vol. 7 no. 1 (Manchester: Arms and Armour Society, 1968)
Adair, John, *Roundhead General, Sir William Waller* (Kineton: Roundwood Press, 1973)
Ashley, Maurice, *George Monck* (London: Jonathan Cape, 1977)
Aylmer, G WE, 'Communication, St Patrick's Day 1628 in Witham, Essex' *Past and Present* vol.61 (London: Chapman and Hall,)
Bas, Francois de et al, *Het Staatsche Leger, 1568-1795* (Breda, 1911)
Beelden, Van een Strizd een strijd: *oorlog en kunst vóór de Vrede van Munster, 1621-1648* (Zwolle: Waanders, 1998)
Beesley, Alfred *The History of Banbury* (London: Nicholas and Son, 1841)
Beller, E A, 'The Military Expedition of Charles Morgan' in *The English Historical Review* Oct 1928
Berg, Holger, *Military Occupation under the Eyes of the Lord* (Gottingen: Vandenboeck & Ruprecht, 2010)
Bezdek, Richard H, *Swords and Sword makers in England and Scotland* (Boulder: Colorado, Paladin Press, 2003)

SELECT BIBLIOGRAPHY

Blackmore, David *Arms and Armour of the English Civil Wars* (London: Royal Armouries, 1990)

Boynton, Lindsay, *The Elizabethan Militia, 1558-1638* (Newton Abbot: David and Charles, 1971)

Bruce, John, *Letters of the Verney family* (London Camden Society, 1858)

Brzezinski, Richard & Hook, Richard, *The Army of Gustavus Adolphus* 1 Infantry, (Cowley: Osprey, 1991)

Chadwyck Healey, C E H, *Sir Ralph Hopton's Narrative* (London: Somerset Record Society, 1902)

Carleton, Charles, *Charles I, The Personal Monarch* (London: Routledge & Kegan Paul, 1983)

Carlton, Charles, *Going to the Wars* (London: Routledge, 1994)

Clarke, Aidan 'Sir Piers Crosby, 1590-1646, Wentworth's Tawney Ribbon' *Irish Historical Studies* (Nov. 1988) vol. 26 no. 102 (Dublin: Hodges, Figgis & Con, 1988)

Clarke, Jack *A Huguenot warrior, the life and times of Henri de Rohan* (The Hague: Martinus Nijhoff, 1936)

Cobbett, William *Parliamentary History of England* vol 2 (London: Longman & Co, 1807)

Cockle, Maurice *A Bibliography of English Books up to 1642* (London: Simpkin, Marshall & Co.,1900)

Cogswell, Thomas, *Home Divisions, aristocracy, the State and Provincial Conflict* (Manchester: Manchester University Press, 1998)

Cogswell, Thomas, *The Blessed Revolution* (Cambridge: Cambridge University Press, 1989)

Cogswell, Thomas, 'Published by Authortie' Newsbooks and the Duke of Buckingham's expedition to the Ile de Rhè' in *Huntington Library Quarterly* vol 67 no. 1

Cogswell, Thomas, 'Popular Political Culture and the assassination of the Duke of Buckingham' *The Historical Journal* (vol. 49 no. 2 Jun 2006) pp.357-385

Coss, Edward J *All for the King's Shilling* (Norman: University of Oklahoma Press, 2010)

Cust, Richard *The Forced Loan and English Politics,* (Oxford: Clarendon, 1987)

Dalton, Charles *Life and Times of General Sir Edward Cecil* (London: Sampson Low, Marston, Searle amd Rivington, 1885)

Doelman, James 'John Earle's Funeral Elegy on Sir John Burroughs' in *English Literary Renaissance,* 2011

Ede-Borrett, Stephen *Flags of the English Civil Wars* (Leeds: Raider Books, 1987)

Eickhoff, Sabine, 'Das Massengrab der Schlacht von Wittstock' in *Militargeschichte* February 2013

Eichoff Sabine and Schopper, Franz *1636 Ihre Letzte Schlacht* (Berlin: Theiss, 2012)

Elliott, J H *Richelieu and Olivares* (Cambridge University Press, 1983)

Eltis, David, *The Military Revolution in the Sixteenth Century* (London: Barnes and Noble Books, 1995)

Firth, C H, *Cromwell's Army* (London: Meuthen & Co. 1912)

Fletcher, Anthony, *Sussex, a county in Peace and War* (Chichester: Philimore, 1980)

Forster, John, *Sir John Eliot: A Biography, 1592-1632* (London: Longmans Green, 1864)

Gardiner, Samuel Rawlinson, *A History of England under the Duke of Buckingham and Charles I* vol. 2 1624-1628 (London: Longman, Green and Co. 1875)

Gardiner, Samuel Rawlinson Gardiner, S R *History of England from the Accession of James I to the Outbreak of the Civil War, 1603-1642,* (London: Longmans & Co.1883)

Genlis, Stephanie F, *The Siege of La Rochelle or misfortune and conscience* (London: C Richards, 1830)

Glozier, Mathew 'Scots in the French and Dutch Armies during the Thirty Years War' in *Scotland and the Thirty Years War* ed. Steve Murdoch (Leiden: Brill, 2001)

Goodwin, Tim, *Dorset in the Civil War* (Tiverton: Dorset Books, 1996)

Guthrie, William, *Battles of the Thirty Years War* vol. 1 From White Mountains to Nordlingen, 1618-1635 (London: Greenwood Press, 2002)

Hall, A R, *Ballistics in the Seventeenth Century* (Cambridge University Press, 1952)

Hibbert, Christopher, *Cavaliers and Roundhead* (Glasgow: Harper Collins, 1993)
Hill, Christopher, *The World Turned upside Down* (London: Penguin Books, 1991)
Hoppe, Isreal Geschichte des Ersten Schwedish-Polnischen Krieges in Preussen nebst Anhang (Leipzig, 1887)
Hubsch, Dr George Das Hochstift Bamberg und seine Politik unmittelbar vor dem ersten Einfalle der Schweden 1631 (Bamberg, 1895)
Krussmann, Walter, *Ernst von Mansfeld*, 1580-1626, (Berlin: Duncker & Humblot, 2010)
Lallemand, Auguste, *Souvenirs de 1814, Les Drapeaux des Invalides* (Paris, 1864)
Langer, Herbert *The Thirty Years War* (Poole: Dorset Press, 1990)
Lawrence, David, *The Complete Soldier* (Leiden: Brill, 2009)
Lediard, Thomas *Naval History of England* vol. 2 (London: Wilcox and Payne, 1735)
Leslie, Lt Col J, 'Survey or muster of Armed and Trayned Companies in London, 1588 and 1599' in *Journal of Army Historical Research* vol. 4
Lockyer, Roger *Buckingham* (London: Longman, 1981)
Lockhart, Paul D *Denmark in the Thirty Years War* (Selinsgrove: Susquehanna University Press, 1996)
Loomie, Albert, 'Gondomar's Selection of English Officers in 1622' in *English Historical Review* vol. Jul 1973
Lynn, John, *Women, Armies and Warfare in Early Modern Europe* (Cambridge: Cambridge University Press, 2008)
MacClean, I J, *De Huwelijksintekenungen van Schotse Militairen in Nedeland* (Zutphen: 1976)
McCreadie, Rory W, *The Barber Surgeon's Mate of the 16th and 17th Century* (Lipton: McCreadie, 2002)
McNair, Don, *The Struggle for Stralsund, 1627-1630* (Farnham: Pike and shot Society, 2012)
Mackay, John *An Old Scots Brigade being the history of Mackay's Regiment* (Tonbridge: Pallas Armata, 1991)
Malden, H E *Victoria County History of Surrey* vol. 2
Mann, Golo, *Wallenstein* (London: Andre Deutsch, 1976)
Manning, Roger B *Swordsmen* (Oxford: Oxford University Press, 2003)
Manning, Roger B *Apprenticeship in Arms* (Oxford: Oxford University Press 2006)
Manning, Roger B, 'Styles of Command in the 17th Century English Armies' in *The Journal of Military History*, Vol. 71, No. 3, July 2007 (Lexington: Va, Society for Military History, 2007)
Mazumder, R N et al, 'Typhus Fever; an overlooked diagnosis' in *Journal of Health, Population and Nutrition* (Dhaka: Centre for Health and Population Research, June 2009)
Mesa, Edward de, *The Irish in the Spanish Armies of the Seventeenth Century* (Woodbridge: Boydell, 2014)
Morris, Robert, *Headwear, Footwear and Trimmings of the Common Man and Woman, 1580-1660* (Bristol: Stuart Press, 2001)
Moxey, Keith, *Peasants, Warriors and Wives* (University of Chicago Press, 2004)
Murdoch, Steve (ed) *Scotland and the Thirty Years War, 1618-1648* (Leiden: Brill, 2001)
Nimwegen, Olaf van *The Dutch Army and the Military Revolutions, 1588-1688* (Woodbridge: Boydell Press, 2010)
Niox, Gustave Leon, *Drapeaux et Trophees, Resume de l'historie militaire* (Ch. Delagrave, 1910)
Norman Wallace, A V B *Wallace Collection Catalogue, European arms and armour supplement* (London: Balding and Mansell, 1986)
Parker, Geoffrey, 'Soldiers of the Thirty Years War' in Repgen, Konrad (ed) *Krieg und Politik* (Munich: R Oldenbourg, 1988)
Parker, Geoffrey, *The Thirty Years War* (London: Routledge, 1997)
Parker, Geoffrey, *Army of Flanders and the Spanish Road* (Cambridge: Cambridge University Press, 1972)
Peachey, Stuart and Turton, Alan, *Old Robin's Foot* (Leigh on Sea: Partizan Press, 1987)

SELECT BIBLIOGRAPHY

Peachey, Stuart and Prince, Les *ECW Flags and Colours, 1: English Foot* (Leigh on Sea: Partizan Press, 1991)

Perjes, G, 'Army provisioning, logistics and strategy in the Second Half of the 17th Century' *Acta Historia Academiae Scientiaruim Hungaricae* vol. 16 (1970)

Questier, Michael, *Stuart Dynastic Policy and Religious Politics, 1621-1625* (Cambridge University Press, 2009)

Reese, Peter, *The Life of General George Monck, for King and Cromwell*, (Barnsley: Pen & Sword, 2008)

Rey, Jean, *Histoire du drapeau, des coleurs et des insignes de la morarchie Francaise* la papier du Richelieu (Paris: Delloye, 1837)

Ringoir, H, *Hoofdofficieren der infantrie van 1568 tot 1813 in Bijdragen van de Sectie Militaire Geschiedenis* (BSMG) nr 9 (S Gravenhage, 1981)

Roberts, Keith and Hook, Adam, *Pike and Shot Tactics, 1590-1660* (Botley: Osprey, 2010)

Sansom, Arthur Ernest, *On the Mortallity after the Operation of Amputation of the Extremities* (London, 1859)

Scott, C L, Turton, A and Arni, E G Von, *Edgehill, the battle reinterpreted* (Barnsley: Pen and Sword, 2005)

Sharpe, Kevin *The Personal Rule of Charles I* (London: Yale University Press, 1992)

Stearns, Stephen 'Conscription and English Society in the 1620s' *Journal of British Studies* May 1972

Stone, Lawrence *Crisis in the Aristocracy* (Oxford University Press, 1965)

Swedish General staff, *Sveriges Krig 1611 – 1632* (Stockholm, 1936-1939)

Terry, Charles S, *The Life and Campaigns of Alexander Leslie, First Earl of Leven* (London: Longman, 1899)

Various, *Oxford Dictionary of National Biography* online http://www.oxforddnb.com/ (accessed between 2014-2015)

Verillon, Pierre *Les Trophes de la France,* (Paris: Leroy, 1907)

Verney, Maria *Memoirs of the Verney Family during the Seventeenth Century* (London: Longman, 1904)

Villemain, Pierre, *Journal des assieges de la Rochelle, 1627-1628* (Paris, 1958)

Webb, Henry J *Elizabethan Military Science* (London: University of Wisconsin Press, 1965)

Wedgewood, C V *The Thirty Years War* (New York: Methuen, 1981)

Welch, Charles, *History of the Cutlers Company of London* (London, Cutlers Company 1923)

White, Jason, *Militant Protestantism and British Identity* (London: Pickering & Chatto, 2012)

Wilson, Peter H *The Thirty Years War, a sourcebook* (Houndsmill: Palsgrave-MacMillan, 2010)